AF245554

COMPUTATIONAL ACOUSTICS

Wave Propagation

1st IMACS Symposium on
Computational Acoustics
New Haven, CT, USA, 6-8 August, 1986

Volume 1: Wave Propagation
Volume 2: Algorithms and Applications

INTERNATIONAL ASSOCIATION FOR MATHEMATICS AND COMPUTERS IN SIMULATION
(Association Internationale pour les Mathématiques et Calculateurs en Simulation)

COMPUTATIONAL ACOUSTICS
Wave Propagation

Proceedings of the 1st IMACS Symposium on
Computational Acoustics
New Haven, CT, USA, 6-8 August, 1986

Volume 1

edited by

Ding LEE

Naval Underwater Systems Center
New London, Connecticut
U.S.A.

Robert L. STERNBERG

Office of Naval Research
Boston, Massachusetts
U.S.A.

Martin H. SCHULTZ

Yale University
New Haven, Connecticut
U.S.A.

1988

NORTH-HOLLAND
AMSTERDAM • NEW YORK • OXFORD • TOKYO

ISBN Volume 1: 0 444 70349 7
ISBN Volume 2: 0 444 70350 0
ISBN SET   : 0 444 70351 9

Published by:

ELSEVIER SCIENCE PUBLISHERS B.V.
P.O. Box 1991
1000 BZ Amsterdam
The Netherlands

Sole distributors for the U.S.A. and Canada:

ELSEVIER SCIENCE PUBLISHING COMPANY, INC.
52 Vanderbilt Avenue
New York, N.Y. 10017
U.S.A.

LIBRARY OF CONGRESS
Library of Congress Cataloging-in-Publication Data

IMACS Symposium on Computational Acoustics (1st : 1986 : New Haven,
Conn.)
    Computational acoustics : proceedings of the 1st IMACS Symposium
on Computational Acoustics, New Haven, CT, USA, 6-8 August, 1986 /
edited by Ding Lee, Robert L. Sternberg, Martin H. Schultz.
        p.   cm.
    Contents: v. 1. Wave propagation -- v. 2. Algorithms and
applications.
    ISBN 0-444-70351-9 (set). ISBN 0-444-70349-7 (v. 1). ISBN
0-444-70350-0 (v. 2)
    1. Sound-waves--Measurement--Congresses. 2. Sound-waves--Computer
simulation--Congresses. 3. Sound-waves--Mathematical models-
-Congresses. 4. Algorithms--Congresses. 5. Numerical analysis-
-Congresses. I. Lee, Ding, 1925-   . II. Sternberg, Robert L.
III. Schultz, Martin H. IV. International Association for
Mathematics and Computers in Simulation. V. Title.
QC242.8.I54 1986
534'.2--dc19   87-30422
                                                          CIP

PRINTED IN THE NETHERLANDS

PREFACE

This book contains invited and contributed lectures presented at the first IMACS (International Association for Mathematics and Computers in Simulation) Symposium on Computational Acoustics held 6-8 August 1986 at Yale University in New Haven, Connecticut, U.S.A. This symposium was sponsored jointly by IMACS, ONR (Office of Naval Research), Yale University, and NUSC (Naval Underwater Systems Center).

The objectives of this symposium were two-fold: (1) to provide a forum for active researchers to report on state-of-the-art research and important contributions in computational acoustics, covering areas of ocean acoustics, seismo-acoustics, and aero-acoustics. Topics included mathematical models, numerical methods, significant developments in the application of effective methods to solve acoustics problems, solutions to acoustics problems by supercomputers, interface acoustics, and other important developments in related areas; and (2) to bring together researchers from different areas of acoustics to exchange new ideas and to stimulate future research.

Participants came from eight countries, representing 29 universities and 27 research laboratories. Their enthusiasms and contributions made this symposium a successful one.

On behalf of the organizing committee, we wish to express our gratitude to Prof. Allan R. Robinson of Harvard University for his effort to give a special lecture on "Nowcasting and Forecasting Ocean Fronts and Eddies," which brought ocean modelers and acoustic modelers closer together. Our gratitude also goes to Prof. Robert Vichnevetsky for his untiring guidance in preparation for this symposium.

This IMACS symposium would not have become possible without the sponsorship of the Office of Naval Research (Drs. Raymond Fitzgerald, Richard L. Lau, and Robert L. Sternberg), Yale University (Prof. Martin H. Schultz), and Underwater Systems Center (Drs. William A. Von Winkle, Kenneth Lima, and Ding Lee).

The continuous diligent efforts of Judith Terrell and Connie Maher (Yale University) in preparing the entire program deserve our particular thanks.

Ding Lee, Robert L. Sternberg, Martin H. Schultz

CONTENTS

SOUND NUMBERS FOR COMPUTING SOUND
by

L. B. Felsen

Two years ago we met at Yale
And heard: Computers must prevail
To tell what's simple and profound
In tracking underwater sound.

This year, we met again at Yale
To update our previous tale.
We heard of earlier schemes improved,
Of instabilities removed.
Problems remain, concerning rigor,
But one thing's plain: the codes got bigger.

The scope is larger than before.
The theme this time is to explore
Acoustics in a broader frame,
Though ocean models drive the game.

As codes enlarge and multiply
The need is strong to verify
That each "exact" code is correct
And does not spew out artifact.

. To come to grips with this concern,
There is a trend one may discern.
Benchmarks computed with great care
May yield the numbers to compare.

Choosing the benchmarks causes grief
Because each has his own belief,
And there are those who, with despair,
Turn thumbs down on the whole affair.
Yet, after all have had their say,
The problem has not gone away.
This is a matter for debate
When we meet on a future date.

Somehow, we have to find a track
That keeps computer codes in check.
And as we plunge into the fray,
Let us reflect, and let us pray:

Give us this day a stable code.
Let it print numbers that don't explode,
Let it solve problems that no one has done
Since computations were first begun.
Grant us belief that the code is exact
And that each digit it prints is correct.
Shield us from benchmarks, lest they spoil the dream
Of our grandiose coding scheme.

COMPUTATIONAL ACOUSTICS: Wave Propagation
D. Lee, R.L. Sternberg, M.H. Schultz (Editors)
Elsevier Science Publishers B.V. (North-Holland)
© IMACS, 1988

INSIGHTS ON MARINE ACOUSTICS
CONCEIVED FROM SPACE (IMACS)

Paul D. Scully-Power

Naval Underwater Systems Center
New London, Connecticut

ABSTRACT

The increasing complexity of underwater acoustic models has, over the years, directly paralleled our increasing knowledge of the complexity of the structure and dynamics of the world's oceans. Simple horizontally stratified ocean models together with the concomitant range independent ray and normal mode acoustic models have gradually been replaced by an ocean with mesoscale perturbations (fronts and eddies) and range-dependent, wide-angle, parabolic equation acoustics. These models have produced a reasonably accurate representation of the two-dimensional acoustic intensity field. At a time when modern numerical techniques coupled with state of the art computer technology now make it feasible to calculate the three-dimensional range-dependent field, another complexity has arisen. Recent space oceanography results show that the ocean is exceedingly complex at the submesoscale—shear currents of width 100 m, horizontal shear fields having an intershear spacing of 1 km, and spiral eddies with a diameter of 10 km. This imposes a coherence constraint in three dimensions which will be strongly dependent on frequency. The challenge of the future, therefore, is to incorporate this new knowledge of the oceans into our acoustic models.

INTRODUCTION

The recent discovery of the ubiquitous nature of submesoscale horizontal shear currents and spiral eddies in the world's oceans (Scully-Power, 1986) poses a number of questions relative to the propagation of sound through such oceanic features. To answer those questions requires a validated, fully three-dimensional, acoustic propagation model. Such models, however, are in their embryonic stages, and one of the pacing issues is the intercomparison of the presently available models. The major reason that this has presented such difficulty is that the input environmental data varies from model to model.

It would thus seem appropriate that a canonical ocean representation be devised, against which each acoustic model could be exercised, and then the differences between the acoustic model predictions could be quantified. This paper is an attempt to do just that.

THE CANONICAL OCEAN

A careful consideration of the type of ocean representations which may be appropriate for three-dimensional acoustic model intercomparisons leads to the delineation of three major requirements for such representations:

    (a)   They should be expressible analytically in three dimensions (range, azimuth, and depth), so that step sizes in the acoustic computer routines can be changed at will without introducing any spurious artifacts,

    (b)   they should have continuous first derivatives in three dimensions (as distinct from being just piecewise continuous) so that false caustics can be avoided,

    (c)   they should be based solidly on ocean physics, so that they are a good approximation to the real ocean.

These representations will consist of two parts: the unperturbed ocean, and a three-dimensional perturbation such as a strong current, a front, or an eddy which is imposed on the unperturbed structure.

THE UNPERTURBED OCEAN

Fortunately, Munk (1974) has solved the problem for the unperturbed ocean. In the open ocean, the most significant physical property is the variation of density with depth. Indeed, the ocean can well be described as a horizontally stratified fluid, i.e., a fluid characterized by horizontal layers, which can be adequately defined by its variation with depth only. Munk's contribution was to note that the real ocean stratification is well represented by an exponential decrease with depth, and then to derive the sound speed structure with depth, based solely on this stratification and the thermodynamic properties of sea water.

In oceanographic parlance, the ocean's stratification is usually introduced by means of the Brunt-Väisälä or buoyancy frequency, $N(z)$, which expresses the natural frequency of a small parcel of water in terms of the vertical density gradient. Munk's exponential form for this is

$$N^2(z) \equiv \frac{g}{\rho}\frac{\partial\rho}{\partial z} = N_0^2 e^{-2z/B} \tag{1}$$

where ρ is the density, g is gravity, N_0 = 3 cycles per hour is
the surface-extrapolated buoyancy frequency, and B = 1.3 km is the
stratification scale which he takes equal to the sound channel
axis depth.

His derived sound speed profile then takes the form

$$c(z) = c_1[1 + \varepsilon(\gamma + e^{-\gamma} - 1)] \tag{2}$$

where $\gamma = 2(z - B)/B$, $\varepsilon = 7.4 \times 10^{-3}$, and c_1 = sound speed (minimum) at the sound channel axis z = B, which Munk takes as 1.492
km/sec. Figure 1 shows equations (1) anu (2) plotted to illustrate this variation of density and sound speed with depth.

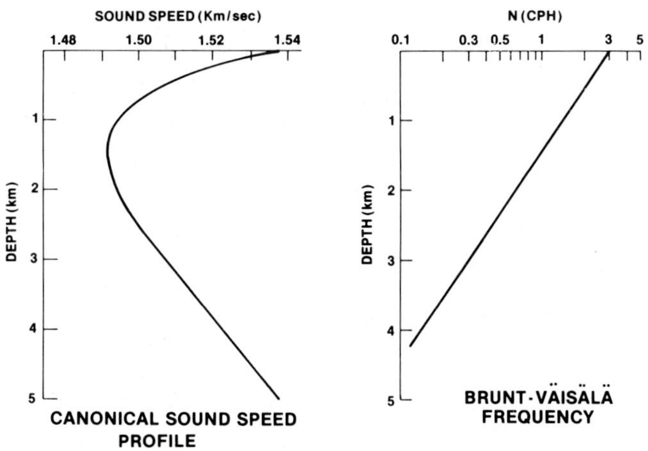

Fig. 1. Brunt-Väisälä (or buoyancy) frequency and canonical sound
speed profile with depth for the unperturbed canonical ocean.

THE PERTURBED OCEAN

Of all the commonly occurring perturbations in the oceans,
such as geostrophic currents, internal waves, thermal fronts,
mesoscale eddies and submesoscale spirals, perhaps the best for
use in three-dimensional acoustic model intercomparisons are the
eddies. This is because they are circularly symmetric (or nearly
so), have a relatively uniform length scale (diameters about 200
km), and perturb the ocean throughout much of the vertical water
column. They are thus ideally suited to demonstrate the three-
dimensional acoustic effects over reasonably long ranges and with

various source/receiver depth combinations. The object, therefore, is to <u>derive</u> the horizontal and vertical structure functions
for such eddies based on the general perturbation theory of the
oceans, to apply Munk's canonical density stratification for the
unperturbed ocean, and thereby calculate the three-dimensional
sound speed variations caused by the eddy perturbation. This will
result in an analytic expression for a three-dimensional ocean
perturbation which can be used directly in the acoustic models and
which will therefore provide a common basis for quantifying three-
dimensional acoustic effects in the ocean and for comparing the
various acoustic models. This canonical perturbed ocean will
therefore provide the stepping stone to validating acoustic models,
which in turn can be used to investigate the acoustic properties
of spirals, which was the original motivation for the work in the
first place.

As a point of departure, we take the nonlinear quasi-geo-
strophic potential vorticity equation of planetary (low frequency)
perturbations in a continuously stratified ocean on a β-plane
($\beta = \partial f/\partial y$, $f = 2\Omega \sin \theta$, where Ω is the magnitude of the planetary rotation and θ is the latitude; i.e., β is assumed constant
and takes the local value of the variation of the Coriolis parameter f with latitude):

$$Dq/Dt = 0$$

where the potential vorticity q is given by

$$q = \nabla_H^2 \Psi + \frac{\partial}{\partial z}\left(\frac{f^2}{N^2}\frac{\partial \Psi}{\partial z}\right) + \beta y$$

Here, $D/Dt \equiv \partial/\partial t + \underset{\sim}{v} \cdot \nabla_H$ is the Lagrangian derivative (following
the motion), ∇_H^2 is the horizontal Laplacian, and Ψ is the geo-
strophic stream function. Written in full, this potential vorticity equation is

$$\left[\frac{\partial}{\partial t} + \frac{\partial \Psi}{\partial x}\frac{\partial}{\partial y} - \frac{\partial \Psi}{\partial y}\frac{\partial}{\partial x}\right]\left[\frac{\partial^2 \Psi}{\partial x^2} + \frac{\partial^2 \Psi}{\partial y^2} + \frac{\partial}{\partial z}\left(\frac{f^2}{N^2}\frac{\partial \Psi}{\partial z}\right) + \beta y\right] = 0 \qquad (3a)$$

or, equivalently,

$$\nabla^2 \Psi_t + [(f^2/N^2)\Psi_z]_{zt} + J[\Psi, \nabla^2 \Psi] + J[\Psi, ((f^2/N^2)\Psi_z)_z]$$
$$+ \beta \Psi_x = 0 \qquad\qquad\qquad (3b)$$

where J is the Jacobian, we have dropped the horizontal designator

on the Laplacian, and subscripts now denote partial derivatives
with respect to that variable.

The derivation of the potential vorticity equation is by no
means trivial, but can be found in standard texts on geophysical
fluid dynamics such as Pedlosky (1979, p. 342) or Gill (1982, p.
530). This equation is usually solved by dropping the nonlinear
terms and then solving for plane harmonic waves (Rossby waves).
We will adopt a different approach here.

First, it should be noted that although the potential vorticity equation is fully three-dimensional (Ψ is a function of
all three space dimensions), it has an overriding two-dimensional
structure since it contains only the <u>horizontal</u> advection of the
potential vorticity. This can be seen most clearly by writing

$$\Psi(x,y,z,t) = \Phi(x,y,t)\psi(z)$$

Then equation (.3b) becomes

$$\nabla^2\Phi_t + \psi^{-1}\Phi_t[(f^2/N^2)\psi_z]_z + \psi J[\Phi,\nabla^2\Phi] + \beta\Phi_x = 0 \tag{4}$$

where we note that the second Jacobian which contained the z-de-
rivatives has dropped out.

The next step is to handle the nonlinear terms which only
appear in the remaining Jacobian. Rather than just simply dropping them, as is the usual practice, we take them to be equal to
the horizontal advection term. Then equation (4) becomes

$$\nabla^2\Phi_t + \psi^{-1}\Phi_t[(f^2/N^2)\psi_z]_z + 2\beta\Phi_x = 0 \tag{5}$$

Although this may seem arbitrary and capricious, it has been
justified at some length in Scully-Power (1980) where it was shown,
by a careful evaluation of the linear and nonlinear effects, that
the most stable circular eddies are those in which this equality
pertains. More will be said on this later in the paper.

Equation (5), which is now linear, can be solved by making
the transformation

$$\Phi(x,y,t) = \mathrm{Re}\, e^{-i[(\beta x/\omega)+\omega t]}\phi(x,y)$$

which is equivalent to a horizontal structure function $\phi(x,y)$ and
a carrier wave of angular frequency ω moving to the west with a
speed ω^2/β. Under this transformation, equation (5) becomes

$$\nabla^2\phi + (\beta^2/\omega^2)\phi + \phi\psi^{-1}[(f^2/N^2)\psi_z]_z = 0 \tag{6}$$

Equation (6) can now be separated into two equations, one in $\phi(x,y)$ and one in $\psi(z)$, in the usual manner:

$$\nabla^2\phi + [(\beta^2/\omega^2) - a_n^2]\phi = 0 \tag{7}$$

and

$$[(f^2/N^2)\psi_z]_z + a_n^2\psi = 0 \tag{8}$$

where a_n^2 is the separation constant. Equation (7) can therefore be thought of as the governing equation for the horizontal structure function $\phi(x,y)$, and equation (8) for the vertical structure function $\psi(z)$. We will investigate each in turn.

THE HORIZONTAL STRUCTURE FUNCTION

The horizontal structure function equation (7) just derived is simply the two-dimensional Helmholtz equation

$$(\nabla^2 + b_n^2)\phi = 0 \tag{9}$$

where

$$b_n^2 = (\beta^2/\omega^2) - a_n^2 \tag{10}$$

Since we are interested in circular eddies, we will transform equation (9) into plane polar coordinates, $\phi = \phi(r,\theta)$, and the equation becomes

$$\frac{\partial^2\phi}{\partial r^2} + \frac{1}{r}\frac{\partial\phi}{\partial r} + \frac{1}{r^2}\frac{\partial^2\phi}{\partial\theta^2} + b_n^2\phi = 0 \tag{11}$$

For ϕ finite at $r = 0$, equation (11) has the well-known Fourier-Bessel series solution

$$\phi = \Sigma_s A_s J_s(b_n r)e^{is\theta} \tag{12}$$

where the J_s are Bessel functions of order s, and the A_s are arbitrary constants.

We now impose the condition that ϕ is circularly symmetric, which is borne out by actual measurements of ocean eddies. In this case ϕ is independent of θ, $s = 0$, and equation (12) reduces to

$$\phi = A_0 J_0(b_n r) \tag{13}$$

Now, the basic ocean perturbation theory that led to the potential vorticity equation also produces the following relationships:

$$\eta(r) \approx \phi(r)$$
$$v(r) \approx \partial\phi/\partial r$$

where $\eta(r)$ is the variation with r of the displacement of a given isopycnal (equal density) surface from its equilibrium position, and $v(r)$ is the concomitant azimuthal current velocity. (Note that there is no radial current velocity in a circularly symmetric eddy.)

Hence the appropriate boundary condition on (13) is

$$\partial\phi/\partial r = 0 \qquad at \ r = R$$

where R is the eddy radius. Physically, this means that the currents are zero at the eddy edge. Hence

$$b_n R = j_{1,n} \tag{14}$$

where the j_1's are the zeros of J_1.

All that is required now for the horizontal structure function of our canonical eddy is to normalize the function. We do this by requiring the structure function to have a value of unity at the eddy center and zero at the edge. This is straightforward, and the result is

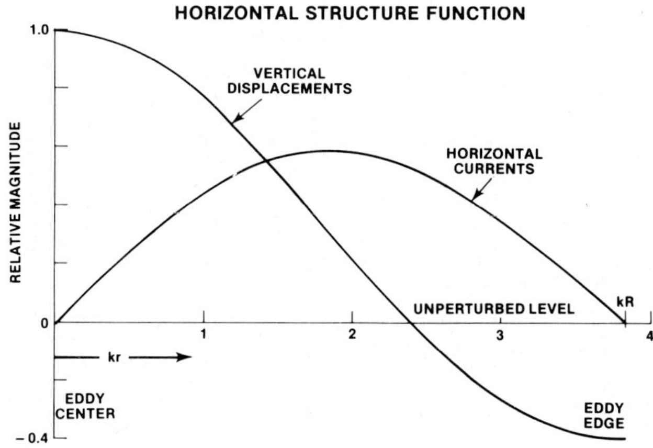

Fig. 2. Relative vertical displacements of isopycnals and horizontal currents across an eddy radius, based on the derived horizontal structure function.

$$\phi(r) = \frac{J_0(b_n r) - J_0(b_n R)}{1 - J_0(b_n R)} \tag{15}$$

to within an arbitrary constant which depends on the strength of the eddy.

Plots of the relative perturbation of the isopycnals and the associated horizontal currents as a function of radial distance are given in Figure 2 to illustrate their variation across the eddy. Note that the currents are zero at both the center and edge of the eddy, as has been noted by direct measurements.

THE VERTICAL STRUCTURE FUNCTION

The vertical structure function equation (8), when used with Munk's exponential form of the stratification (Eq. (1)), leads to the following eigenfunction equation:

$$[e^{2z/B}\psi_z]_z + (a_n^2 N_0^2/f^2)\psi = 0 \tag{16}$$

The solution of this equation becomes straightforward upon using the following transformations:

$$\zeta(z) = e^{2z/B}\psi \tag{17a}$$

$$\xi = e^{-z/B} \tag{17b}$$

Then Eq. (16) becomes

$$\zeta_{\xi\xi} + \xi^{-1}\zeta_\xi + \alpha_n^2\zeta = 0 \tag{18}$$

where

$$\alpha_n^2 = a_n^2 N_0^2 B^2/f^2 \tag{19}$$

Equation (18) is simply Bessel's equation of zero order, whose solution is

$$\zeta = A_0 J_0(\alpha_n\xi)$$

where A_0 is an arbitrary constant. Then, backtransforming and noting that $\psi = -(B^2/\alpha_n^2)\zeta_z$ [from Eq. (16)] we have

$$\psi(z) = Ae^{-z/B}J_1(\alpha_n e^{-z/B}) \tag{20}$$

where again, A is an arbitrary constant.

Now, as before, the basic ocean perturbation theory provides the following relationships:

$$v(z) \approx \psi(z)$$
$$\eta(z) \approx \partial\psi/\partial z$$

where $v(z)$ is the variation with z of the horizontal currents and $\eta(z)$ is the concomitant displacement of a given isopycnal from its equilibrium position.

Hence the appropriate boundary condition on (20) is

$$\partial\psi/\partial z = 0 \qquad \text{at } z = 0$$

which, upon making use of Eq. (17a) gives

$$\alpha_n = j_{0,n}$$

where the j_0's are the zeros of J_0. Thus, using Eq. (19)

$$a_n = (f/N_0 B)\, j_{0,n} \tag{21}$$

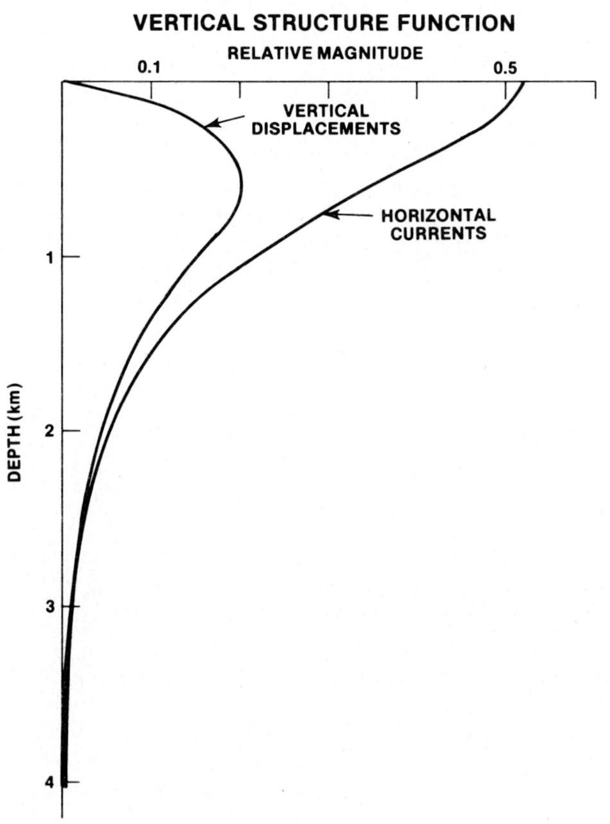

Fig. 3. Relative vertical displacements of isopycnals and horizontal currents as a function of depth, based on the derived vertical structure function.

Plots of the relative perturbation of the isopycnals and the associated horizontal currents as a function of depth are given in Figure 3 to illustrate their functional form.

THE SOUND SPEED PERTURBATION

For small perturbations, the change in sound speed Δc due to the eddy perturbation is given by

$$\Delta c(r,z) = (\partial c/\partial z)\eta(r,z) = (\partial c/\partial z)\eta(r)\eta(z) \tag{22}$$

where $c(z)$ is given by Eq. (2).

Now Munk (op. cit.) has shown that, based on the thermodynamic relations for seawater,

$$(\partial c/\partial z) \approx c(z)N^2(z)$$

and so we have for the fractional change in sound speed $\nu(r,z) = \Delta c(r,z)/c(z)$ due to the eddy perturbation

$$\nu(r,z) \approx \eta(r)\eta(z)N^2(z) \tag{23}$$

where

FRACTIONAL VARIATION IN SOUND SPEED

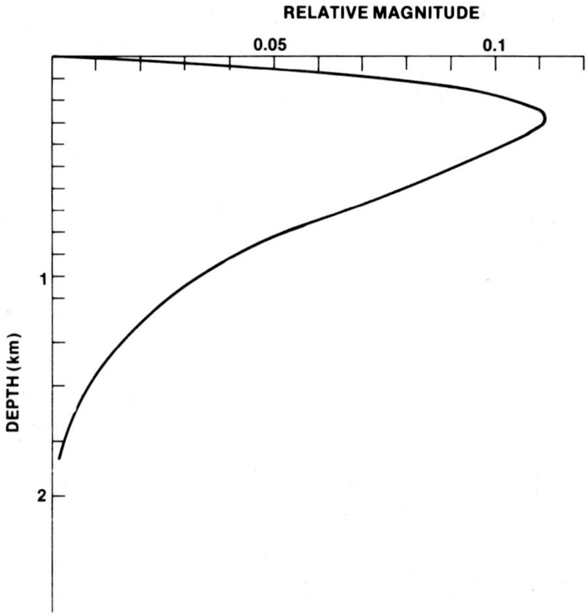

Fig. 4. Relative magnitude of the fractional variation in sound speed with depth due to the eddy.

$$\eta(r) \approx J_0(b_n r) \tag{24}$$

$$\eta(z) \approx e^{-2z/B} J_0(\alpha_n e^{-z/B}) \tag{25}$$

and $N^2(z)$ is given by Eq. (1).

The purely vertical component of $v(r,z)$ is plotted in Figure 4. Note that it reaches a maximum around 300 meters, which is in conformance with measured results.

THE FINAL CANONICAL PERTURBED OCEAN

In order to obtain the final form to be used in acoustic model validation, some of the parameters used in the derivations need to be evaluated.

First, the separation constants a_n which are the eigenvalues of the vertical structure function given by Eq. (8) and whose reciprocals are known as the Rossby deformation radii of the corresponding eigenfunctions. These are given by Eq. (21). Although there are an infinite number of them, we will take for our purposes only the prime value given by the first zero of J_0. This is not as restrictive as it would seem, since it gives the largest possible Rossby deformation radius $\mu = 1/a_1$, and Lighthill (1969) has shown that this introduces only small errors.

Numerically, taking $f = 8.37 \times 10^{-5}$ rads/sec (at 35° latitude), $N_0 = 3$ cph $= 5.24 \times 10^{-3}$ rads/sec as proposed by Munk, and the stratification scale $B = 1$ km following the more recent work of Flatte et al. (1979), gives for the Rossby deformation radius

$$\mu = 26.1 \text{ km}$$

which fits the real ocean extremely well.

We also have to evaluate the b_n, which are the eigenvalues of the horizontal structure function given by Eq. (7). These are given by Eq. (14), where again we take only the prime value given by the first zero of J_1.

At this point, however, b_1 is not fully specified since it also depends on the eddy radius R via Eq. (14). We could now simply assign a value to R. However, R can be derived from ocean perturbation theory.

Scully-Power (1980) has shown that the phase velocity associated with the most stable circular eddies is $1/2 \, \beta\mu^2$ directed toward the west. Equating this with the westward phase velocity ω^2/β of the carrier wave associated with the horizontal structure

function as previously derived, gives for the frequency of the
carrier wave

$$\omega = (1/\sqrt{2})\beta\mu \tag{26}$$

Then, since b_1 and a_1 are associated via Eq. (10) in terms of ω,
we have by combining Eqs. (10) and (26),

$$b_1^2 = 2/\mu^2 - a_1^2 = a_1^2$$

This is an interesting result and is a direct consequence of equating the nonlinear and horizontal advection terms in the original derivation.

Moreover, since a_1 is fully specified, we now have a means of calculating R via Eq. (14). Numerically, this gives for the eddy radius

$$R = 99.9 \text{ km}$$

which is in excellent agreement with measured ocean values. Moreover, since the westward phase velocity of the carrier wave

$$\omega^2/\beta = \frac{1}{2}\beta\mu^2$$

has a typical value of 0.005 meters per second, which is five orders of magnitude smaller than the sound speed $c(z)$, we are justified in making the frozen ocean approximation and neglecting the effects of the carrier wave in acoustic calculations.

The canonical perturbed ocean is now fully specified and takes the form:

$$\Delta c(r,z) = \pm A \, c(z) e^{-4z/B} J_0(\alpha e^{-z/B}) \, \frac{J_0(r/\mu) - J_0(R/\mu)}{1 - J_0(R/\mu)} \tag{27}$$

where
      $R = 100$ km is the eddy radius

      $\mu = 26$ km is the Rossby deformation radius

      $B = 1$ km is the stratification scale

      $\alpha = N_0 B/\mu f = j_{0,1} = 2.40$

      $c(z)$ = the canonical sound speed profile given by Eq. (2)

      A = the strength of the eddy which takes typical values of
            0.05 to 0.25 in the actual ocean

    + = anticyclonic (warm core) eddy

    − = cyclonic (cold core) eddy

REFERENCES

1. Scully-Power, P. D., Navy Oceanographer Shuttle Observations, STS 41-G: Mission Report, Naval Underwater Systems Center Tech. Document 7611, March 1986.

2. Munk, W. H., Sound channel in an exponentially stratified ocean, with application to SOFAR, JOURNAL OF THE ACOUSTICAL SOCIETY OF AMERICA, 55 (1974), 220-226.

3. Pedlosky, J., GEOPHYSICAL FLUID DYNAMICS, Springer-Verlag, New York, 1979.

4. Gill, A. E., ATMOSPHERE-OCEAN DYNAMICS, Academic Press, New York, 1982.

5. Scully-Power, P. D., Mesoscale inhomogeneities and turbulence in ocean acoustics, CAVITATION AND INHOMOGENEITIES IN UNDER-WATER ACOUSTICS, W. Lauterborn, ed., Springer Series in Elec-trophysics, Springer-Verlag, New York, Vol. 4, 1980, 294-307.

6. Lighthill, M. J., Dynamic response of the Indian Ocean to on-set of the southwest monsoon, PHILOSOPHICAL TRANSACTIONS OF THE ROYAL SOCIETY, A, 265 (1969), 45-92.

7. Flatte, S., ed., SOUND TRANSMISSION THROUGH A FLUCTUATING OCEAN, Cambridge University Press, Cambridge, 1979.

COMPUTATIONAL ACOUSTICS: Wave Propagation
D. Lee, R.L. Sternberg, M.H. Schultz (Editors)
Elsevier Science Publishers B.V. (North-Holland)
© IMACS, 1988

AN ANALYTICAL SOLUTION FOR THE THREE-DIMENSIONAL
ACOUSTIC FIELD IN A PENETRABLE OCEAN WEDGE

Michael J. Buckingham

Radio and Navigation Department
Royal Aircraft Establishment
Farnborough, Hampshire, England

and

Institute of Sound and Vibration Research
The University
Southhampton, England

ABSTRACT

A new solution is introduced for the 3-D acoustic field due
to a point source in an ocean wedge with a penetrable bottom. The
field consists of a superposition of range-dependent normal modes.
These modes are matched to range-independent modes in a notional
"effective" wedge with a perfectly reflecting bottom whose apex is
offset from that of the penetrable wedge. A known solution for
the "perfect" wedge is modified on the basis of the effective wedge
model to provide a relatively straightforward expression for the
full spatial dependence of the field in the penetrable wedge. This
solution satisfies the boundary conditions and exhibits features
which conform with ray-theoretic predictions. Examples of the
field computed from the new theory are presented for directions
perpendicular to and parallel to the apex of the wedge.

I. INTRODUCTION

In ocean channels of variable depth the field is three-dimensional, that is, it depends on three spatial coordinates. Three-
dimensional fields in general exhibit several distinctive features,
including <u>ray-path curvature</u> in the horizontal, <u>shadow zones</u> in
the horizontal, and <u>intra-mode interference</u> exemplified by very
rapid spatial oscillations within each mode. These features all
originate in the mechanism of horizontal refraction [1], whereby
wavefronts are twisted in the horizontal as they undergo reflections
from the inclined boundaries of the channel.

Any satisfactory solution of the Helmholtz equation for a variable depth channel must incorporate horizontal refraction, implicitly or explicitly, in order to represent correctly the essential

3-D nature of the acoustic field. Such solutions can be achieved analytically for a certain class of ocean domains, viz., isovelocity channels whose boundaries are perfectly reflecting surfaces corresponding to the coordinate surfaces in one of the separable coordinate systems. An exact solution for the field in a channel satisfying these conditions may be derived by applying a sequence of three integral transforms (corresponding to the three spatial coordinates) to both sides of the inhomogeneous Helmholtz equation, followed by the associated sequence of inversion integrals. The appropriate choice of transforms is dictated by the geometry of the channel and the boundary conditions.

This theoretical approach has been applied to a wedge-shaped ocean with perfectly reflecting boundaries [2], to yield a solution for the 3-D field in the channel. In view of the cylindrical symmetry of the wedge, a cylindrical coordinate system with the axis horizontal and running along the apex is the natural choice for the problem. The solution obtained is exact, consisting of a sum of uncoupled normal modes, each represented by a trigonometric eigenfunction in which the variable is angular depth measured about the apex. The mode coefficients, containing the range (out from the apex) and z (distance parallel to the apex) dependence of the field, are given by an integral over finite limits in which the integrand is highly oscillatory. Embedded in this integral are all the features arising from horizontal refraction which are characteristic of 3-D fields in general.

An analogous solution has been derived for the field in an ocean channel around a perfectly reflecting conical seamount [3,4] whose apex just touches the surface of the ocean. In this case, the channel boundaries correspond with the coordinate surfaces of a spherical polar coordinate system with origin at the apex. The normal modes in the solution are functions of angular depth, as in the wedge, but the eigenfunctions are no longer trigonometric functions. Instead, they are associated Legendre functions, reflecting the spherical symmetry of the channel. However, for the small slopes typical of seamounts, the associated Legendre functions approximate closely to the more familiar trigonometric forms. The mode coefficients give the range (out from the apex) and azimuthal dependence of the field. In particular, they show a shadow zone behind the seamount (i.e., on the side remote from the source) whose angular width expands with increasing mode number. Thus,

the angular width of the shadow in the total field is determined
by that of the lowest order mode.

A serious limitation inherent in these two canonical models
of 3-D ocean acoustics (the wedge and the seamount) lies in the
boundary conditions: the bottom of the real ocean is not usually
a perfect reflector. In many situations a better approximation is
a fast bottom, that is, a water-sediment interface which shows a
critical grazing angle. A ray travelling in the ocean which is
incident upon this type of bottom will either be totally internal-
ly reflected—a process accompanied by a phase shift—or it will
penetrate, depending on whether the grazing angle is less than or
greater than the critical value. Such an interface is therefore
said to be penetrable.

No exact solution is available for the field in a variable
depth channel with a penetrable bottom boundary. Even though the
geometry of the channel may be such that the boundary surfaces are
matched by the coordinate surfaces of one of the separable coordinate systems, the penetrable boundary condition ensures that the
Helmholtz equation is not separable. Thus, in such problems the
integral transform technique cannot be applied, at least, not
directly.

An obvious alternative is to turn to numerical methods of solution. Jensen and Kuperman [5] used the parabolic equation (p.e.)
approximation to evaluate the <u>two-dimensional</u> field in a penetrable wedge, and demonstrated the existence of energy propagating
down into the bottom from around the penetration points of the
modes. This result stimulated several analytical investigations
[6-10] of the 2-D penetrable wedge, which achieved reasonable agreement with the original p.e. computation. However, no approximate
solution, analytical or numerical, is known to have been published
for any three-dimensional penetrable channel, although a 3-D p.e.
algorithm [11] has been developed which should, in principle, be
capable of handling penetrable domains. This approach is almost
certain to be computationally demanding, perhaps prohibitively so.

The purpose of this paper is to show that, with appropriate
modifications, the analytical integral-transform techniques normally considered to work only when the Helmholtz equation is separable,
can yield useful approximate solutions for the 3-D field in certain
variable-depth ocean channels with a penetrable bottom. This is
exemplified in the following discussion, which concentrates exclusively on a new analytical solution for the 3-D field in a

M.J. Buckingham

penetrable wedge. The analysis is remarkably similar to the theory [2] for the field in a "perfect" wedge, although two additional features appear in the new model: the phase change on reflection from the penetrable bottom gives rise to range dependent eigenfunctions (indicative of the nonseparability of the Helmholtz equation); and any ray whose grazing angle exceeds the critical grazing angle at any point along its path is excluded from the field.

This new 3-D solution is achieved by a simple coordinate transformation, based on the concept of a notional "effective" perfect wedge in which both boundaries are pressure-release surfaces. The effective bottom boundary is parallel to but displaced below the actual penetrable bottom by a distance calculated to provide the appropriate phase change on reflection. The introduction of this effective wedge means that much of the groundwork laid in the perfect wedge theory can be transferred directly into the penetrable wedge analysis. The mathematical details of the new solution are given elsewhere [12], but it should be noted that the expression for the field satisfies the boundary conditions and exhibits features which are consistent with ray-theoretic arguments. These features include well-defined shadow zones, certain regions of strong intra-mode interference and other regions of none, where the field is relatively smooth. A discussion of the physics underlying the analysis is given here, in geometrical terms where possible, and some results are presented which illustrate the differences and similarities between the field in the penetrable wedge and that in the perfect wedge.

2. THE DISPERSION RELATION

The geometry of the penetrable wedge is illustrated in Fig. 1, which shows the water column overlying a fluid (i.e., no-shear) sediment. Across the boundary between these two regions the acoustic field must show continuity of both pressure and normal component of velocity, and at the pressure-release sea surface the velocity potential of the field is zero. These three requirements constitute the boundary conditions of the penetrable wedge problem.

On examining the (homogeneous) Helmholtz equation for the field in the ocean/sediment regions, in conjunction with the boundary conditions, it is possible to determine the mode shapes in the water column. As indicated below, these mode shapes are given as the solution of a dispersion relation.

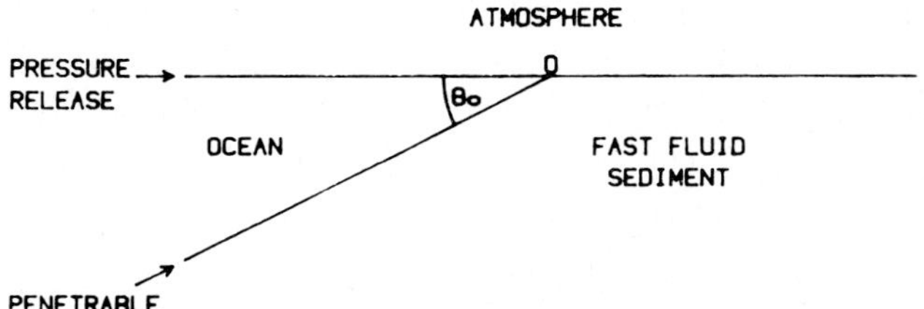

Fig. 1. Geometry of the penetrable wedge.
O is the apex, running normal to the plane of
the paper.

Strictly the modes are not separable, since they show a range
dependence, but by assuming this dependence to be weak they may be
treated as quasi-separable, allowing a similar formalism to be followed in achieving a solution for the field as in the case of the
perfect wedge [2]. This includes the use of a cylindrical coordinate system.

To begin, the procedure is to represent each mode in the water
column by a sinusoidal function of the form

$$\sin(v\theta) \tag{1}$$

where θ is the angular depth measured down from the surface about
the apex O, and

$$v \equiv v(r) = m\pi/\theta_1 \qquad m = 1, 2, \ldots \tag{2}$$

In the perfect wedge, θ_1 would be a constant equal to the wedge
angle θ_0, but here it depends weakly on range r (measured out from
the apex). The parameter m in equation (2) is the mode number.
The eigenfunction in equation (1) is now used as the kernel in a
finite Fourier sine transform in the variable θ taken over the
interval $[0,\theta_1]$, and this transform is applied to the Helmholtz
equation for the field in the water column. The range dependence
of v is neglected everywhere in this process except in the second
derivative of the field with respect to θ. A similar quasi-separ-
ation is performed for the field in the sediment and, by employing
the boundary conditions at the penetrable bottom, the fields above
and below the interface are matched to yield the following equation
for v:

$$\tan(v\theta_0) = -gv/\{[k_1 r \sin(\alpha_c)]^2 - v^2\}^{1/2} \tag{3}$$

Here θ_0 is the wedge angle, $g > 1$ is the sediment/seawater density
ratio and k_1 is the wavenumber in the water column. α_c is the

critical grazing angle of the bottom, defined by

$$\alpha_c = \cos^{-1}(c_1/c_2) \tag{4}$$

where $c_1/c_2 < 1$ is the seawater/sediment sound speed ratio.

Equation (3) is the dispersion relation which must be solved for the eigenvalues v in the penetrable wedge. It may look familiar, since it has exactly the same form as the dispersion relation obtained by Pekeris [13] for a channel of uniform depth overlying a fast fluid sediment. It is a transcendental equation with real roots for the mth mode lying in the interval $(m - (1/2))\pi/\theta_0 < v < m\pi/\theta_0$. At ranges where v falls within these limits the mode propagates through the wedge-like channel, undergoing total internal reflection at the bottom boundary. The region where such behavior occurs is established from two limiting cases. First, in the limit of very long range it follows from equation (3) that

$$\lim_{r \to \infty} \tan(v\theta_0) = 0 \tag{5a}$$

giving

$$\lim_{r \to \infty} v = m\pi/\theta_0 \qquad m = 1, 2, \ldots \tag{5b}$$

These are the same eigenvalues as obtain in the perfect wedge, signifying that far from the apex the penetrable bottom has a negligible effect on the mode shapes. The second limiting case is when r takes a value such that the denominator on the right of equation (3) is zero. Then

$$r \equiv r_{cm} = v/[k_1 \sin(\alpha_c)] \tag{6}$$

and

$$v\Big|_{r=r_{cm}} = (m - \tfrac{1}{2})\pi/\theta_0 \tag{7}$$

On substituting this value for v into equation (6) the cut-off range, r_{cm}, for the mth mode can be expressed as

$$r_{cm} = (m - \tfrac{1}{2})\pi/[k_1\theta_0 \sin(\alpha_c)] \tag{8}$$

This corresponds exactly with the cut-off depth for a uniform-depth channel, as obtained by Pekeris [13].

When $r < r_{cm}$, equation (3) has no real solution, total internal reflection no longer occurs, and the energy in the mode penetrates the bottom to radiate away into the sediment. This is the cut-off region of the mode. At longer ranges, where $r > r_{cm}$, the mode remains trapped in the water column with an eigenvalue lying between the two limiting values in equations (5b) and (7).

An approximate explicit expression for the eigenvalue of the mth mode over the propagation region can be derived by writing equation (3) in the form

$$v\theta_0 = m\pi - \tan^{-1}\{gv/\{[k_1 r \sin(\alpha_c)]^2 - v^2\}^{1/2}\} \qquad (9)$$

On expanding the arctan function in a Taylor series to first order in v, this becomes

$$v\theta_0 \simeq m\pi - \{gv/[k_1 r \sin(\alpha_c)]\} \qquad (10)$$

which has the solution

$$v \simeq (m\pi/\theta_0)\{1 + \{g/[k_1 r\theta_0 \sin(\alpha_c)]\}\}^{-1} \qquad (11)$$

In the long-range limit this approximation reduces correctly to the result in equation (5b). At the other range limit, where $r = r_{cm}$, equation (11) may be matched to the eigenvalue in equation (7) by setting $g = \pi/2$, thus:

$$v \simeq (m\pi/\theta_0)\{1 + \{\pi/[2k_1 r\theta_0 \sin(\alpha_c)]\}\}^{-1} \qquad (12)$$

This final expression for the range dependence of the eigenvalues is independent of the density ratio g, which is reasonable since a numerical solution of the full dispersion relation shows v to be insensitive to g in the realistic range $1 < g < 2$.

Equation (12) is a very useful formula for the mode shapes. Moreover, it has a simple geometrical interpretation which ultimately leads to a solution for the r/z dependence of the field in the penetrable wedge.

3.   THE EFFECTIVE WEDGE

The expression for v in equation (12) suggests that a straightforward coordinate transformation arising from the construction of an effective perfect wedge can simplify enormously the penetrable wedge problem. The effective wedge, with apex $\tilde{O}$, is illustrated in Fig. 2. It shows the same wedge angle, θ_0, as the actual wedge, but the lower boundary is a fictitious pressure-release surface displaced below the penetrable bottom by a distance Δ.

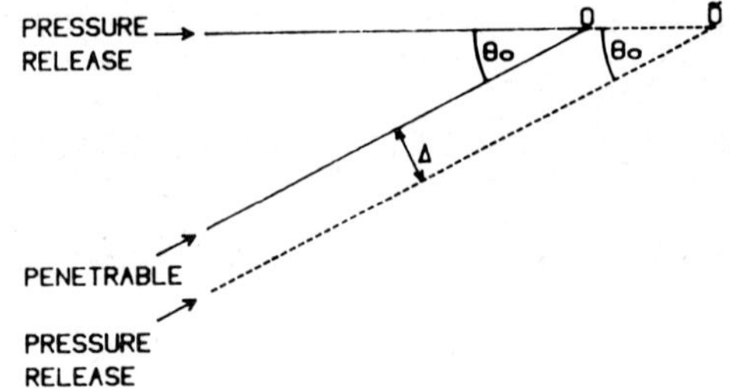

Fig. 2. The penetrable wedge with apex O and
effective perfect wedge with apex Õ.

The range $\tilde{r}$ and angular depth $\tilde{\theta}$ coordinates in the new geometry are related to their original counterparts through the equa-
tions

$$\tilde{r} = r + (\Delta/\theta_0) \tag{13a}$$

and

$$\tilde{\theta} = \theta/[1 + (\Delta/r\theta_0)] \tag{13b}$$

where it is assumed that the wedge angle is small, as is the case
in most ocean wedges. On setting

$$\Delta = \pi/[2k_1 \sin(\alpha_c)] \tag{14}$$

these expressions become

$$\tilde{r} = r + \pi/[2k_1\theta_0 \sin(\alpha_c)] \tag{15a}$$

and

$$\tilde{\theta} = \theta/\{1 + \pi/[2k_1 r\theta_0 \sin(\alpha_c)]\} \tag{15b}$$

By examining the Rayleigh laws of reflection, it is easy to show
that Δ in equation (14) is just the displacement required to accommodate the phase shift experienced by a ray undergoing total internal reflection from the penetrable bottom.

The significance of the transformation should now be apparent.
In the effective wedge the modes are given by the eigenfunctions

$$\sin(m\pi\tilde{\theta}/\theta_0) \qquad m = 1, 2, \ldots \tag{16}$$

which are independent of the new range coordinate $\tilde{r}$. On substituting for $\tilde{\theta}$ from equation (15b), however, these same modes can be
seen to depend on the old range r and, what is critically important,
this range dependence is exactly the same as that in equation (12)
for the modes in the penetrable wedge. This means that by working

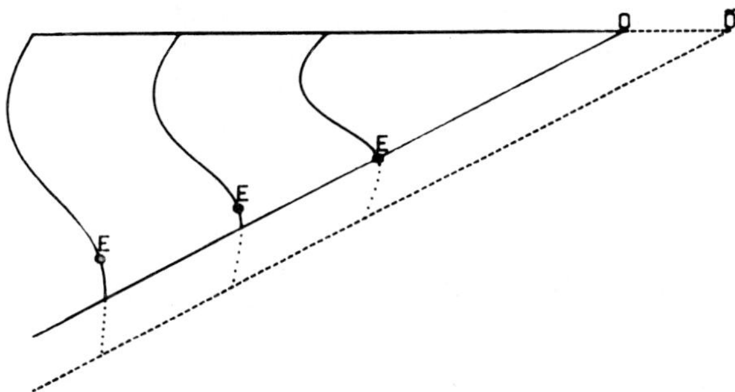

Fig. 3. Schematic showing the second mode apparently sinking down
into the penetrable bottom as the apex O is approached. Penetration occurs when the lowest extremum in the mode (labeled E) just
touches the bottom.

with the effective perfect wedge, the boundary conditions in the
penetrable wedge are automatically satisfied (within the degree of
approximation indicated here). Thus the effective wedge provides
a natural basis for deriving a solution for the r/z dependence of
the field in the penetrable wedge which is convenient since the
solution for the perfect wedge is known [2].

     According to equation (16), the modes in the penetrable wedge
appear to sink down into the bottom as the apex is approached.
This is illustrated for the second mode in Fig. 3. Penetration of
the bottom occurs when the lowest extremum in the eigenfunction
(the point E in the figure) just touches the penetrable interface.
Equations (6) and (7) have already indicated that this happens at
the cut-off range of the mode.

4.   THE r/z DEPENDENCE OF THE FIELD

     The field in the penetrable wedge created by a point source
in the channel can now be determined from a solution of the inhomogeneous Helmholtz equation expressed in the new cylindrical coordinates. The boundary conditions are simply those of the effective perfect wedge, since these ensure that the actual boundary
conditions are satisfied on the penetrable water/sediment interface. Now, the solution [2] for the field in a wedge with pressure-release boundaries is obtained by applying a sequence of
three integral transforms to both sides of the Helmholtz equation:
a finite Fourier sine transform over angular depth $\tilde{\theta}$, a generalized Hankel transform over range $\tilde{r}$, and a unilateral Laplace transform over z, the direction parallel to the apex. When the corresponding inverse transforms are applied to the triply transformed

result, an expression for the field is obtained of the form

$$\phi = (Q/\theta_0) \sum_{m=1}^{\infty} I_\mu(\tilde{r},\tilde{r}',z)\,\sin(\mu\tilde{\theta})\sin\tilde{\theta}') \tag{17}$$

where the primes denote source coordinates, Q is the source strength, and μ is an abbreviation for the limit in equation (5b):

$$\mu = m\pi/\theta_0 \tag{18}$$

The complex mode coefficients, I_μ, give the amplitude and phase of the field in the $\tilde{r}/z$ plane.

In the perfect wedge the mode coefficients are given by an integral [2] over the interval $[0,\pi]$:

$$(1/\pi\tilde{R}_0)\int_0^\pi \cos(\mu\sigma)\,\{\exp\{-jk_1 R_0[1 - 2\tilde{a}\cos(\sigma)]^{1/2}\}$$

$$\times\,[1 - 2\tilde{a}\cos(\sigma)]^{-1/2}\,d\sigma \tag{19}$$

where

$$\tilde{R}_0 = (\tilde{r}^2 + \tilde{r}'^2 + z^2)^{1/2} \tag{20a}$$

and

$$\tilde{a} = \tilde{r}\tilde{r}'/\tilde{R}_0^2 \tag{20b}$$

This integral must be modified to apply to the penetrable wedge, for in its present form it does not allow for the energy lost to the bottom when penetration occurs. In other words, according to equation (19), all the acoustic energy remains trapped in the water column, as indeed it does in a perfect wedge. The penetrable wedge, however, is a leaky waveguide, which significantly alters the physics of the situation.

The required mathematical amendment, however, is very simple. All it involves is a change in the upper limit on the integral in equation (19). Thus, for the penetrable wedge the mode coefficients are given by

$$I_\mu(\tilde{r},\tilde{r}',z) = (1/\pi)\int_0^{\sigma_1} \cos(\mu\sigma)\,[\exp(-jk_1\tilde{R})]/\tilde{R}\,d\sigma \tag{21}$$

where

$$\tilde{R} = \tilde{R}_0[1 - 2\tilde{a}\cos(\sigma)]^{1/2}$$

$$= \{[\tilde{r} - \tilde{r}'\cos(\sigma)]^2 + \tilde{r}'^2\sin^2(\sigma) + z^2\}^{1/2} \tag{22}$$

All that remains to be done now is to identify the upper limit σ_1 and justify the formulation of I_μ in equation (21).

The integrand in equation (21) contains the function

$$[\exp(-jk_1\tilde{R})]/\tilde{R} \qquad (23)$$

This represents spherical waves emanating from an image point at $\tilde{R} = 0$. On referring to equation (22), it is apparent that this image point lies in the $z = 0$ plane (i.e., the plane perpendicular to the line of the apex, $\tilde{O}$, and passing through the source) at a position which depends on σ. The locus of image points associated with values of σ in the interval $[0,\pi]$ (corresponding to the integration interval in equation (19) for the effective perfect wedge) is a semi-circle in the $z = 0$ plane with radius $\tilde{r}'$ and centered on $\tilde{O}$. Thus, according to this representation, the modal field at any point in the channel is a superposition of (straight-line) rays originating in a continuous, semi-circular distribution of virtual images. Now, the criterion we require in connection with the penetrable wedge is that none of these rays should cross any of the image planes with a grazing angle greater than the critical grazing angle, α_c. (It is implicit in this statement that a zero-wedge-angle approximation is being adopted.) This, however, sets an upper limit on the value σ can take. From three-dimensional geometry [12] this upper limit is

$$\sigma_1 = \cos^{-1}\{(\tilde{r}/\tilde{r}')\sin^2(\alpha_c) + \cos(\alpha_c)\{1 - (r/r')^2\sin^2(\alpha_c)$$
$$- (z/\tilde{r}')^2\tan^2(\alpha_c)\}^{1/2}\} \qquad (24a)$$

when $|z| \le z_1$,

$$\sigma_1 = \cos^{-1}\{[\tilde{r}^2 + \tilde{r}'^2 - z^2\tan^2(\alpha_c)]/2\tilde{r}\tilde{r}'\} \qquad (24b)$$

when $z_1 < |z| \le z_2$, and

$$\sigma_1 = \pi \qquad (24c)$$

when $|z| > z_2$, where

$$z_1 = (\tilde{r}'^2 - \tilde{r}^2)^{1/2}\cot(\alpha_c) \qquad (24d)$$

and

$$z_2 = (\tilde{r} + \tilde{r}')\cot(\alpha_c) \qquad (24e)$$

Equations (24) are valid when $\tilde{r} < \tilde{r}'$, and remain so when $\tilde{r} > \tilde{r}'$ provided $\tilde{r}$ and $\tilde{r}'$ are interchanged.

By restricting the integration range in equation (21) to the interval $[0,\sigma_1]$, all rays that penetrate the bottom before reaching the receiver are eliminated from the mode coefficients. Note that when $|z|$ is sufficiently large (i.e., $|z| > z_2$) all the virtual image sources contribute to the field. In these regions of the channel the only difference between the field in the penetrable wedge and that in the perfect wedge is due to the phase shift on reflection from the penetrable interface.

5. RESULTS FOR THE FIELD

Figure 4a shows an example of the first mode coefficient for the penetrable wedge plotted as a function of range with $z = 0$. This was computed from equations (21) and (24). For comparison

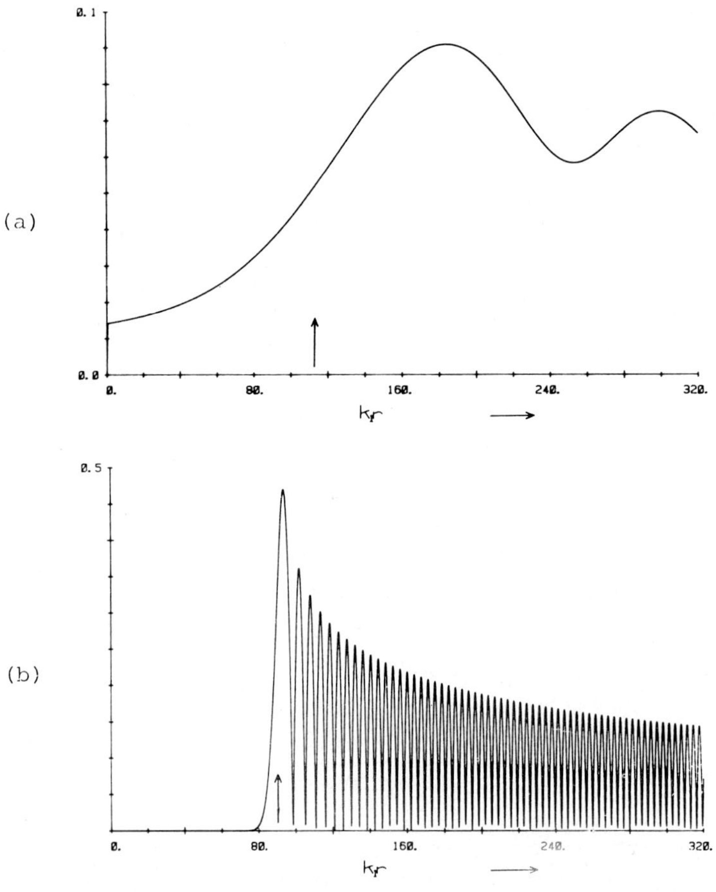

Fig. 4. Modulus of the first mode coefficient as a function of range in the $z = 0$ plane ($k_1r' = 10^4$, $\theta_0 = 2°$): (a) penetrable wedge ($\alpha_c = 25°$); (b) perfect wedge.

the corresponding mode coefficient for the perfect wedge is shown
in Fig. 4b. The two vertical arrows in the figure indicate the
cut-off ranges estimated from the related normal mode theories of
uniform-depth channels. Although cut-off in the penetrable wedge
occurs around the correct range, the roll-off rate is much gentler
than in the perfect wedge. This is consistent with a ray picture
of the cut-off process: in the penetrable wedge cut-off is a result of ray penetration, which takes place over an extended range
region, whereas in the perfect wedge, cut-off is associated with
the much more abrupt phenomenon of ray turn-round, and so the
field falls rapidly to zero. At ranges down-slope of cut-off, the
field in the penetrable wedge is much smoother than that in the
perfect wedge. In the latter case, strong intra-mode interference
is evident, interpreted as being due to mutual interference between modal rays travelling up and down slope. In contrast, the
amount of energy travelling down slope in the penetrable wedge is
negligible (when $r < r'$) and hence rapid spatial oscillations in
the field do not occur.

Perhaps more interesting than the range dependence is the z
dependence of the field, for this shows features which are unique-
ly three-dimensional in origin. Figure 5 shows the first modal
coefficient as a function of $k_1 z$, at fixed range $k_1 r = 400$, for
both penetrable and perfect wedges. Particularly noticeable in
Fig. 5a is the absence of intra-mode interference throughout the
central part of the ensonified region. This behavior, and the onset of spatial oscillations toward the outer edges of the field
region, are in accord with ray-theoretic arguments [12]. Intra-
mode interference is evident, of course, throughout the whole of
the field in the perfect wedge (Fig. 5b). The unlabeled vertical
arrows in the figure depict the edge of the shadow zone, and the
arrows labeled 1 and 2, respectively, in Fig. 5b indicate the
boundaries $z = z_1$ and $z = z_2$, as given by equations (24d) and
(24e).

6. CONCLUDING REMARKS

The analytical solution introduced here for the acoustic
field in a penetrable ocean wedge provides a useful insight into
the nature of three-dimensional fields in depth-varying ocean channels. Inevitably it involves certain approximations, but for the
very small bottom slopes typical of ocean wedges ($\theta_0 \simeq 2°$) these
are all remarkably good. This becomes apparent when the solution

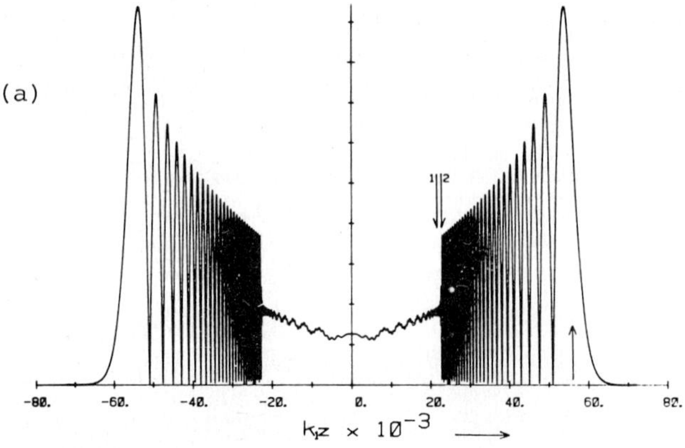

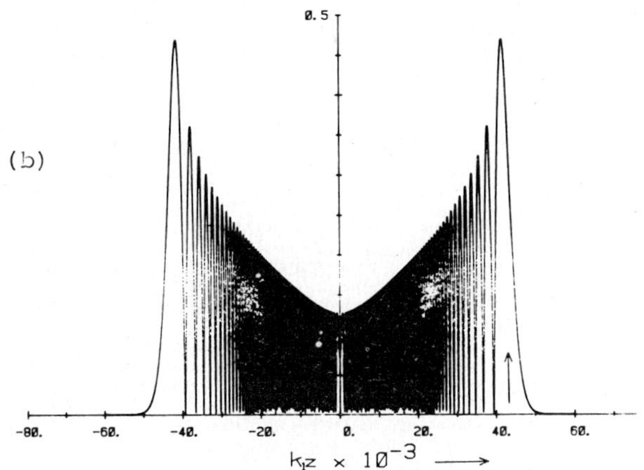

Fig. 5. Modulus of the first mode coefficient as a function of z at a fixed range $k_1 r = 400$ ($k_1 r' = 10^4$, $\theta_0 = 2°$):   (a) penetrable wedge ($\alpha_c = 25°$) and (b) perfect wedge.

is compared with ray-theoretic results [12] for the edge of the shadow zone and other large-scale features of the field.

The theory has been formulated in such a way that it gives the field in the water column but not in the fast sediment lying below the bottom. This omission is not too serious because in practice it is usually (though not always) the field in the ocean that is of interest rather than the field in the seabed. A possible use of the new theory is as a reliable reference for testing the versatile, 3-D numerical algorithms that are currently under development [11]. The theory of the perfect wedge, although it is exact, is not as useful as a test case because the more promising algorithms have an inherent grazing angle limitation and cannot therefore be expected to give the field correctly in a domain with perfectly reflecting boundaries.

ACKNOWLEDGMENT

I am indebted to Dr. S. A. S. Jones for producing the diagrams in this paper.

REFERENCES

1. D. E. Weston, Horizontal refraction in a three-dimensional medium of variable stratification, PROCEEDING OF THE PHYSICAL SOCIETY 78 (1961), 46-52.

2. M. J. Buckingham, Acoustic propagation in a wedge-shaped ocean with perfectly reflecting boundaries, in Proceedings of the NATO Advanced Research Workshop on Hybrid Formulation of Wave Propagation and Scattering, IAFE, Castel Gandolph (Rome), Italy, August 30-Sept. 3, 1983 (Martinus Nijhoff, Dordrecht), pp. 77-105; and NRL Report 8793 (March 1984).

3. M. J. Buckingham, Theory of acoustic propagation around a conical seamount, JOURNAL OF THE ACOUSTICAL SOCIETY OF AMERICA 80 (1) (1986), 265-277.

4. M. J. Buckingham, S. A. S. Jones, and P. N. Harriman, Stationary phase evaluation of the integral for the acoustic field around a conical seamount, JOURNAL OF THE ACOUSTICAL SOCIETY OF AMERICA, 80 (1) (1986), 278-281.

5. F. B. Jensen and W. A. Kuperman, Sound propagation in a wedge-shaped ocean with a penetrable bottom, JOURNAL OF THE ACOUSTICAL SOCIETY OF AMERICA 67 (5) (1980), 1564-1566.

6. A. D. Pierce, Guided mode disappearance during upslope propagation in variable depth shallow water overlying a fluid bottom, JOURNAL OF THE ACOUSTICAL SOCIETY OF AMERICA 72 (2) (1982), 523-531.

7. A. D. Pierce, Augmented adiabatic mode theory for upslope propagation from a point source in variable depth shallow water overlying a fluid bottom, JOURNAL OF THE ACOUSTICAL SOCIETY OF AMERICA 74 (6) (1983), 1837-1847.

8. A. Kamel and L. B. Felsen, Spectral theory of sound propagation in an ocean channel with weakly sloping bottom, JOURNAL OF THE ACOUSTICAL SOCIETY OF AMERICA 73 (4) (1983), 1120-1130.

9.	J. M. Arnold and L. B. Felsen, Rays and local modes in a wedge-shaped ocean, JOURNAL OF THE ACOUSTICAL SOCIETY OF AMERICA 73 (4) (1983), 1105-1119.

10.	J. M. Arnold and L. B. Felsen, Intrinsic modes in a nonseparable ocean waveguide, JOURNAL OF THE ACOUSTICAL SOCIETY OF AMERICA 76 (3) (1984), 850-860.

11.	W. L. Siegmann, G. A. Kriegsmann, and Ding Lee, A wide-angle three-dimensional parabolic wave equation, JOURNAL OF THE ACOUSTICAL SOCIETY OF AMERICA 78 (2) (1985), 659-664.

12.	M. J. Buckingham, Theory of three-dimensional acoustic propagation in a wedge-like ocean with a penetrable bottom, to be submitted for publication in the JOURNAL OF THE ACOUSTICAL SOCIETY OF AMERICA.

13.	C. L. Pekeris, Theory of propagation of explosive sound in shallow water, GEOLOGICAL SOCIETY OF AMERICA, Memo 27, 1948, pp. 1-117.

COMPUTATIONAL ACOUSTICS: Wave Propagation
D. Lee, R.L. Sternberg, M.H. Schultz (Editors)
Elsevier Science Publishers B.V. (North-Holland)
© IMACS, 1988

RANGE-DEPENDENT CANONICAL SOLUTIONS
FOR SHALLOW OCEAN ACOUSTICS

I. T. Lu and L. B. Felsen

Department of Electrical Engineering/Computer Science
Weber Research Institute
Polytechnic University
Farmingdale, New York

ABSTRACT

Numerical and analytical techniques are now being developed
for successively more general models of acoustic propagation in
range-dependent shallow ocean environments. To assess the quality
of such schemes, it is necessary to test them against reliable reference data, usually referred to as benchmarks. Exactly solvable
canonical problems provide an important potential source of benchmarks because the burden of generating the reference data is then
entirely on the numerical implementation of rigorous analytical
forms. A group of coordinate separable two-dimensional and three-
dimensional classical canonical problems is examined here from the
perspective of shallow ocean acoustics with range-dependent bottom
topography and bottom surface impedance; generalizing the usually
assumed rigid bottom in these coordinate separable canonical test
configurations, the surface impedance feature permits modeling of
partial reflection from, and absorption through, the bottom boundary, and also allows for inclusion of bottom guided interface waves.
The problems are formulated in terms of general spectral integrals
over the constituent separable one-dimensional "characteristic"
Green's functions in a curvilinear coordinate frame, and rigorous
alternative forms are then developed which emphasize rays, normal
modes, adiabatic modes, horizontally refracted modes, and other
propagation mechanisms relevant to shallow ocean acoustics. Specific applications include the wedge-shaped ocean, a ridge modeled
as a hyperbolic cylinder, a conical ridge and a hyperbolic spheroidal ridge. Numerical results are shown for the special case of
horizontal refraction in the presence of a weakly sloping rigid
conical seamount.

I. INTRODUCTION

    With the increasing complexity of analytical and numerical
models for underwater acoustic propagation in range-dependent shallow oceans, there is a correspondingly urgent need for testing
these models against benchmark reference solutions. Such reference
solutions and their numerical implementation must be _reliable_ and
they should be _relevant_; meeting these requirements makes their
selection difficult. One potential class of canonical test solutions is provided by classical coordinate separable configurations,
which can be solved exactly. To make them "relevant" to shallow
ocean range-dependent propagation, the solutions must be presented
in a form that permits identification of various wave phenomena
known to be important in a more general context. The building
blocks in this hierarchy include spectral representations, rays,
normal modes, adiabatic modes, hybrid combinations, horizontal refraction, etc. With this in mind, we analyze a class of two-dimensional and three-dimensional coordinate separable curvilinear structures which, though highly idealized, incorporate a range-dependent
bottom topography with a particular range-dependent surface impedance. Inclusion of a surface impedance condition (though of a
special form) is a nontrivial departure from the usual rigid bottom
assumption in such cases because it can account phenomenologically
for partial transmission (with or without absorption) into the bottom, and also for the existence of an interface surface mode guided
along the bottom boundary.

    The formulation of the problem in separable curvilinear coor-
dinates, as carried out in Section II, leads to a general representation of the time harmonic Green's function for the pressure field
as complex spectral integrals over products of individual one-dimensional characteristic Green's functions. Alternative representations derived from this general form lead to spectral objects that
can be interpreted in terms of the basic wave phenomena listed
above. Some remarks are added about extension of these prototype
solutions to weakly nonseparable structures. The general constructs
are made specific in Section III for a slanted planar (ocean wedge)
and hyperbolic cylinder (ocean ridge) bottom configuration, and for
a three-dimensional conical and hyperbolic spheroidal seamount.
Numerical results for a rigid conical seamount, generated from horizontal ray asymptotics based on the exact solution, are presented
in Section IV. These results compare favorably with those produced
by Buckingham [1,2] from an entirely different integral represent-

ation, and they thereby dispute his assertion that "it would be
difficult to make quantitative comparisons because the ray description does not readily yield amplitude or phase information" [1].
Concluding remarks are made in Section V.

II.   RIGOROUS FORMULATION

A.   Statement of the Problem
      We consider the Green's function problem in two dimensions

$$(\nabla_\rho^2 + k^2)G_2(\rho,\rho') = -\delta(\rho - \rho') \tag{1a}$$

and in three dimensions

$$(\nabla_r^2 + k^2)G(r,r') = -\delta(r - r') \tag{1b}$$

where $\rho = (u,v)$ and $r = (\rho,w)$ are two and three-dimensional orthogonal coordinates, respectively. In configurations with coordinate separable boundary conditions, general solutions for these
two and three-dimensional Green's functions can be expressed as
quadratures over the constituent one-dimensional characteristic
Green's functions [3, Sec. 3.3].

B.   Alternative Representations
1.   One-Dimensional Characteristic Green's Function
     (a)   Exact formulation: The typical one-dimensional charac-
teristic Green's function problem is summarized below. It yields
representation theorems in integral or eigenfunction form, thereby
providing a variety of options for alternative representations of
the multidimensional Green's function in (1a,b). The characteristic Green's function is defined as follows:

$$\left[\frac{d}{dx}\, p(x)\, \frac{d}{dx} + \zeta q(x) + \lambda_x t(x)\right] g_x(x,x';\zeta;\lambda_x) = -\delta(x - x') \tag{2}$$

with boundary conditions,

$$\left(p(x)\, \frac{d}{dx} + \alpha_{1,2}\right) g_x(x,x';\zeta;\lambda_x) = 0 \qquad x = x_{1,2} \tag{2a}$$

ζ is a fixed parameter and λ_x is the characteristic spectral variable whose domain in the complex λ_x-plane is restricted so as to
ensure a unique solution for g_x. The variable x stands for any of
the individual (u,v) or (u,v,w) coordinates that appear in the two
or three-dimensional problem, respectively. Let $\overleftarrow{g}_x$ and $\overrightarrow{g}_x$ be two
independent solutions of (2) which satisfy the boundary conditions
(2a) at x_1 and x_2, respectively. Then,

$$g_x = \frac{\overleftarrow{g}_x(x_<)\,\overrightarrow{g}_x(x_>)}{-pW(\overleftarrow{g}_x,\overrightarrow{g}_x)} \tag{3}$$

where the Wronskian is

$$W(\overleftarrow{g}_x,\overrightarrow{g}_x) = \overleftarrow{g}_x \frac{d}{dx}\,\overrightarrow{g}_x - \frac{d\overleftarrow{g}_x}{dx}\,\overrightarrow{g}_x \tag{3a}$$

and $x_<$ and $x_>$ represent the smaller and the greater, respectively, of x and x'. The corresponding eigenvalue problem is:

$$\left[\frac{d}{dx}\,p(x)\,\frac{d}{dx} + \zeta q(x) + \lambda_{x_m} t(x)\right]\Phi_{x_m}(x,\zeta) = 0 \tag{4}$$

with the same boundary conditions as those on g_x in (2a). Here Φ_{x_m} is the eigenfunction and λ_{x_m} the eigenvalue. The completeness relation for the one-dimensional x-domain can be stated in the form of a weighted delta function representation that involves either a general spectral integration or an orthonormal eigenfunction expansion (with $\overline{\Phi}_{x_m}$ denoting the adjoint eigenfunction):

$$\frac{\delta(x - x')}{t(x')} = -\frac{1}{2\pi i} \oint_{C_x} g_x(x,x';\zeta;\lambda_x)\,d\lambda_x$$

$$= \sum_m \Phi_{x_m}(x,\zeta)\,\overline{\Phi}_{x_m}(x',\zeta) \tag{5}$$

The contour C_x in (5) encircles all singularities (poles at λ_{x_m} and (or) branch points) of g_x in the complex λ_x-plane in the positive sense, and the discrete and (or) continuous eigenfunction expansion in (5) is obtained therefrom by residue and (or) branch cut evaluation. In the latter instance, the discrete summation index is replaced by a continuous (integration) index.

(b) Asymptotic approximation: The solutions of the homogeneous equation (2), which synthesize the characteristic Green's function g_x in (3), generally are in the form of transcendental functions. For ease of computation as well as physical interpretation of the wave phenomena described by these functions, it is desirable to explore asymptotic techniques emphasizing "high frequency." This can be done systematically by first expressing the defining equation in standard form via the transformation

$$g_x = \hat{g}_x/\sqrt{p(x)} \tag{6}$$

which leads to

$$\left[\frac{d^2}{dx^2} + \gamma^2(x)\right]\hat{g}_x = \frac{-1}{\sqrt{p(x')}}\,\delta(x - x')$$ (7)

where

$$\gamma^2(x) = \frac{\zeta q(x) + \lambda_x t(x)}{p(x)} - \frac{1}{\sqrt{p(x)}}\,\frac{d^2}{dx^2}\sqrt{p(x)}$$ (7a)

For large values of the parameters ζ and (or) λ_x appearing in $\gamma(x)$, one may employ asymptotic procedures for approximating $\hat{g}_x$. These depend critically on whether, or not, the coordinate variables (x,x') are located near zeros (if any) of $\gamma(x)$. Away from such zeros, $\hat{g}_x$ can be expressed by standard propagating or decaying WKB local plane wave functions. In the transition regions near such zeros (turning points), uniform approximations are required whose detailed behavior depends on whether the turning points are isolated or clustered (if there is a multiplicity). Details of the general procedure may be found in the literature [3, Sec. 3.5], and specific examples are given in the context of special configurations treated later on.

2. Multidimensional Green's Functions

The multidimensional Green's functions can be synthesized in terms of the one-dimensional characteristic solutions. In the two-dimensional (u,v) space,

$$G_2(\underset{\sim}{\rho},\underset{\sim}{\rho}') = \frac{-1}{2\pi i}\oint_{C_v} g_v(v,v';\lambda_v)g_u(u,u';\lambda_u)d\lambda_v$$ (8a)

where the contour C_v in the complex λ_v plane encloses in the positive sense all singularities of g_v but no others. Additional singularities in the λ_v plane arise due to $g_u(u,u';\lambda_u)$ where λ_u is a function of λ_v. From (8a), alternative representations can be derived by contour deformation and residue or branch cut evaluation. For example,

$$G_2 = \sum_m G_{2m} \qquad G_{2m}(\underset{\sim}{\rho},\underset{\sim}{\rho}') = \sum_m \Phi_{v_m}(v)\overline{\Phi}_{v_m}(v')g_u(u,u';\lambda_{u_m})$$ (8b)

results from residue evaluation around the g_v singularities and emphasizes propagation (transmission) along the u coordinate. Similarly, in the three-dimensional (u,v,w) space,

$$G(\underset{\sim}{r},\underset{\sim}{r}') = \frac{1}{(2\pi i)^2}\oint_{C_w}\oint_{C_v} g_u(u,u';\lambda_u)g_v(v,v';\lambda_v)$$

$$\times\, g_w(w,w';\lambda_w)d\lambda_v d\lambda_w$$ (9a)

and alternatively,

$$G(\underset{\sim}{r},\underset{\sim}{r}') = \sum_m G_m(\underset{\sim}{r},\underset{\sim}{r}')$$

$$G_m(\underset{\sim}{r},\underset{\sim}{r}') = \frac{1}{2\pi i}\, \Phi_{v_m}(v)\bar{\Phi}_{v_m}(v') \int_{C_w} g_u(u,u';\lambda_{u_m})g_w(w,w';\lambda_w)d\lambda_w \tag{9b}$$

with the further reduction

$$G_m(\underset{\sim}{r},\underset{\sim}{r}') = \sum_p G_{mp}(\underset{\sim}{r},\underset{\sim}{r}')$$

$$G_{mp}(\underset{\sim}{r},\underset{\sim}{r}') = [\Phi_{v_m}(v)\Phi_{w_p}(u)][\bar{\Phi}_{v_m}(v')\bar{\Phi}_{w_p}(u')]g_u(u,u';\lambda_{u_{mp}}) \tag{9c}$$

The contour $C_w(C_v)$ in the complex $\lambda_w(\lambda_v)$ plane encloses in the positive sense all singularities of $g_w(g_v)$ but no others. Additional singularities in the λ_w or λ_v planes arise from $g_u(u,u';\lambda_u)$. The functional dependence $\lambda_u = \lambda_u(\lambda_v,\lambda_w)$ of λ_u on λ_v and λ_w is dictated by the nature of the coordinate representation in the (u,v,w) domains. As before, (9a) can be employed to derive alternative representations. Propagation along the u coordinate is emphasized by G_{mp} in (9c) while propagation in the two-dimensional u-w plane is emphasized by G_m in (9b). The latter representation is useful for charting the lateral "modal rays" that will be explored in detail for shallow ocean channels with lateral inhomogeneities. Comparing (8a) and (9a), one concludes that the generalization from the two-dimensional to the three-dimensional problem or, conversely, the reduction from three dimensions to two dimensions is straightforward in concept. In what follows, we shall deal primarily with the three-dimensional case.

3. Ray and Hybrid Ray-Mode Representations

When the spectrum of the operator in the v coordinate or in both the v and w coordinates is discrete, the resulting eigenfunction representation, which is in the form of a series of "guided modes" G_m or G_{mp} that propagate in the u-w plane or only along the u coordinate, respectively, may be slowly convergent. It is possible to derive an alternative series based on power series expansion of the "resonant denominators" in g_v or g_w, whose zeros at the eigenvalues λ_{v_m} and λ_{w_p} generate the spectral poles. After substituting into the remaining spectral integral (9a) or (9b), the elements in the single or double series can be interpreted as generalized ray integrals because their asymptotic approximation at high frequencies is structured around ray trajectories. Decomposing G_m in (9b) in this manner leads to a generalized lateral ray series representation of the mth "guided mode," propagating in the u-w plane, in the form

$$G_m(\underset{\sim}{r}, \underset{\sim}{r}') = \sum_{\ell} G_m^{(\ell)}$$

$$G_m^{(\ell)} = \oint_{C_w} A_m^{(\ell)}(\lambda_w) \, \exp[i\Psi_m^{(\ell)}(\lambda_w)] d\lambda_w \tag{10}$$

where the integrand has been grouped so that $A_m^{(\ell)}$ and $\Psi_m^{(\ell)}$ express a slowly varying amplitude and rapidly varying phase, respectively. Such a grouping can be effected, for example, by employing WKB approximations [see Section II.B.1(b)] for the functions that comprise G_m, regarding $\Psi_m^{(\ell)}$ as the composite WKB phase, and the exact integrand multiplied by $\exp[-i\Psi_m^{(\ell)}]$ as the amplitude $A_m^{(\ell)}$. By the stationary phase approximation,

$$G_m^{(\ell)} \sim \left[\left(\frac{-2\pi}{i \dfrac{d^2}{d\lambda_w^2} \Psi_m^{(\ell)}} \right)^{1/2} A_m^{(\ell)} \, \exp(i\Psi_m^{(\ell)}) \right]_{\lambda_w = \lambda_{ws}} \tag{11}$$

where λ_{ws} is the stationary phase point derived from

$$\left[\frac{d}{d\lambda_w} \Psi_m^{(\ell)} \right]_{\lambda_w = \lambda_{ws}} = 0 \tag{11a}$$

The stationary phase condition in (11a) determines the lateral ray trajectory traversed by the composite field $G_m^{(\ell)}$ in the (u,w) plane. The expansions of G_m in (9c) and (10) afford alternative options that emphasize via the elementary solutions G_{mp} and $G_m^{(\ell)}$ guided (modal) or freely propagating (ray) features of the u-w plane field, respectively. Rigorously constructed hybrid combinations of some of each are likewise possible [4]. For many given ranges of physical parameters, the choice can be made so as to achieve favorable overall convergence and a cogent description of the propagation phenomenology.

C. Adiabatic Extension to Weak Nonseparability

When a general physical environment can be modeled as a weak deviation from a strictly separable canonical configuration, one may employ spectral scalings and adiabatic invariants to account approximately for the weak nonseparability, generally manifested in the lateral range dependence [5]. Within the present format, this implies assumption of <u>local separability</u> between the depth coordinate v and range coordinate $\underset{\sim}{\xi} = (u,w)$. The constituent separate equations become accordingly,

$$[\nabla_v^2 + \lambda_v(\xi)]g_v(v,v';\lambda_v(\xi)) = -\delta(v - v') \tag{12a}$$

$$[\nabla_\xi^2 + k^2 - \lambda_v(\xi)]g_\xi(\xi,\xi';\lambda_v(\xi)) = -\delta(\xi - \xi') \tag{12b}$$

where the spectral separation parameter λ_v is no longer a constant but a function of the range coordinate ξ. The dependence of λ_v on ξ is determined from an adiabatic invariant. If the problem is two-dimensional, $\xi = u$ and (12b) reduces to a one-dimensional equation as in (2). The spectral value λ_v in (12a) associated with the depth coordinate and the eigenvalue λ_{v_m} are dependent on range. To satisfy reciprocity in the overall Green's function, additional symmetrizing factors are introduced to furnish the following expression [5],

$$G_2(\rho,\rho') \approx \frac{1}{2\pi i} \oint g_v(v,v';\lambda_v(u))g_u(u,u';\lambda_u)[d\lambda_v(u)d\lambda_v(u')]^{1/2} \tag{13}$$

where the symmetrized symbolic derivative $[d\lambda_v(u)d\lambda_v(u')]^{1/2}$ is defined as

$$[d\lambda_v(u)d\lambda_v(u')]^{1/2} = \left[\frac{\partial F(q(u))/\partial q(u)}{\partial F(q(u'))/\partial q(u')}\right]^{1/2} dq(u) \tag{13a}$$

or

$$[d\lambda_v(u)d\lambda_v(u')]^{1/2} = \left[\frac{\partial F(q(u'))/\partial q(u')}{\partial F(q(u))/\partial q(u)}\right]^{1/2} dq(u') \tag{13b}$$

Here, $q = (\lambda_v)^{1/2}$, and the formulations in (13a) and (13b) are based on using the receiver range u and the source range u', respectively, as the reference coordinate. The function F, descriptive of a plane wave propagating through a complete cycle along the depth coordinate, represents essentially the denominator in (3) with $x \equiv v$, and the condition

$$F(q(u)) = F(q(u')) = \text{constant} \tag{14}$$

scales the spectrum with range so that it adapts locally to the changing guiding environment. Similar to (8b), the resonances generated by the zeros of F yield as residues in (13),

$$G_2 = \sum_m G_{2m} \tag{15}$$

$$G_2(\rho,\rho') = \sum_m \Phi_{v_m}(v,q(u))\bar{\Phi}_{v_m}(v',q(u'))g_u(u,u';\lambda_{u_m})$$

Near the cutoff range of one of the modes, where its resonance pole approaches another singularity in the intergrand, that portion of

the spectrum must be retained in "intrinsic mode" integral form
[5-7] and thereby describes the uniform transition through cutoff.

In three dimensions, if the two-dimensional wave equation
(12b) in the lateral domain is separable (e.g., through translational or rotational symmetry in the medium), then the integral
representation of the three-dimensional Green's function becomes

$$
G(\underset{\sim}{r},\underset{\sim}{r}') = \frac{1}{(-2\pi i)^2} \oint_{C_w} \oint_{C_v} g_u(u,u';\lambda_u) g_w(w,w';\lambda_w)
$$

$$
\times\, g_v(v,v';\lambda_v(u,w))\, d\lambda_w [d\lambda_v(u,w)\, d\lambda_v(u',w')]^{1/2} \tag{16}
$$

The integral (16) can then be treated in a manner similar to that
for the two-dimensional case, combined with the options in Section
II.B.2. If the two-dimensional wave equation (12b) for the given
configuration is not separable, i.e., in an environment with truly
three-dimensional heterogeneity, one may construct the approximate
Green's function with the use of adiabatic transforms [5].

III. APPLICATIONS

The general considerations in Section II are now applied to
various separable geometrical configurations, which can serve as
prototype models for range-dependent acoustic propagation in a shallow ocean with a sloping bottom described by a variable (separable)
surface impedance. Typically, these examples model a coastal slope
(wedge) (Fig. 1), a smooth ridge (hyperbolic cylinder) (Fig. 2),
and rotationally symmetric peaked (conical) and smooth (spheroidal)
seamounts (Figs. 3 and 4, respectively). As stated in the Introduction, the main purpose here is to construct exact canonical reference solutions that could serve as benchmarks for more general
numerical codes, and also to gain an understanding of range-dependent propagation mechanisms in three-dimensionally varying bottom
environments. The results are listed in a tabular format since
they follow the pattern established in Section II.

The curvilinear coordinates, three-dimensional wave equations
and boundary conditions for various configurations are shown in
Table I. The three one-dimensional characteristic problems and
their solutions (exact or asymptotic) are summarized in Tables II-
IV. In the range coordinate u (see Table II), the domain extends
from the origin to infinity. In the high frequency regime, the
Green's function can be approximated by Airy type solutions [3,
Sec. 3.5]. In essence, the domain of interest is then bounded below by the vicinity of the turning point u_t farthest from the origin.

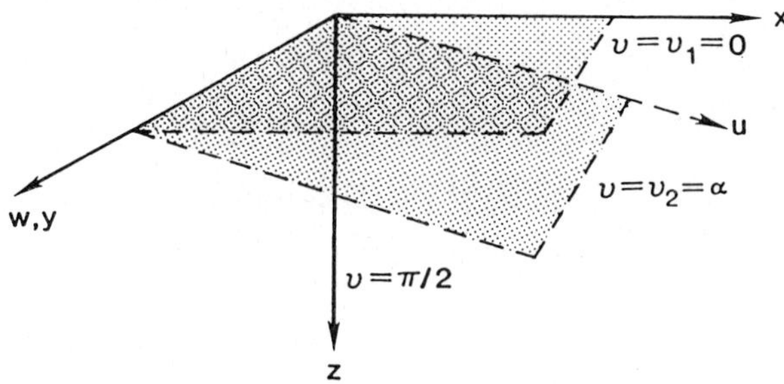

Fig. 1. Wedge configuration.

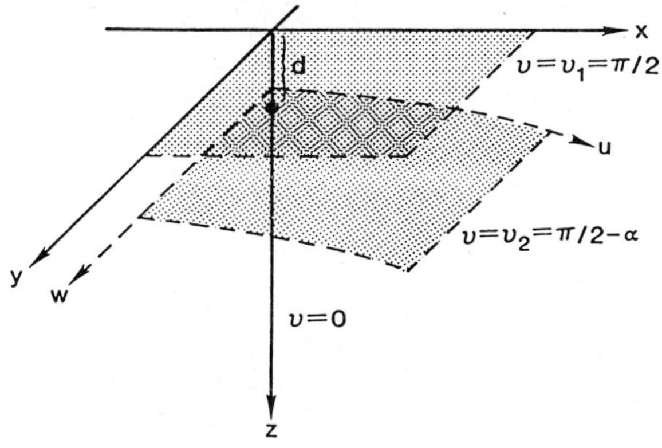

Fig. 2. Hyperbolic cylinder configuration.

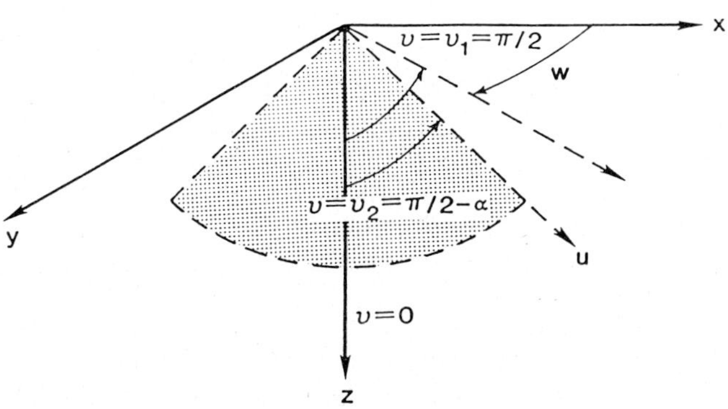

Fig. 3. Cone configuration.

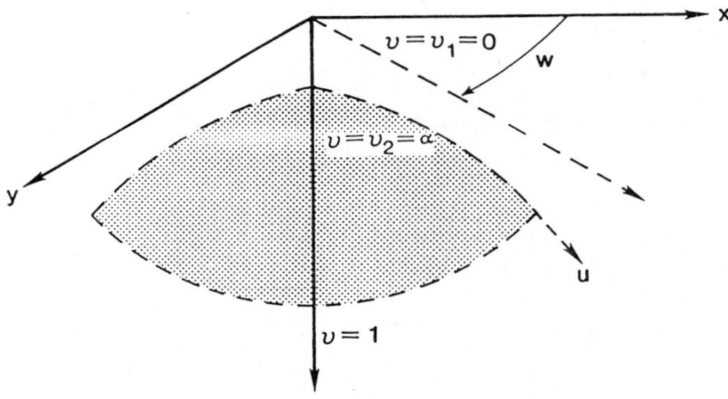

Fig. 4.   Hyperbolic spheroid configuration.

When both source and receiver are far from the turning point, one
can effect further simplification via the asymptotic (WKB) form of
the Airy functions with negative large argument [3, Sec. 3.5].   In
the depth coordinate v (see Table III), the relevant domain is limited to small angles and likewise allows use of the WKB approximation.   This approximation and restriction is unnecessary for the
wedge configuration, for which the asymptotic solution is exact.
Note that all examples are based on the same procedure for solving
the range and depth Green's functions.   In the lateral coordinate
w (see Table IV), translational symmetry gives rise to a continuous
spectrum whereas rotational symmetry gives rise to a discrete spectrum.   As mentioned in Section II.B.3, in the latter instance, one
may employ power series expansion of the resonance denominator to
generate a more rapidly convergent and physically incisive alternative representation.

From Sections II.B.2, II.B.3, and Tables I-IV, the three-dimensional Green's function is given in the form of (9b) and (10),

$$G = \sum_m G_m(\underset{\sim}{r},\underset{\sim}{r}')$$

$$G_m(\underset{\sim}{r},\underset{\sim}{r}') \approx \sum_{\ell=1}^{2} \int_{-\infty}^{\infty} A_m^{(\ell)}(\beta)\ \exp[i\Psi_m^{(\ell)}]\ d\beta \qquad \beta = \sqrt{\lambda_w} \qquad (17)$$

with (slowly varying) amplitude terms

$$A_m^{(1)}(\beta) = \frac{\Phi_{v_m}(v)\,\overline{\Phi}_{v_m}(v')}{4\pi(2/3)^{1/3}i\,[\gamma_u\gamma_{u'}\,p(u)p(u')]^{1/2}}$$

$$\tag{17a}$$

$$A_m^{(2)} = -iA_m^{(1)}$$

Table I - The Curvilinear Coordinates, Wave Equations and
Boundary Conditions for Various Bottom Configurations

Configuration	Translational Symmetry		Rotational Symmetry	
	Wedge	Hyperbolic Cylinder	Cone	Hyperbolic Spheroid
Curvilinear Coordinates (u,v,w) and Rectangular Coordinates (x,y,z)	Cylindrical $x=u\cos v,\ 0\le u<\infty$ $y=w,\quad 0\le v\le 2\pi$ $z=u\sin v,\ -\infty\le w\le\infty$ (see Fig. 1)	Elliptic $x=d\sinh u\sin v,\ 0\le u<\infty$ $y=w,\quad 0\le v\le\pi$ $z=d\cosh u\cos v,\ -\infty\le w\le\infty$ (see Fig. 2)	Spherical $x=u\sin v\cos w,\ 0\le u<\infty$ $y=u\sin v\sin w,\ 0\le v\le\pi$ $z=u\cos v,\quad 0\le w\le 2\pi$ (see Fig. 3)	Hyperbolic Spheroidal $x=d[1-v^2)(u^2-1)]^{1/2}\cos w,\ 1\le u<\infty$ $y=d[(1-v^2)(u^2-1)]^{1/2}\sin w,\ -1\le v\le 1$ $z=duv,\quad 0\le w\le 2\pi$ (see Fig. 4)
Metric Coefficients	$h_u=1,\ h_v=u,\ h_w=1$	$h_u=h_v=d[\sinh^2u+\sin^2v]^{1/2},\ h_w=1$	$h_u=1,\ h_v=u,\ h_w=u\sin v$	$h_u=d\left(\dfrac{u^2-v^2}{u^2-1}\right)^{1/2},\ h_v=d\left(\dfrac{u^2-v^2}{1-v^2}\right)^{1/2}$ $h_w=d[(1-v^2)(u^2-1)]^{1/2}$
3-Dim Wave Equations	$\left[\dfrac{1}{u}\dfrac{\partial}{\partial u}u\dfrac{\partial}{\partial u}+\dfrac{1}{u^2}\dfrac{\partial^2}{\partial v^2}+\dfrac{\partial^2}{\partial w^2}+k^2\right]G$ $=-\dfrac{1}{u}\delta(u-u')\delta(v-v')\delta(w-w')$	$\left[\dfrac{1}{d^2(\sinh^2u+\sin^2v)}\left(\dfrac{\partial^2}{\partial u^2}+\dfrac{\partial^2}{\partial v^2}\right)+\dfrac{\partial^2}{\partial w^2}+k^2\right]G$ $=-\dfrac{\delta(u-u')\delta(v-v')\delta(w-w')}{d^2(\sinh^2u+\sin^2v)}$	$\left[\dfrac{1}{u^2}\dfrac{\partial}{\partial u}u^2\dfrac{\partial}{\partial u}+\dfrac{1}{u^2\sin v}\dfrac{\partial}{\partial v}\sin v\dfrac{\partial}{\partial v}\right.$ $\left.+\dfrac{1}{u^2\sin^2v}\dfrac{\partial^2}{\partial w^2}\right]G$ $+k^2G=-\dfrac{\delta(u-u')\delta(v-v')\delta(w-w')}{u^2\sin v}$	$\left[\dfrac{\partial}{\partial u}(u^2-1)\dfrac{\partial}{\partial u}+\dfrac{\partial}{\partial v}(1-v^2)\dfrac{\partial}{\partial v}\right.$ $\left.+\dfrac{(u^2-v^2)}{(1-v^2)(u^2-1)}\dfrac{\partial^2}{\partial w^2}\right]G$ $+k^2G=\dfrac{-\delta(u-u')\delta(v-v')\delta(w-w')}{d}$

Boundary Conditions					
	u	Radiation Condition at $u=\infty$; Finiteness Condition at u_{min}			
		$u_{min}=0$	$u_{min}=0$	$u_{min}=0$	$u_{min}=1$
	v	$G=0$ at $v=v_1$: $\dfrac{1}{h_v}\dfrac{\partial G}{\partial v}=aG$, $a=\dfrac{b}{h_v}$, at $v=v_2$			
		$v_1=0,\ v_2=\alpha$	$v_1=\dfrac{\pi}{2},\ v_2=\dfrac{\pi}{2}-\alpha$	$v_1=\dfrac{\pi}{2},\ v_2=\dfrac{\pi}{2}-\alpha$	$v_1=0,\ v_2=\alpha$
	w	Radiation Condition at $w=\pm\infty$		Periodicity Condition: $\dfrac{\partial G}{\partial w}=0$, $|w-w'|=\pi$	

Table II - One-Dimensional Characteristic Problem

In Range Domain u

Configuration	Wedge	Hyperbolic Cylinder	Cone	Hyperbolic Spheroid
Characteristic Equation	$\left[\dfrac{d}{du}u\dfrac{d}{du}+(k^2-\lambda_w)u-\dfrac{\lambda_v}{u}\right]g_u$ $=-\delta(u-u')$	$\left[\dfrac{d^2}{du^2}+d^2(k^2-\lambda_w)\sinh^2 u-\lambda_v\right]g_u$ $=-\delta(u-u')$	$\left[\dfrac{d}{du}u^2\dfrac{d}{du}+k^2u^2-\lambda_v\right]g_u$ $=-\delta(u-u')$	$\left[\dfrac{d}{du}(u^2-1)\dfrac{d}{du}-\dfrac{\lambda_w}{(u^2-1)}+d^2k^2u^2-\lambda_v\right]g_u$ $=-\delta(u-u')$
Transformation	(from (6))	$\mathfrak{g}_u=\sqrt{p(u)}\,g_u$		
Standard Form	(from (7))	$\left[\dfrac{d^2}{du^2}+\gamma_u^2\right]\mathfrak{g}_u=\dfrac{-1}{\sqrt{p(u')}}\delta(u-u')$		
$p(u)$	u	1	u^2	u^2-1
γ_u^2	$(k^2-\lambda_w)-\dfrac{\lambda_v-\frac{1}{4}}{u^2}$	$d^2(k^2-\lambda_w)\sinh^2 u-\lambda_v$	$k^2-\dfrac{\lambda_v}{u^2}$	$\dfrac{k^2d^2u^4-(k^2d^2+\lambda_v)u^2+(1+\lambda_v-\lambda_w)}{(u^2-1)^2}$
Boundary Condition for $\mathfrak{g}_u$ and g_u	Radiation Condition at $u=\infty$; Finiteness Condition at u_{min}			
	$u_{min}=0$	$u_{min}=0$	$u_{min}=0$	$u_{min}=1$
Green's Function (from (3))[*]	$g_u=\dfrac{\mathfrak{g}_u}{\sqrt{p(u)}}$, $\mathfrak{g}_u=\dfrac{-1}{\sqrt{p(u)}}\dfrac{\mathfrak{g}_u(u_<)\mathfrak{g}_u(u_>)}{W(\mathfrak{g}_u,\mathfrak{g}_u)}$, $\mathfrak{g}_u\approx\dfrac{Y^{1/6}}{\gamma_u^{1/2}}\,Ai(-Y^{2/3})$, $\mathfrak{g}_u\approx\dfrac{Y^{1/6}}{\gamma_u^{1/2}}\left[Ai(-Y^{2/3})-iBi(-Y^{2/3})\right]$, $Y=\dfrac{3}{2}\displaystyle\int_{u_t}^{u}\gamma_u du$, $W=\dfrac{1}{\pi}(\dfrac{2}{3})^{1/3}$			
Largest Turning Point u_t	$\left[(\lambda_v-\tfrac{1}{4})/(k^2-\lambda_w)\right]^{1/2}$	$\dfrac{1}{2}\ell n\left\{\left[2\dfrac{\lambda_v}{(k^2-\lambda_w)d^2}+1\right]+\left[\left[2\dfrac{\lambda_v}{(k^2-\lambda_w)d^2}+1\right]^2-1\right]^{1/2}\right\}$	$\dfrac{\sqrt{\lambda_v}}{k}$	$\dfrac{1}{2}\left[(1+\dfrac{\lambda_v}{k^2d^2})+\sqrt{(1+\dfrac{\lambda_v}{k^2d^2})^2-\dfrac{4-(\lambda_v-\lambda_w+1)}{k^2d^2}}\right]^{1/2}$
Away from turning point $Y^{2/3}\to\infty$	Airy Functions $Ai(-Y^{2/3})\sim\dfrac{1}{\sqrt{\pi}\,Y^{1/6}}\sin\left(\dfrac{2}{3}Y+\dfrac{\pi}{4}\right)$, $Bi(-Y^{2/3})\sim\dfrac{1}{\sqrt{\pi}\,Y^{1/6}}\cos\left(\dfrac{2}{3}Y+\dfrac{\pi}{4}\right)$, $g_u\sim\dfrac{1}{2\left(\frac{2}{3}\right)^{1/3}}\dfrac{-i\exp\left[i\int_{u_t}^{u}\gamma_u du+i\int_{u_t}^{u'}\gamma_u du\right]+\exp\left[i\int_{u_<}^{u_>}\gamma_u du\right]}{\sqrt{p(u)p(u')}\gamma_u\gamma_{u'}}$			

[*]Contour integral form of completeness relation in terms of the Green's function is analogous to that shown in Table IV

Table III - One-Dimensional Characteristic Problem

In Depth Domain v

Configuration	Wedge	Hyperbolic Cylinder	Cone	Hyperbolic Spheroid
Characteristic Equation	$\left(\dfrac{d^2}{dv^2}+\lambda_v\right)g_v=-\delta(v-v')$	$\left[\dfrac{d^2}{dv^2}+d^2(k^2-\lambda_w)\sin^2 v+\lambda_v\right]g_v$ $=-\delta(v-v')$	$\left[\dfrac{d}{dv}\sin v\dfrac{d}{dv}-\dfrac{\lambda_w}{\sin v}+\lambda_v\sin v\right]g_v$ $=-\delta(v-v')$	$\left[\dfrac{d}{dv}(1-v^2)\dfrac{d}{dv}-\dfrac{\lambda_w}{1-v^2}-d^2k^2v^2+\lambda_v\right]g_v$ $=-\delta(v-v')$
Boundary Condition	$g_v=0$ at $v=v_1$; $\dfrac{dg_v}{dv}=bg_v$ at $v=v_2$			
	$v_1=0$, $v_2=\alpha$	$v_1=\dfrac{\pi}{2}$, $v_2=\dfrac{\pi}{2}-\alpha$	$v_1=\dfrac{\pi}{2}$, $v_2=\dfrac{\pi}{2}-\alpha$	$v_1=0$, $v_2=\alpha$
Transformation	(from (6))	$g_v=\sqrt{p(u)}\,g_v$		
Standard Form	(from (7))	$\left[\dfrac{d^2}{dv^2}+\gamma_v^2\right]g_v=\dfrac{-1}{\sqrt{p(v')}}\delta(v-v')$		
$p(v)$	1	1	$\sin v$	$(1-v^2)$
γ_v^2	λ_v	$d^2(k^2-\lambda_w)\sin^2 v+\lambda_v$	$\lambda_v+\dfrac{1}{4}-\dfrac{\lambda_w-\frac{1}{4}}{\sin^2 v}$	$\dfrac{k^2d^2v^4-(k^2d^2+\lambda_v)v^2+(\lambda_v-\lambda_w+1)}{(1-v^2)^2}$
Boundary Condition	$g_v=0$ at $v=v_1$; $\dfrac{dg_v}{dv}=\dot{b}\,g_v$ at $v=v_2$			
	$\dot{b}=b$	$\dot{b}=b$	$\dot{b}=b+\dfrac{1}{2}\cot v$	$\dot{b}=b-\dfrac{v}{1-v^2}$
Green's Function (from (3))*	$g_v=\dfrac{g_v}{\sqrt{p(v)}}$; $g_v=\dfrac{-1}{\sqrt{p(v')}}\dfrac{\mathcal{Z}_v(v_<)\mathcal{Z}_v(v_>)}{W(\mathcal{Z}_v,\mathcal{Z}_v)}$; $\mathcal{Z}_v=R_1^{-1/2}g_v^-+R_1^{1/2}g_v^+$, $\mathcal{Z}_v=R_2^{1/2}g_v^-+R_2^{-1/2}g_v^+$, $W=2i(R_1^{-1/2}R_2^{-1/2}-R_1^{1/2}R_2^{1/2})$			
Up and down going waves	$g_v^\pm=\dfrac{\exp[\pm i\gamma_v v]}{\sqrt{\gamma_v}}$	$g_v^\pm\approx\dfrac{\exp\left[\pm i\int_{v_1}^{v}\gamma_v dv\right]}{\sqrt{\gamma_v}}$ (WKB approximation)		
Reflection Coefficients	$R_1=-1$, $R_2=\dfrac{i\gamma_v-\dot{b}}{i\gamma_v+\dot{b}}\exp[2i\gamma_v v_2]$	$R_1=-1$, $R_2\approx\dfrac{i\gamma_v-\dot{b}}{i\gamma_v+\dot{b}}\exp\left[2i\int_{v_1}^{v_2}\gamma_v dv\right]$		
Eigenfunction Eigenvalue	Eigenfunction $\Phi_{vm}=\left[\dfrac{g}{-2ip\frac{\partial F}{\partial v}}\right]^{1/2}=\left[\dfrac{g}{-2ip\frac{\partial F}{\partial v}}\right]^{1/2}$ at eigenvalue $\lambda_v=\lambda_{v_m}$ which satisfies resonance condition $F=R_1R_2=1$			

*Contour integral form of completeness relation in terms of the Green's function is analogous to that shown in Table IV

Table IV - One-Dimensional Characteristic Problem
In Lateral Domain w

Configuration	Wedge and Hyperbolic Cylinder (Translational Symmetry)	Cone and Hyperbolic Spheroid (Rotational Symmetry)
Characteristic Equation	$\left(\dfrac{d^2}{dw^2}+\lambda_w\right)g_w=-\delta(w-w')$	
Boundary Condition	radiation condition at $w=\pm\infty$	periodicity $\dfrac{\partial g_w}{\partial w}=0$, $\quad |w-w'|=0$
Green's Function	$g_w=\dfrac{-e^{\,i\sqrt{\lambda_w}\,|w-w'|}}{2i\sqrt{\lambda_w}}\qquad \mathrm{Im}(\sqrt{\lambda_w})>0$	$g_w=-\dfrac{\cos\sqrt{\lambda_w}(\pi-|w-w'|)}{2\sqrt{\lambda_w}\sin\sqrt{\lambda_w}\pi}$
Completeness Relation	$\delta(w-w')=-\dfrac{1}{2\pi i}\displaystyle\oint_{C_w} g_w d\lambda_w$	
Contour C_w		
Power Series Expansion	Not Applicable	$-(\sin\sqrt{\lambda_w}\pi)^{-1}=2i\displaystyle\sum_{n=0}^{\infty}\exp[(2n+1)i\sqrt{\lambda_w}\pi]$

and (rapidly varying) phase terms

$$\psi_m^{(1)} = \beta |w - w'| + \int_{u_<}^{u_>} \gamma_u du$$

$$\psi_m^{(2)} = \beta |w - w'| + \left[\int_{u_t}^{u} + \int_{u_t}^{u'} \right] \gamma_u du$$

(17b)

These equations are valid only when u and u' are away from the turn-ing point u_t. Here, the higher order (n > 0) terms pertaining to configurations with rotational symmetry in Table IV have been ig-nored because their asymptotic contributions are of higher order (integrands have no real saddle points). The dominant asymptotic result obtained from (17) for isolated saddle points has been given in (11). Near a caustic [3, Secs. 4.5 and 5.8],

$$G_m \sim 2\pi A_m^{(2)} (\beta_{m_0}) \left[\frac{1}{d^3 \psi_m^{(2)} / d\beta^3} \right]_{\beta_{m_0}}^{1/3}$$

$$\times \exp \left[\frac{i\psi_m^{(2)} (\beta_{m_1}) + i\psi_m^{(2)} (\beta_{m_2})}{2} \right] Ai(X)$$

(18)

where

$$\beta_{m_0} = (\beta_{m_1} + \beta_{m_2})/2$$

(18a)

and

$$X = \left\{ \frac{3i}{4} [\psi_m^{(2)} (\beta_{m_1}) - \psi_m^{(2)} (\beta_{m_2})] \right\}^{2/3}$$

(18b)

Here, β_{m_1} and β_{m_2} are the two nearby saddle points which satisfy

$$\left[\frac{d}{d\beta} \psi_m^{(2)} \right]_{\beta=\beta_{m_{1,2}}} = 0 \quad \text{with} \quad \left[\frac{d^2}{d\beta^2} \psi_m^{(2)} \right]_{\beta=\beta_{m_{1,2}}} \begin{array}{c} < \\ > \end{array} 0$$

(19)

and the caustic is determined by

$$\frac{d}{d\beta} \psi_m^{(2)} = \frac{d^2}{d\beta^2} \psi_m^{(2)} = 0$$

(20)

Uniform asymptotics valid both near and away from a caustic can be found in the literature [3, Sec. 4.5].

IV. NUMERICAL EXAMPLE: CONICAL SEAMOUNT

The implications of the analytical forms presented in Sections II and III are now illustrated on a special example, the rigid cone, which has been treated recently by an alternative method [1] that is, however, less suitable for generalization to the weakly non-separable case as discussed in Section II.C. Referring to [1], we choose the following parameters:

$$\alpha = \frac{\pi}{12} \qquad b = 0 \qquad k = 1m^{-1} \tag{21}$$

and make the small angle approximation

$$1 \approx \sin\left(\frac{\pi}{2} + \alpha\right) \le \sin v \le 1 \qquad \frac{\pi}{2} \le v \le \frac{\pi}{2} + \alpha \;. \tag{22}$$

Then from Table III,

$$\gamma_v^2 \approx \lambda_v + \frac{1}{2} - \beta^2 \qquad \lambda_w = \beta^2 \tag{23a}$$

$$g_v^{\pm} \approx \frac{\exp[\pm i\gamma_v(v - \pi/2)]}{\sqrt{\gamma_v}} \tag{23b}$$

$$R_1 = -1 \qquad R_2 \approx \exp[2i\gamma_v \alpha] \tag{23c}$$

$$\Phi_{v_m} = i \sin \gamma_v v / (\gamma_v \sin^{1/2} v) \tag{23d}$$

yielding the approximate resonance condition

$$\exp[2i\gamma_{v_m} 2] \approx -1 \tag{24}$$

with the eigenvalue [see (23a) and (24)]

$$\lambda_{v_m} \approx \gamma_{v_m}^2 + \beta^2 - \frac{1}{2} \qquad \gamma_{v_m} \approx (2m - 1) \frac{\pi}{2\alpha} \tag{25}$$

From Table II,

$$\gamma_u^2 = k^2 - \frac{\gamma_v^2 - \frac{1}{2}}{u^2} - \frac{\beta^2}{u^2} \tag{26}$$

whence the phase terms for the mth mode in (17b) are

$$\psi_m^{(1,2)} = \beta|w - w'| + \left\{ [k^2 u_>^2 - \lambda_{v_m}]^{1/2} - \sqrt{\lambda_{v_m}} \cos^{-1}\left|\frac{\sqrt{\lambda_{v_m}}}{ku_>}\right| \right\}$$

$$\mp \left\{ [k^2 u_<^2 - \lambda_{v_m}]^{1/2} - \sqrt{\lambda_{v_m}} \cos^{-1}\left|\sqrt{\frac{\lambda_{v_m}}{ku_<}}\right| \right\} \tag{27}$$

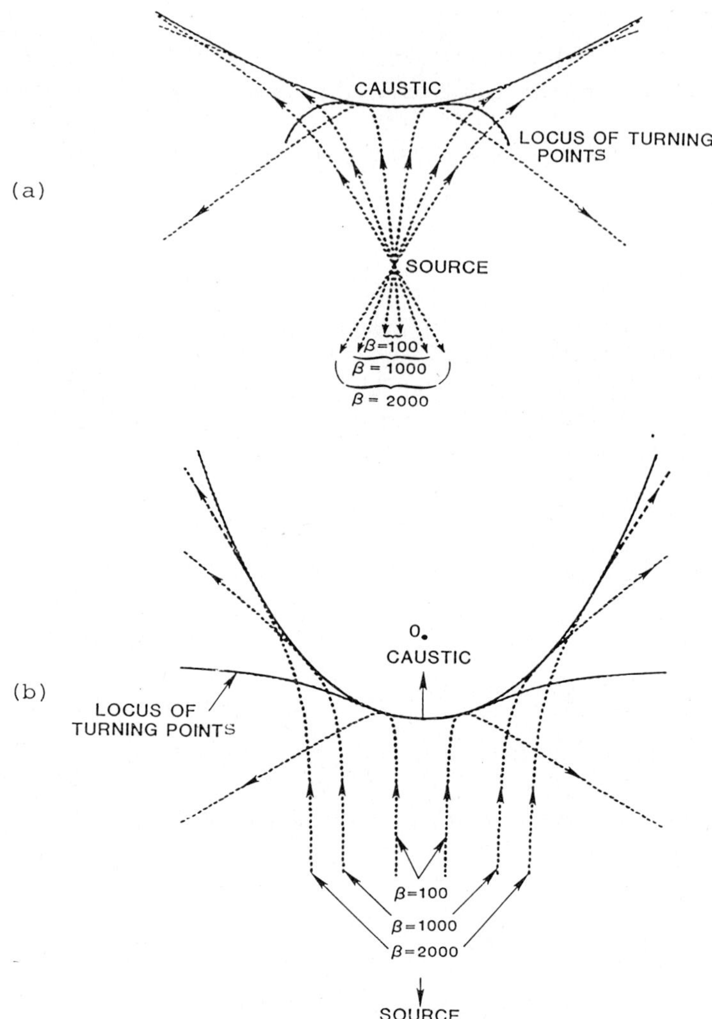

Fig. 5. Lateral rays, caustic, and turning point locus for rigid cone configuration. Cone parameters: $k = 1$, $\alpha = \pi/12$, mode $m = 3$. Three rays with parameters $\beta = 100$, 1000, and 2000 are shown. (a) $u' = 100$ m; (b) $u' = 10,000$ m.

The lateral ray trajectories traversed by the mth local mode are determined by the stationary phase conditions [see (11a)]

$$|w - w'| = \left| \frac{\beta}{\sqrt{\frac{\lambda}{v_m}}} \left\{ \cos^{-1}\left| \sqrt{\frac{\frac{\lambda}{v_m}}{ku_>}} \right| \mp \cos^{-1}\left| \sqrt{\frac{\frac{\lambda}{v_m}}{ku_<}} \right| \right\} \right| = 0 \qquad (28)$$

which are equivalent to the ray equation in [8] by noting that the constant γ_{v_m} in (25) and the constant ray parameter β in (28) are equivalent to Weston's invariant [9] and the cone invariant in [8], respectively. The caustic of the lateral rays satisfies the equation [see (20)],

$$\frac{\lambda_{v_m}^2 - \frac{1}{2}}{\lambda_{v_m}^{3/2}} \left\{ \cos^{-1}\left|\frac{\sqrt{\lambda_{v_m}}}{ku_>}\right| + \cos^{-1}\left|\frac{\sqrt{\lambda_{v_m}}}{ku_<}\right| \right\} - \frac{\beta^2}{\lambda_{v_m}} \left(\frac{1}{\sqrt{k^2 u_>^2 - \lambda_{v_m}}} + \frac{1}{\sqrt{k^2 u_<^2 - \lambda_{v_m}}} \right) = 0 \tag{29}$$

and the turning point, i.e., the location of the closest approach to the origin of a ray with ray parameter β, is given by (see Table II),

$$u_t^2 = \left[\gamma_{v_m}^2 + \beta^2 - \frac{1}{2} \right]/k^2 = \lambda_{v_m}/k^2 \tag{30}$$

Figures 5a and 5b show typical lateral ray trajectories for mode m = 3 emanating from a source located at u' = 100 m and u' = 10,000 m, respectively, with w' = 0. The corresponding caustic curves and turning point loci are shown as well. The fields for modes m = 1 through 8, observed at u = 100 m with source location at u' = 10,000 m, are plotted in Fig. 6. Equation (11) or (18) has been used in the computation when the observation point is far from or near the caustic, respectively. The two results coincide in an overlapping region near the caustic for m = 1 through 4 but not for m = 5 through 8. The lack of coincidence of (11) and (18) in a common boundary layer in the latter instance is due to the proximity at the u = 100 m range of the ray turning point locus and the caustic. Since $k^2 u^2 \approx \lambda_{v_m}$ there [see (30)], the amplitude factor in front of the exponential in (18) tends to infinity, in view of the form in (27). More sophisticated asymptotic analysis is required here. The patterns in Fig. 6 for the lower order modes agree well with those obtained in [1] by an entirely different approach except that they do not exhibit the small modulation effect that occurs in Fig. 5 of [1]. It remains to be explored, by numerical evaluation of the exact integral in (9), whether this is due to the asymptotic approximations leading to (11). In any event, our ability to derive quantitative mode fields by lateral ray asymptotics contradicts the statement in [1] that this cannot readily be done by ray techniques. Actually, our asymptotic formulas could have been developed directly by lateral ray theory [10].

I.T. Lu and L.B. Felsen

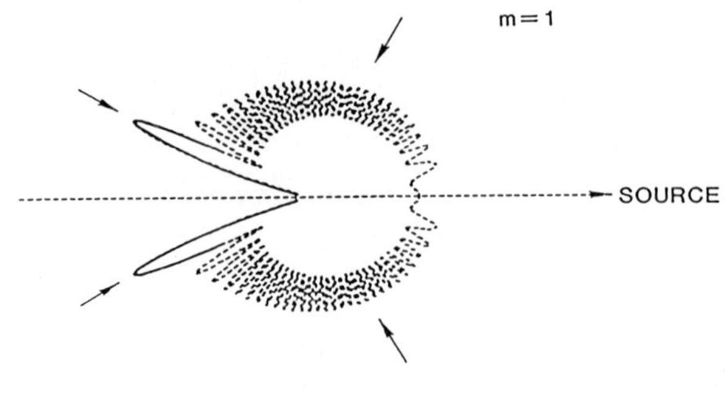

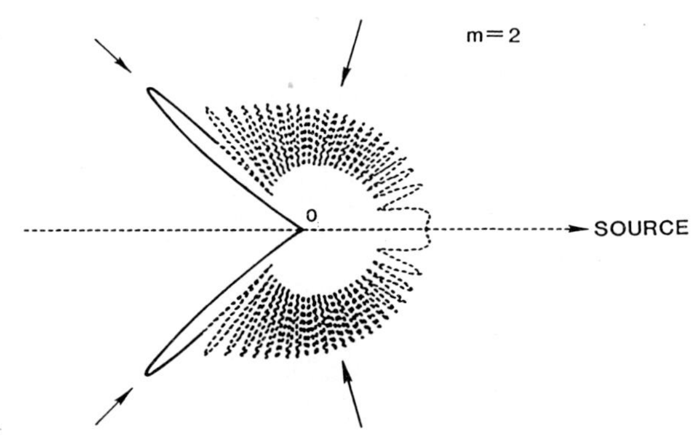

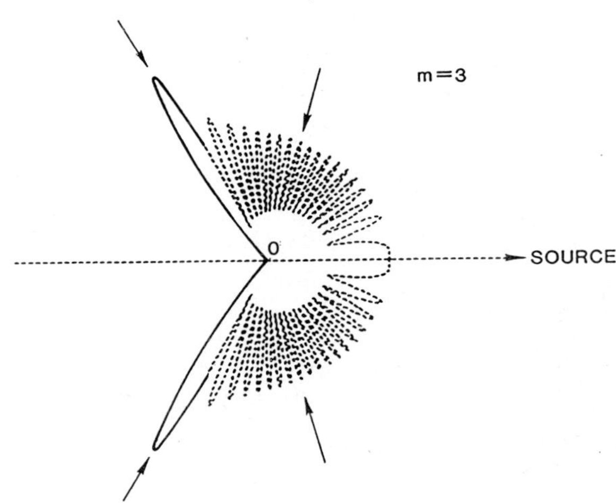

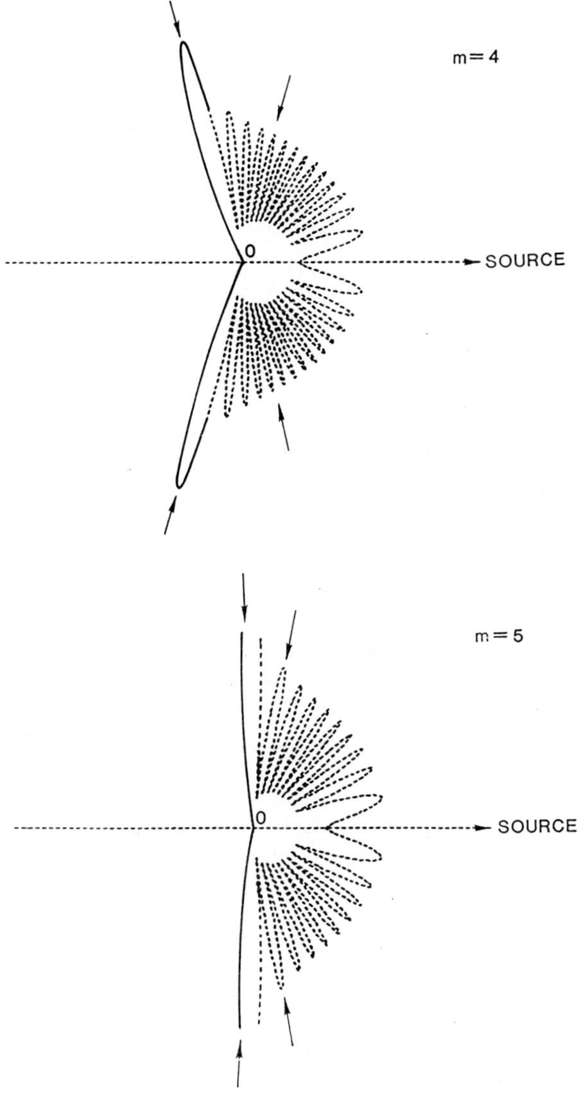

Fig. 6. Azimuthal patterns of mode fields m = 1 through 8 for rigid cone configuration, with source located at u' = 10,000 m, w' = 0, and observations made at u = 100 m. The arrows at the edges of the patterns denote the location of the caustics; the other arrows denote the locations of ray turning points at the 100 m range. Solutions computed from (11) and (18) are denoted by dashed lines and solid lines, respectively. Overlapping domains for these approximate asymptotic forms exist for m = 1-4 but not for m = 5-8.

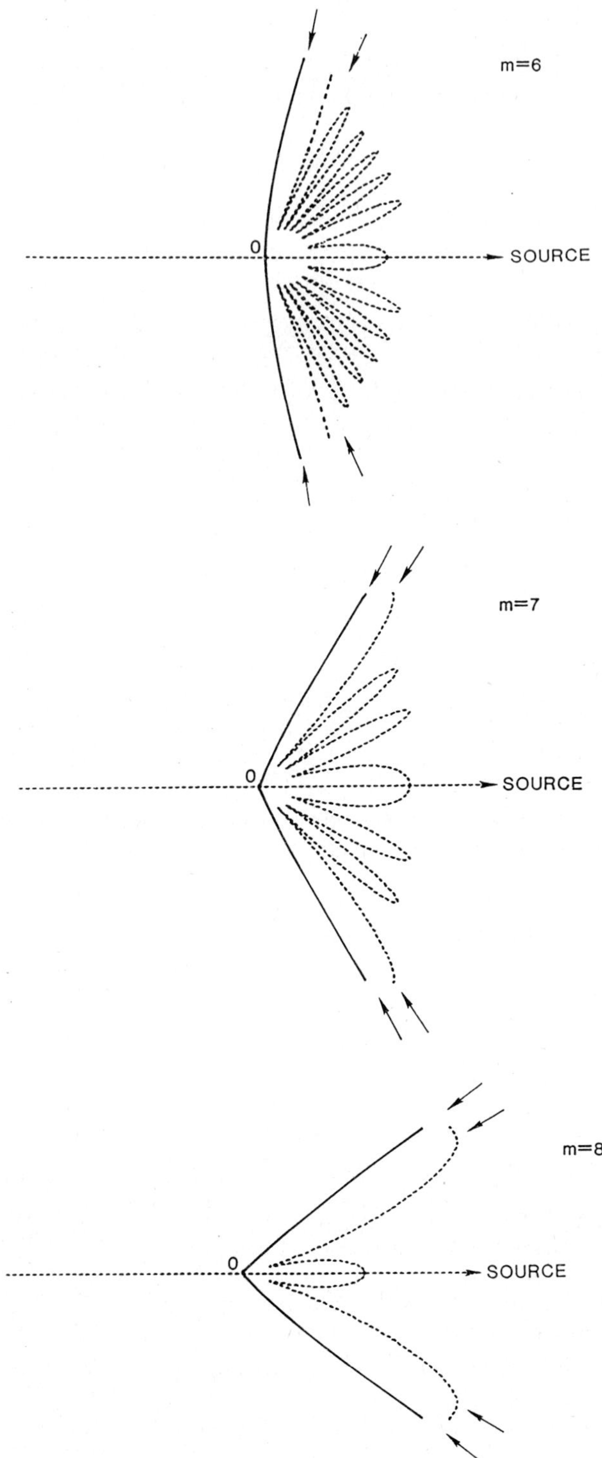

m=6
SOURCE
0
m=7
SOURCE
0
m=8
SOURCE
0

V. CONCLUDING REMARKS

We have presented here exact solutions for a class of coordinate separable two-dimensional and three-dimensional prototype configurations that can serve as idealized models for acoustic propagation in a homogeneous shallow ocean with range dependent bottom topography and bottom surface impedance. Careful numerical implementation of these exact formal solutions can produce benchmark data, with which results from more generally applicable numerical or approximate analytical schemes could be compared. Our solutions have been cast in alternative forms, from which one may extract interpretations in terms of various physical wave processes that play an important role in underwater acoustic propagation. These phenomena become explicit when our exact spectral objects are reduced by asymptotic approximations; again, the quality of such approximations can be assessed by comparison with the rigorous generating prototype. Special attention has been given to formulations that highlight horizontal refraction in three-dimensional problems, and to their asymptotic reduction, which describes the phase and amplitude of local modes (in depth) that travel along curved horizontal (lateral) ray trajectories. A comparison of our lateral ray asymptotic results with those of an independent study based on an entirely different approach [1,2] confirms that ray asymptotics works and is an appealing option that can also be generalized. In fact, all of the prototype constructs can be extended approximately to include weakly nonseparable configurations. This has been a further motivation for dealing with the problems at hand.

ACKNOWLEDGMENT

This work was supported by the Office of Naval Research under Contract No. N-00014-79-C-0013.

REFERENCES

1. Buckingham, M. J., Theory of acoustic propagation around a conical seamount, JOURNAL OF THE ACOUSTICAL SOCIETY OF AMERICA, 80 (1986), 265-277.

2. Buckingham, M. J., S. A. S. Jones, and P. N. Harriman, Stationary phase evaluation of the integral for the acoustic field around a conical seamount, JOURNAL OF THE ACOUSTICAL SOCIETY OF AMERICA, 80 (1986), 278-281.

3. Felsen, L. B., and N. Marcuvitz, RADIATION AND SCATTERING OF
    WAVES, Prentice Hall, Inc., Englewood Cliffs, N.J., 1973.

4. Felsen, L. B., Progressing and oscillatory waves for hybrid
    synthesis of source excited propagation and diffraction, IEEE
    TRANS. ON ANTENNAS AND PROPAGATION, AP-32 (1984), 775-796.

5. Lu, I. T., and L. B. Felsen, Adiabatic transforms for spec-
    tral analysis and synthesis of weakly range dependent shallow
    ocean Green's functions, JOURNAL OF THE ACOUSTICAL SOCIETY OF
    AMERICA (to appear).

6. Arnold, J. M., and L. B. Felsen, Local intrinsic modes: Layer
    with nonplanar interface, WAVE MOTION, 8 (1986), 1-14.

7. Lu, I. T., and L. B. Felsen, Canonical propagation problems
    for a wedge-shaped ocean: I. Layered fluid-solid bottom;
    II. Bottom with linear surface impedance variation, OCEAN
    ACOUSTICS, Proceedings of the Symposium on Underwater Acous-
    tics, Halifax, Nova Scotia, Canada, 1986 (to appear).

8. Harrison, C. H., Three-dimensional ray paths in basins,
    troughts, and near seamounts by use of ray invariants, JOUR-
    NAL OF THE ACOUSTICAL SOCIETY OF AMERICA, 62 (1977), 1382-
    1388.

9. Weston, D. E., Guided propagation in a slowly varying medium,
    PROCEEDINGS OF THE PHYSICAL SOCIETY OF LONDON, 73 (1959),
    365-384.

10. Weinberg, H., and R. Burridge, Horizontal ray theory for
    ocean acoustics, JOURNAL OF THE ACOUSTICAL SOCIETY OF AMERICA,
    55 (1974), 63-78.

COMPUTATIONAL ACOUSTICS: Wave Propagation
D. Lee, R.L. Sternberg, M.H. Schultz (Editors)
Elsevier Science Publishers B.V. (North-Holland)
© IMACS, 1988

MULTIGRID PRECONDITIONERS APPLIED TO
THREE-DIMENSIONAL PARABOLIC EQUATION TYPE MODELS

C. I. Goldstein

Applied Mathematics Department
Brookhaven National Laboratory
Upton, New York

ABSTRACT

The numerical solution of three dimensional standard parabolic equation (SPE) and high angle parabolic equation (HAPE) approximations of the Helmholtz equation is investigated. The model is discretized using an unconditionally stable, implicit finite difference scheme based on the Crank-Nicolson method in the range direction. The discretization in the transverse direction is based on a standard finite difference or finite element method. The use of an implicit scheme in range requires the solution of a large, sparse, nonselfadjoint system of equations at each range step of the marching algorithm. This system is solved by preconditioning the matrix and then applying the conjugate gradient method to the normal equations (PCGN).

The main result is the development of a preconditioner that significantly reduces the condition number of the original matrix and is efficiently implemented into the marching algorithm. The preconditioner is based on a multigrid cycle with a limited number of grid levels applied to the discrete Laplacian. The optimal number of grid levels is conveniently expressed in terms of the grid sizes, the wave number, and the form of the parabolic approximation. It is shown that PCGN is particularly well suited for implementation in a marching algorithm since the solution computed at each range step may be used to furnish a good starting guess at the following range step, thereby greatly reducing the number of iterations. Furthermore, it is shown that the system of equations to be solved at each range step is better conditioned than the system corresponding to the discretized Helmholtz equation. Consequently, an extremely small number of iterations suffices at each range step. This is validated by a variety of numerical experiments for both SPE and HAPE.

1. INTRODUCTION

In this paper we investigate multigrid preconditioners in connection with the iterative solution of the system of equations arising at each range step of a marching algorithm obtained by discretizing a three dimensional parabolic approximation of the Helmholtz equation with a variable index of refraction. We consider both the standard parabolic equation (SPE) and a high angle parabolic equation (HAPE). See [1] and [2] as well as references cited there for detailed discussions and applications of SPE and HAPE. We employ an implicit finite difference discretization based on the Crank-Nicolson method in the range direction and a standard finite difference or finite element scheme in the transverse direction (see [2] or [3]). This implicit scheme is unconditionally stable, so that it is not necessary to employ an excessive number of range steps. (Explicit schemes may not even be conditionally stable for HAPE.) However, the use of an implicit scheme in range requires the solution of a large, sparse, nonselfadjoint two dimensional system of linear equations at each range step.

It was proposed by Schultz, Lee, and Jackson in [3] (Section 4) for a three dimensional HAPE to apply the conjugate gradient iterative method to the normal equations. This iterative algorithm, referred to as CGN, can be efficiently implemented and is guaranteed to converge. However, the convergence rate is prohibitively slow for large systems since the condition number of the resulting matrix is quite large. For a comprehensive discussion of the conjugate gradient and related iterative methods, see [4] and references cited there. The use of a preconditioner before applying CGN to reduce the condition number (i.e., compress the eigenvalues) of the coefficient matrix can significantly reduce the number of iterations and hence the computational work (see, e.g., [4] and [5]).

The solution of the two dimensional system of equations at each range step comprises most of the computational work. Hence the development of an effective preconditioner is extremely important. Our preconditioner is based on a multigrid cycle with a limited number of grid levels applied to the discrete transverse Laplacian. See [6]-[8] for detailed discussions of the multigrid method. We shall see in Sections 4 and 5 that the multigrid preconditioner is extremely effective provided the number of grid levels is properly chosen. The main result of the paper shows

how the optimal number of grid levels depends on the form of the
parabolic approximation, as well as the grid sizes, wave number,
and reference wave number. Furthermore, under suitable assumptions, the optimal number of iterations remains bounded as these
parameters vary. Hence there is a close connection between the
physical characteristics of the problem and the appropriate choice
of grid sizes and preconditioner.

The multigrid preconditioner was previously employed in [9]
in connection with the Helmholtz equation. Its effectiveness was
demonstrated in [9] by means of several numerical examples. This
preconditioner was rigorously analyzed in [10] and extended to
singular perturbation problems. Furthermore, a practical method
for determining the optimal number of grid levels was described in
[10]. In the present paper, it will be seen that CGN with this
multigrid preconditioner is particularly well suited for implementation in a marching algorithm since the solution computed at each
range step may be used to furnish a good starting guess at the following range step. This considerably reduces the number of iterations. Furthermore, it will be seen that the system of equations
considered here is better conditioned than the system corresponding to the discretized Helmholtz equation.

We close this section by outlining the remainder of the paper.
In Section 2 we briefly review the parabolic approximations and
their discretizations. In Section 3 we discuss the preconditioned
conjugate gradient iterative method. In Section 4 we describe the
multigrid preconditioner and obtain our main theoretical results.
In Section 5 we present numerical results validating the effectiveness of the preconditioner for both SPE and HAPE. In particular,
we shall see that CGN with this preconditioner results in a substantial reduction in computational work compared to CGN with no
preconditioning (typically between one and two orders of magnitude) and that this reduction in cost increases with the problem
size. Furthermore, the algorithm is to a large extent vectorizable. We close Section 5 with several observations concerning both
the discretization error and the error due to approximating the
Helmholtz model by SPE and HAPE.

2. CONTINUOUS AND DISCRETE PARABOLIC APPROXIMATIONS

To simplify the exposition we consider the Helmholtz equation
with constant index of refraction in Cartesian coordinates, although
the following discussion can be readily generalized to variable

index of refraction and other coordinate systems (see Remark 2.1 below). Let D denote a three dimensional waveguide with uniform cross-section S. We consider the following Helmholtz boundary value problem:

$$(-\Delta - K^2)u = 0 \text{ in } D \qquad u(r_0,x,y) = g(x,y) \tag{2.1}$$

u satisfies an outgoing radiation condition at $r = \infty$ and u satisfies prescribed boundary conditions on ∂S, where $D = \{(r,x,y):$ $r_0 \le r < \infty,\ (x,y) \in S\}$, K is real, ∂S denotes the boundary of S, and Δ denotes the three dimensional Laplacian in Cartesian coordinates. The boundary conditions on ∂S can be combinations of Dirichlet, Neumann, impedance, and periodic boundary conditions.

    We next approximate (2.1) using a parabolic approximation.
Let K_o denote a reference wave number and set

$$U(r,x,y) = e^{-iK_o r} u(r,x,y) \tag{2.2}$$

We assume that K_o is chosen so that U is a smooth function of r.
Methods for obtaining K_o will be discussed in Section 5. In complete analogy with the method described in [1] and [2] for the two dimensional case, we first factor the Helmholtz operator in (2.1) and retain the "outgoing" part of this factorization to obtain

$$\frac{\partial u}{\partial r} = iK_o Qu \tag{2.3}$$

with the pseudo-differential operator Q given by

$$Q = \left(\frac{1}{K_o^2} \Delta^T + \frac{K^2}{K_o^2}\right)^{1/2} \tag{2.4}$$

where Δ^T denotes the transverse Laplacian on S corresponding to the prescribed boundary conditions on ∂S. Now define the following operators:

$$q_1 = \frac{K^2 - K_o^2}{K_o^2} + \frac{1}{K_o^2}\frac{\partial^2}{\partial y^2} \qquad q_2 = \frac{1}{K_o^2}\frac{\partial^2}{\partial x^2} \tag{2.5}$$

Combining (2.4) and (2.5), we obtain

$$Q = (1 + q_1 + q_2)^{1/2} \tag{2.6}$$

    We may obtain various approximations to (2.1) by approximating the square root in (2.6) by rational linear functions of q_1 and q_2

(or other convenient approximations such as continued fractions) and then substituting this in (2.3). This is analogous to the derivation in [2] and references cited there, where K^2 is typically replaced by $K_o^2 n^2$ with n a spatially dependent index of refraction. In our case, $n = K/K_o$. If we employ (2.2), (2.3), and a linear approximation, we obtain the standard three dimensional parabolic equation (SPE):

$$\frac{\partial U}{\partial r} = \frac{iK_o}{2} (q_1 + q_2) U \tag{2.7a}$$

This was obtained in [1] and solved in [11] for the three dimensional case using a Fourier method. If we combine (2.2) and (2.3) with the rational linear approximation

$$Q \approx \frac{1 + \frac{3}{4}(q_1 + q_2)}{1 + \frac{1}{4}(q_1 + q_2)}$$

we obtain the three dimensional Claerbout high angle parabolic equation (HAPE):

$$\left(1 + \frac{q_1 + q_2}{4}\right) \frac{\partial U}{\partial r} = \frac{iK_o}{2} (q_1 + q_2) U \tag{2.7b}$$

We next discretize (2.7a) and (2.7b). Let Δ^h denote a prescribed finite difference or finite element discretization of the transverse Laplacian, Δ^T, with h denoting the maximum grid size in the x and y directions. To be specific, we employ a second order finite difference scheme using central differences for $\partial^2/\partial x^2$ and $\partial^2/\partial y^2$ with uniform grid size, h, in both the x and y directions. We employ a Crank-Nicolson scheme in the range (r) direction as in [2] and [3]. Let h_r denote the range step size, $N_r + 1$ the number of range grid points, and set $r_j = r_0 + jh_r$, $j = 0, \ldots, N_r$. Let N_x and N_y denote the number of grid points, x_ℓ and y_m, in the x and y directions, respectively, and denote the grid points on S at range $r = r_j$ by

$$P_{\ell,m}^j = (r_j, x_\ell, y_m) \qquad j = 0, \ldots, N_r, \quad \ell = 1, \ldots, N_x,$$
$$m = 1, \ldots, N_y$$

Set $U_{\ell,m}^j = U(P_{\ell,m}^j)$ and

$$\Delta^h U_{\ell,m}^j = \frac{(U_{\ell+1,m}^j + U_{\ell-1,m}^j + U_{\ell,m+1}^j + U_{\ell,m-1}^j - 4U_{\ell,m}^j)}{h^2}$$

with suitable modifications in Δ^h at grid points on ∂S. Employing
(2.5) and (2.7a) and (2.7b), we now readily obtain the following
finite difference equations:

$$\frac{U_{\ell,m}^{j+1} - U_{\ell,m}^{j}}{h_r} = \frac{i}{4K_o}\,(\Delta^h U_{\ell,m}^{j} + \Delta^h U_{\ell,m}^{j+1} + (K^2 - K_o^2)\,(U_{\ell,m}^{j} + U_{\ell,m}^{j+1}))$$

$$\tag{2.8a}$$

for SPE, $j = 0,\ \ldots,\ N_{r-1}$, $\ell = 1,\ \ldots,\ N_x$, $m = 1,\ \ldots,\ N_y$, and

$$\left[1 + \frac{1}{4K_o^2}(\Delta^h + (K^2 - K_o^2))\right]\left(\frac{U_{\ell,m}^{j+1} - U_{\ell,m}^{j}}{h_r}\right)$$

$$= \frac{i}{4K_o}(\Delta^h U_{\ell,m}^{j} + \Delta^h U_{\ell,m}^{j+1} + (K^2 - K_o^2)\,(U_{\ell,m}^{j} + U_{\ell,m}^{j+1})) \tag{2.8b}$$

for HAPE, $j = 0,\ \ldots,\ N_{r-1}$, $\ell = 1,\ \ldots,\ N_x$, $m = 1,\ \ldots,\ N_y$.

<u>Remark 2.1</u>: For a more detailed and more general derivation of
the analogous equations in connection with a three dimensional
model of underwater acoustics, see [3]. The index of refraction
in [3] is spatially dependent and a cylindrical coordinate system
is employed, so that x is replaced by the azimuthal angle, θ, and
the operator q_2 in (2.5) is replaced by $(1/(K_o r)^2)(\partial^2/\partial\theta^2)$. A periodic boundary condition is specified for θ in a full annular cylindrical region. Our multigrid preconditioner described in Section 4 is applicable to more general models than (2.1), including
the underwater acoustics model in [3].

Equation (2.8a) or (2.8b) defines a marching algorithm in
which it is necessary to solve a system of equations for $U_{\ell,m}^{j+1}$ in
terms of $U_{\ell,m}^{j}$, $\ell = 1,\ \ldots,\ N_x$, $m = 1,\ \ldots,\ N_y$, for each range step,
r_j, starting with prescribed values of $U_{\ell,m}^{0}$. The solution of this
system at each step comprises most of the computational work and
is the subject of the remainder of this paper. In order to most
efficiently solve this system using an iterative method at each
step of the marching algorithm, it is useful to use the solution
computed at the previous step. To this end, we wish to solve for

$$\delta U_{\ell,m}^{j+1} = U_{\ell,m}^{j+1} - U_{\ell,m}^{j} \qquad \ell = 1,\ \ldots,\ N_x,\ m = 1,\ \ldots,\ N_y$$

instead of $U_{\ell,m}^{j+1}$ (starting with zero initial guess for each j).
Multiplying both sides of (2.8a) by $-4iK_o$ [of (2.8b) by $-4iK_o/$
$(1 + i/K_o h_r)$] and rearranging terms, we readily obtain

$$\begin{cases} (-\Delta^h - \gamma_S(K,K_o,h_r))\delta U^{j+1}_{\ell,m} = 2(\Delta^h + (K^2 - K_o^2))U^j_{\ell,m} \\ \qquad \text{for SPE with } j = 0, \ldots, N_{r-1}, \; \ell = 1, \ldots, N_x, \\ \qquad\qquad m = 1, \ldots, N_y \\ \text{and} \\ \gamma_S(K,K_o,h_r) = (K^2 - K_o^2) + \dfrac{4iK_o}{h_r} \end{cases} \qquad (2.9a)$$

$$\begin{cases} (-\Delta^h - \gamma_H(K,K_o,h_r))\delta U^{j+1}_{\ell,m} = \dfrac{2}{1 + \dfrac{i}{K_o h_r}} (\Delta^h + (K^2 - K_o^2))U^j_{\ell,m} \\ \qquad \text{for HAPE with } j = 0, \ldots, N_{r-1}, \; \ell = 1, \ldots, N_x, \\ \qquad\qquad m = 1, \ldots, N_y \\ \text{and} \\ \gamma_H(K,K_o,h_r) = (K^2 - K_o^2) + \dfrac{4iK_o}{h_r\left(1 + \dfrac{i}{K_o h_r}\right)} \end{cases} \qquad (2.9b)$$

3. PRECONDITIONED CONJUGATE GRADIENT METHOD

The system of linear equations given by (2.9a) or (2.9b) may
be expressed in matrix form at each range step by

$$AX = F \qquad\qquad (3.1)$$

where X and F are vectors with $N = N_x N_y$ elements and A is a sparse
$N \times N$ matrix. X, F, and A depend on r in general and the coefficient matrix A has the form

$$A = A_o + \gamma I \qquad\qquad (3.2)$$

where A_o corresponds to the operator $-\Delta^h$, I is the identity matrix,
and γ is a complex constant [$\gamma = \gamma_S$ in (2.9a) for SPE and $\gamma = \gamma_H$
in (2.9b) for HAPE]. When A is large and sparse, it is generally
desirable to employ iterative solution methods such as the conjugate gradient (CG) method to solve (3.1), since the matrix A need
not be inverted or stored. See [4] for a comprehensive discussion
of CG and related iterative methods.

In order to guarantee convergence of CG and other standard
iterative methods, the matrix A must be positive definite symmetric
[4] (i.e., the eigenvalues of A must be positive real numbers).
This is not the case for A in (3.2) since γ is complex [see (2.9a)
or (2.9b)]. To overcome this difficulty, we may apply CG to the
normal equations (CGN). Thus (3.1) is replaced by the equivalent
system

$$A*AX = A*F \qquad\qquad (3.3)$$

where A* is the adjoint (conjugate transpose) of A. Since A*A is
positive definite symmetric, CGN is guaranteed to converge and was
proposed in [3] for use in connection with a three dimensional
HAPE.

The main practical difficulty with using CGN is that the number of iterations required for convergence is typically quite large.
The reason is that the number of iterations is proportional to
$\Lambda = \sqrt{\kappa(A^*A)}$, where

$$\kappa(A^*A) = \frac{\lambda_{max}(A^*A)}{\lambda_{min}(A^*A)}$$

is the condition number of A*A and λ_{max} (λ_{min}) denotes the largest
(smallest) eigenvalue of A*A [4]. $\kappa(A^*A)$ will grow quite rapidly
with N and hence the convergence will be prohibitively slow for N
large. An effective method for reducing the condition number and
hence the computational cost is to first precondition A by a "partial inverse," M^{-1}, of A_O. CGN is then applied to the resulting
system with coefficient matrix A'. This was proposed in [5] in
connection with the Helmholtz equation. The two main requirements
are that $M^{-1}V$ be inexpensive to compute for each vector V and that
A' have significantly smaller condition number than A. The resulting algorithm, referred to as PCGN, requires 5N + 2 multiplications
and divisions for the five-point finite difference scheme, two
matrix vector products (AV and A*V) and two preconditioning solves
($M^{-1}V$ and $M^{-1^*}V$) for each iteration.

4. MULTIGRID PRECONDITIONER

We begin by briefly describing a multigrid cycle applied to
the matrix A_O in (3.2) corresponding to the discrete transverse
Laplacian, $-\Delta^h$, with uniform grid size h on the transverse plane.
Such a cycle, denoted by M^{-1}, acts as a "fast Laplace solver" in
the sense that a finite number of such cycles, independent of the
number of unknowns, N, solves the system of equations

$$A_O U = F \quad \text{for each vector F} \tag{4.1}$$

in the asymptotically optimal 0(N) number of operations. To define M^{-1}, consider a sequence of grids $G^0, \ldots, G^L$, with uniform
grid size h_j on grid G^j given by $h_j = 2^{L-j}h$, j = 0, $\ldots$, L, where
h_0 is independent of h. We choose some relaxation scheme for each
grid level. We also choose an interpolation operator I_{j-1}^j from
G^{j-1} to G^j and a weighting operator I_j^{j-1} from G^j to G^{j-1}.

To begin the cycle, we choose an initial guess and make r_L relaxation sweeps on the finest grid level G^L to obtain an approximate solution U^L. We then transfer the residual to G^{L-1} using I_L^{L-1} and obtain an equation analogous to (4.1) for the correction, V^{L-1}, on G^{L-1}. This process is repeated using r_j relaxation sweeps on G^j until we get to the coarsest grid G^0. To return to the next finer grid, we make s_0 relaxation sweeps on G^0 to improve the coarse grid correction V^0. We then recalculate V^1 by $V^1_{new} = V^1_{old} + I_0^1 V^0$. We continue this process using s_j relaxation sweeps on each successive grid level G^j up to G^L. The entire process defines one complete multigrid cycle, M^{-1}. For various specific choices of the relaxation sweeps and operators, I_i^j, as well as a detailed description and analysis of various multigrid cycles, see [6]-[8].

In this paper we employ M^{-1} as a preconditioner for CGN instead of a fast solver. Hence we want the multigrid cycle to be positive definite symmetric. To this end, we use linear interpolation for I_{j-1}^j, and a full weighting for I_j^{j-1} as in [9], [10], and [12]. For our relaxation sweeps, we can use Weighted Simultaneous Displacement (WSD) as in [10]. Alternatively, we can use Gauss-Seidel with red-black (black-red) ordering when going to coarser (finer) grids, as in [9], [12], and the numerical examples of the next section. These operations also make the multigrid cycle vectorizable.

We next employ M^{-1} as a preconditioner for CGN to solve (3.1) with A given by (3.2). Hence we replace A by the preconditioned matrix $A' = M^{-1}A$. In view of the discussion in Section 3, we wish to estimate the condition number of A'^*A'. Fundamental to the following discussion is a result rigorously proved in [10] which states that "a finite number, J, of multigrid cycles with a limited number of grid levels acts as a fast Laplace solver on the subspace of eigenvectors corresponding to the larger eigenvalues of A_0 and scales the remaining eigenvectors by h_0^2, where h_0 is the coarsest grid size in the multigrid cycle." Thus instead of completing the cycle, we stop (possibly after only one or a few grid levels) and do not invert the Laplacian on the coarse grid levels. This result, which extends the concept of a fast Laplace solver, is stated precisely in [10] and proved using Fourier analysis. For simplicity, we assume here that J = 1.

Our goal is to choose the optimal number of grid levels, or optimal coarse grid size, h_0, so as to minimize the condition number

$$\kappa(A'^*A') = \frac{\lambda_{max}(A'^*A')}{\lambda_{min}(A'^*A')} \tag{4.2}$$

It follows from results in [10] that the optimal choice of h_0 is given by

$$h_0 \approx |\gamma|^{-1/2} \tag{4.3}$$

and in this case

$$\lambda_{max}(A'^*A') = 0(1) \qquad \text{as } |\gamma| \to \infty . \tag{4.4}$$

It also follows from [10] that

$$\lambda_{min}(A'^*A') \geq C_0 \frac{|\gamma^I|^2}{(1 + |\gamma|)^2} \tag{4.5}$$

where C_0 is independent of h, K, and K_0, and γ^I denotes the imaginary part of γ. Combining (4.2), (4.4), and (4.5), we obtain

$$\kappa(A'^*A') = 0\left(\frac{|\gamma|^2}{|\gamma^I|^2}\right) \qquad \text{as } |\gamma| \to \infty \tag{4.6}$$

provided h_0 is given by (4.3).

We next briefly explore some implications of this optimal choice of coarse grid size, h_0. It follows from (2.9) that γ is given by γ_S or γ_H, defined by

$$\gamma_S = (K^2 - K_0^2) + 4iK_0/hr \tag{4.7a}$$

$$\gamma_H = (K^2 - K_0^2) + 4iK_0/h_r(1 + i/K_0h_r) \tag{4.7b}$$

The crucial parameter to be selected when implementing the algorithm is the number of grid levels, L, in the multigrid cycle, defined by

$$h_0 = 2^L h \tag{4.8}$$

It is easily seen from (4.3), (4.7), and (4.8) how the optimal choice of L depends on each of the parameters K, h_r, and h with the remaining parameters fixed. For example, when h is fixed, L must be decreased as h_0 decreases or equivalently $|\gamma|$ increases. We see from (4.7) that both $|\gamma_S|$ and $|\gamma_H|$ increase with K. Furthermore, it is readily seen from (4.7) that

$$|\gamma_S| \to \infty \qquad \text{as } h_r \to 0 \tag{4.9a}$$

and

$$|\gamma_H| \rightarrow K^2 + 3K_o^2 \qquad \text{as } h_r \rightarrow 0 \qquad\qquad (4.9b)$$

Hence PCGN behaves differently for SPE and HAPE when h_r is small, as will be seen in the next section. When $|\gamma|$ (and hence h_0) is fixed, it follows from (4.8) that L must be increased as h decreases. Finally, we note that some or all of these parameters are often related to each other in view of accuracy considerations (see [13] in connection with the Helmholtz equation).

In order to show that $\kappa(A'^*A')$ is uniformly bounded, we apply (4.6) and make two further assumptions. First of all, suppose that

$$|K^2 - K_o^2| = 0(K_o^2) \qquad \text{as } K_o \rightarrow \infty \qquad\qquad (4.10)$$

It will be seen from the numerical examples in the next section that this is frequently the case. (If $K = K_o n$ with n a variable index of refraction, (4.10) holds provided $|n|$ is bounded.) We also assume that the range step size h_r and wave number K_o are constrained by

$$K_o h_r = 0(1) \qquad \text{as } K_o \rightarrow \infty \text{ or } h_r \rightarrow 0 \qquad\qquad (4.11)$$

Combining (4.6), (4.7), (4.10), and (4.11), we obtain

THEOREM 4.1: Assume that (4.10) and (4.11) hold. Then for SPE (HAPE), the condition number, $\kappa(A'^*A')$, is uniformly bounded as $h \rightarrow 0$ and K or $K_o \rightarrow \infty$, provided the number of grid levels in the multigrid preconditioner is chosen such that (4.3) holds with $\gamma = \gamma_S$ ($\gamma = \gamma_H$).

This is the optimal choice of grid levels.

Remark 4.1: It can be readily seen using the preceding analysis that the condition number (and hence the number of iterations) is not uniformly bounded as $h \rightarrow 0$ or $|\gamma| \rightarrow \infty$ when condition (4.3) does not hold. In particular, this is the case when M^{-1} is given by a complete multigrid cycle (or any other fast Laplace solver). Furthermore, when M^{-1} is the identity matrix (i.e., no preconditioning is employed), it is known that $\kappa(A^*A) = 0(h^{-4})$ as $h \rightarrow 0$ with γ fixed. Hence the number of iterations increases like $0(h^{-2})$.

Remark 4.2: It was shown in [9] and [10] that for the discrete Helmholtz equation (with coefficient matrix A_H), the optimal preconditioner, M^{-1}, considered in Theorem 4.1 yields the condition number

$$\kappa(A_H'^*A_H') = 0(K) \qquad \text{as } K \to \infty$$

Hence the number of iterations is not bounded as $K \to \infty$ and is much larger for the Helmholtz equation than for SPE and HAPE, particularly for K large. This is consistent with our numerical experiments.

5.   NUMERICAL RESULTS

In this section we validate the theoretical results of Section 4 concerning the multigrid preconditioner by means of numerical examples for which the solution of the Helmholtz equation is known in closed form. We consider a three dimensional waveguide D in Cartesian coordinates with uniform cross section S given by the square

$$S = \{(x,y): 0 \le x \le \pi, 0 \le y \le \pi\} \qquad (5.1)$$

We begin with the following Helmholtz boundary value problem:

$$(-\Delta - K^2)u = 0 \text{ in } D \qquad u(0,x,y) = g(x,y)$$
$$u \text{ satisfies an outgoing radiation condition at } r = \infty$$

and   (5.2)

$$\frac{\partial u(r,0,y)}{\partial x} = \frac{\partial u(r,\pi,y)}{\partial x} = \frac{\partial u(r,x,0)}{\partial y} = u(r,x,\pi) = 0$$

Note that we have obtained numerical results analogous to those described below for other boundary conditions on S of the form discussed in Section 2. The data g in (5.2) is determined by the solution. We consider solutions of the form

$$u = \sum_{m=1}^{M_x} \sum_{n=1}^{M_y} a_{mn} e^{iK_{mn}r} \cos mx \cos \left(n - \frac{1}{2}\right)y \qquad (5.3)$$

where $M_x, M_y \ge 1$, the a_{mn} are complex constants, and the modal wave numbers K_{mn} are defined by

$$K_{mn} = \sqrt{K^2 - \Lambda_{mn}} \qquad \text{with } \Lambda_{mn} = m^2 + \left(n - \frac{1}{2}\right)^2 \qquad (5.4)$$

As described in Section 2, we approximate the Helmholtz model (5.2) by the two parabolic approximations SPE and HAPE given by (2.7a) and (2.7b), respectively, where U is defined in terms of u by means of (2.2). The reference wave number, K_o, appearing in

(2.2) will be defined in connection with the numerical examples below. For a more detailed discussion of suitable choices for K_o, see [14] and [15]. As in Section 2, we now obtain equations (2.9a) or (2.9b) at each range step, where h_r is the uniform range step size and Δ^h corresponds to a discretization of the transverse Laplacian obtained using central differences with uniform grid size h in both directions. We employ the iterative method PCGN discussed in Section 3 using the multigrid preconditioner, M^{-1}, described in Section 4 to solve (2.9a) or (2.9b). We use two relaxation sweeps at each grid level except the coarsest, where we use four. The main goal of our numerical experiments is to investigate the behavior of this preconditioner. We define convergence of the iterative method to mean that the normalized mean square norm of $\bar{r} = A*M^{-1}r$ is less than 10^{-6}, where r is the residual. As described in [9], $\bar{r}$ is naturally produced by the implementation of the algorithm and the magnitude of $\bar{r}$ is about the same as that of r. Note that our stopping criteria is more stringent than might be required in practice.

For our first two examples, we consider the solution (5.3) with only one mode, corresponding to $m = m_x$ and $n = n_y$. Thus $a_{m_x n_y} = 1$ and all other $a_{mn} = 0$ in (5.3). In the first example, we choose $m_x = n_y = 1$. We choose $K_o = K_{11} \cos \alpha$, $0 \le \alpha \le \pi/2$. The r-dependence of U in (2.2) is given by

$$e^{iK_{11}(1-\cos\alpha)r}$$

In order for SPE and HAPE to be accurate approximations to the Helmholtz equation, we must have $\alpha \le \pi/12$ and $\alpha < \pi/4$, respectively [2]. Let N denote the number of grid intervals in the x and y directions. We express the range step size h_r in terms of the wavelength $\lambda = 2\pi/K$ as follows:

$$h_r = \delta\lambda = \frac{2\pi\delta}{K} \qquad \delta > 0 \tag{5.5}$$

In Table 1 we exhibit the optimal number of grid levels, NLEV, in the multigrid preconditioner and the corresponding number of iterations, NIT, required for convergence for various choices of K, N, δ, and α. Although the results in Table 1 are for HAPE, we have obtained virtually the same results for SPE. The results in Table 1 (as well as in Tables 2 and 3) are typical of our results for various solutions of the form (5.3).

C.I. Goldstein

TABLE 1   ($m_x = m_y = 1$; HAPE)

α	K	N	δ	NLEV	NIT
$\pi/12$	5	16	1	1 or 2	5 (121↑123)*
$\pi/12$	5	32	1	2	3 (416↑418)*
$\pi/12$	5	64	1	3	5 (1394↑1395)*
$\pi/12$	15	32	1	1	2
$\pi/12$	15	64	1	2	3
$\pi/12$	5	32	1/10	1	2
$\pi/12$	5	64	1/10	2	3
$\pi/5$	5	32	1/10	1	2
$\pi/5$	5	64	1/10	2	4

*No preconditioning, first five range steps.

In the first three entries of Table 1, we increase N (decrease h) keeping the other parameters fixed. We see that the optimal number of grid levels increases with N. The runs were carried out for several range steps (typically between 25 and 100) and the number of iterations was observed to be essentially constant over all of the range steps when the optimal number of grid levels was employed. When a suboptimal number of levels was employed, the number of iterations generally increased as we marched out in range. The results obtained using any number of grid levels were far superior to those obtained without preconditioning (shown in parenthesis in the first three entries for the first five range steps). The computation time for each iteration is about twice as long when the preconditioning is employed. Note that in the case of no preconditioning, the number of iterations increases like $0(h^{-2})$ as $h \to 0$ (see Remark 4.1).

In the remaining entries of Table 1, we see that the dependence of NLEV on N is analogous for different values of K, δ, and α. We also see that NLEV decreases as K increases or δ (i.e., h_r) decreases. Finally, observe that the optimal number of iterations is very nearly independent of the parameters K, N, δ, and α. All of the observed results are consistent with and explained by the theory of the previous section. See [10] for a practical method for determining the optimal number of grid levels.

For our second example, we assume that $m_x = 4$ and $n_y = 2$. Hence $a_{42} = 1$ and all other $a_{mn} = 0$ in (5.3). In this case, we set $K_o = K_{42} \cos \alpha$. In Table 2, we illustrate the behavior of PCGN for both SPE and HAPE for the first 25 range steps with K = 5,

TABLE 2   ($m_x = 4$, $n_y = 2$, $K = 5$, $\alpha = \pi/12$)

Model	N	δ	NLEV	NIT
HAPE	32	1	2	5↑8
HAPE	32	1/10	2	5
SPE	32	1	2	5↑7
SPE	32	1/10	1	2↑3
HAPE	64	1	3	8
HAPE	64	1/10	2	5
HAPE	64	1/100	2	5
SPE	64	1	3	8
SPE	64	1/10	2	4
SPE	64	1/100	1	2

$\alpha = \pi/12$, and different values of N and δ. As before, we see that the optimal number of grid levels is nondecreasing as N or δ increases with the other parameters fixed. Furthermore, the results are similar for SPE and HAPE, except when $\delta \ll 1$. In this case, we observe that NLEV$\downarrow 1$ as $\delta \downarrow 0$ for SPE. On the other hand, we have observed that the optimal number of grid levels for HAPE is independent of δ for δ small and varies inversely with h^2 and $K^2 + 3K_o^2$. These observations follow readily from (4.9), (5.5), and the theory in Section 4.

For our last example, we consider the solution given by (5.3) with $a_{mn} = 1$ for $m = 1, \ldots, M_x$ and $n = 1, \ldots, M_y$. In this case, we define K_o by the average of the modal wave numbers

$$K_o = \frac{1}{M_x M_y} \sum_{M=1}^{M_x} \sum_{n=1}^{M_y} K_{mn}$$

We illustrate typical results in Table 3 with $M_x = 1$, $M_y = 4$, $K = 10$, and different values of δ and N for both SPE and HAPE over 25 range steps. We see that the dependence of NLEV and NIT on N and δ is analogous to that in the first two examples. The behavior of PCGN is the same for SPE and HAPE except when δ is small. Furthermore, the number of iterations is greatly increased when no preconditioning is used.

We close this section with some observations concerning the discretization error and the modeling error due to SPE and HAPE. We measured the error, E_2, using the normalized discrete mean-square norm in the transverse plane at each range step. Note that

TABLE 3   ($M_x = 1$, $M_y = 4$, $K = 10$)

Model	N	δ	NLEV	NIT
HAPE	32	1	2	7 (121↑130)*
HAPE	32	1/10	1	6
SPE	32	1	2	7
SPE	32	1/10	1	7↑11
HAPE	64	1	2 or 3	8 (404↑431)*
HAPE	64	1/10	1 or 2	7
SPE	64	1	2 or 3	8
SPE	64	1/10	1	5

*No preconditioning, first five range steps.

there are a variety of other error measures that can be employed
in practice, such as propagation loss commonly used in connection
with underwater acoustics problems [1]. We have observed from our
numerical experiments that the error, E_2, is additive over all the
range steps provided h_r is chosen sufficiently small. Hence, in
order to march out as far as possible in the range direction,
while maintaining accuracy, it is desirable to choose sufficiently
small mesh sizes that the error is primarily due to the parabolic
approximation. We have determined whether the error is mainly a
discretization or modeling error by decreasing the mesh sizes and
observing the behavior of the error.

   To describe results of typical numerical experiments, let
E_2/λ denote the percent error accumulated over one wavelength,
$\lambda = 2\pi/K$, in the range direction. Consider the first example discussed earlier (see Table 1). Using SPE and $\alpha = \pi/12$, we observed
that $E_2/\lambda = 0.3\%$ and that it suffices, as in other problems for
which SPE is a valid approximation, to use $h_r = \lambda$ (or even larger).
Furthermore, E_2/λ is independent of $h_r \leq \lambda$ and $N = 16$, 32, or 64,
indicating that this error is due to the SPE approximation. When
HAPE was employed, we observed that $E_2/\lambda = 0.004\%$ with $h_r = \lambda/10$
and 0.07% with $h_r = \lambda$. Thus accuracy is lost for HAPE by choosing
$h_r = \lambda$. When $\alpha = \pi/5$ in this example and $h_r \leq 1/10$, we observed
that $E_2/\lambda = 1.6\%$ for HAPE and is independent of $N = 32$ or 64.
Hence this is the modeling error using HAPE. Furthermore, the
computed solution is inaccurate for $\alpha = \pi/5$ when $h_r = \lambda$ or SPE is
employed.

   For the second example discussed above (Table 2), we observed
an error of about 1.25% with $N = 64$ and 5% with $N = 32$ for both

SPE and HAPE, where $\alpha = \pi/12$ and $h_r \leq \lambda$. This indicates that the error is predominantly due to the $0(h^2)$ discretization error in the transverse direction and is explained by the fact that higher modes have more of an oscillatory behavior in this direction. For the last example above (Table 3), we observed for HAPE and $h_r = \lambda/10$ that the error is dominated by discretization error in the transverse direction. For SPE, most of the error appears to be due to the parabolic approximation. In view of the preceding observations and the theory in Section 4, we conclude that there is a very close relationship between the physical characteristics of the problem (determining the oscillatory behavior of the solution) and the appropriate choice of grid sizes and preconditioner.

ACKNOWLEDGMENTS

This work was supported (in part) by the Applied Mathematical Sciences subprogram of the Office of Energy Research, U.S. Department of Energy, under Contract No. DE-AC02-76CH00016.

The author wishes to express his gratitude to Alvin Bayliss for several useful discussions concerning this work.

REFERENCES

1. F. D. Tappert, The Parabolic Approximation Method, in WAVE PROPAGATION AND UNDERWATER ACOUSTICS, J. B. Keller and J. S. Papadakis, eds., Springer-Verlag, New York, 1977.

2. NORDA Parabolic Equation Workshop, NORDA Tech. Note 143, J. Davis, D. White, and R. Cavanagh, eds., 1981.

3. Recent Progress in the Development and Application of the Parabolic Equation, NUSC Technical Document 7145, D. Lee, ed., 1984.

4. H. C. Ellman, Iterative methods for large, sparse, nonsymmetric systems of linear equations, Yale University, Dept. Comp. Science, Research Report 229, 1981.

5. A. Bayliss, C. I. Goldstein, and E. Turkel, An iterative method for the Helmholtz equation, JOURNAL OF COMPUTATIONAL PHYSICS, 49 (1983), 443-457.

6. A. Brandt, Multi-level adaptive solution to boundary value problems, MATHEMATICS OF COMPUTATION 31 (1977), 333-391.

7.   R. A. Nicolaides, On the ℓ^2 convergence of an algorithm for solving finite element equations, MATHEMATICS OF COMPUTATION 31 (1977), 892-906.

8.   R. E. Bank and T. Dupont, An optimal order process for solving finite element equations, MATHEMATICS OF COMPUTATION 36 (1981), 35-51.

9.   A. Bayliss, C. I. Goldstein, and E. Turkel, The numerical solution of the Helmholtz equation for wave propagation problems in underwater acoustics, COMPUTERS AND MATHEMATICS WITH APPLICATIONS 11 (1985), 655-665.

10. C. I. Goldstein, Multigrid preconditioners applied to the iterative solution of singularly perturbed elliptic boundary value problems, BNL Report 51916.

11. J. S. Perkins and R. N. Baer, An approximation to the three-dimensional parabolic equation method for acoustic propagation, JOURNAL OF THE ACOUSTICAL SOCIETY OF AMERICA 72 (1982), 515-522.

12. J. Gozani, A. Nachson, and E. Turkel, Conjugate gradient coupled with multigrid for an indefinite problem, ADVANCES IN COMPUTER METHODS FOR PARTIAL DIFFERENTIAL EQUATIONS, R. Vichnevetsky and R. Stepleman, eds., IMACS (1984), 425-427.

13. A. Bayliss, C. I. Goldstein, and E. Turkel, On accuracy conditions for the numerical computation of waves, JOURNAL OF COMPUTATIONAL PHYSICS 59 (1985), 396-404.

14. A. D. Pierce, The natural reference wavenumber for parabolic approximations in ocean acoustics, COMPUTERS AND MATHEMATICS WITH APPLICATIONS 11 (1985), 831-841.

15. C. I. Goldstein, Finite element methods applied to nearly one-way propagation, JOURNAL OF COMPUTATIONAL PHYSICS 64 (1986), 56-81.

COMPUTATIONAL ACOUSTICS: Wave Propagation
D. Lee, R.L. Sternberg, M.H. Schultz (Editors)
Elsevier Science Publishers B.V. (North-Holland)
© IMACS, 1988

AN EFFICIENT METHOD FOR SOLVING THE
THREE-DIMENSIONAL WIDE ANGLE WAVE EQUATION

Ding Lee

Naval Underwater Systems Center
New London, Connecticut

Youcef Saad

Department of Computer Science
Yale University
New Haven, Connecticut

Martin H. Schultz

Department of Computer Science
Yale University
New Haven, Connecticut

ABSTRACT

We propose a new method for the solution of the wide angle
wave equation in three dimensions. In contrast with standard
techniques, our approach requires only solutions of successive
tridiagonal systems in the resulting finite difference parabolic
equations. The method is based on a simple approximation to the
square-root operator written formally as $\sqrt{I + X + Y}$ where X is a
partial differential operator with respect to the depth z and Y is
a partial differential operator with respect to the azimuthal
angle θ. We exploit the fact that the partial derivative term Y
with respect to azimuthal angle is small, but not negligible, as
compared with other terms. It is then natural to replace the
square-root operator by an expansion which is of order 2 with respect to the X operator, and of order 1 with respect to the Y operator. An important feature of this approach is that it is then
possible to derive a rational function approximation to the exponential of the square-root operator which has the property of being stable, and accurate. Moreover, the approximation decouples
naturally as a product of a (1,1) rational function of X times a
(1,1) rational function of Y. As a consequence, this will result
in a solution technique that requires only two tridiagonal system
solutions per step, namely, one for the X operator and one for the
Y operator. Numerical examples are reported that show the wide
angle capability of this method.

1. INTRODUCTION

The wide angle three-dimensional parabolic approximation
technique developed by Siegmann, Kriegsmann, and Lee [5] has been
proven capable of handling wide angle propagation in the vertical
plane. This technique is based on a pseudo-differential three-
dimensional parabolic wave equation of which the 3-D parabolic
approximation introduced by Tappert [7] is a special case. Methods for solving this equation have been developed by Baer and
Perkins [1], using the fast Fourier transform, but are only applicable to small angles of propagation. Moreover, these methods do
not extend easily to equations with variable coefficients and do
not handle rigid boundary conditions.

In order to be able to handle wide angle propagation in the
general variable coefficient case and to easily treat rigid boundary conditions, we choose the three-dimensional parabolic wave
equation as the representative equation. The main contribution of
this paper is a method for efficiently solving this equation for
wide angle propagations. A solution technique for the wide angle
3-D equation has been previously developed by Schultz, Lee, and
Jackson [4] using the Crank-Nicolson scheme in conjunction with a
preconditioned conjugate gradient method. Due to the resulting
properties of the discretized finite difference equations, the operator is neither Hermitian nor positive real. Therefore no effective preconditioners are known for this case and the solution
adopted by Schultz, Lee, and Jackson [4] was to precondition the
normal equations. This squares the condition number of the initial matrix but gives satisfactory results as far as accuracy is
concerned.

In this paper we propose a new method to solve the same 3-D
wide angle parabolic approximation. What makes our technique
attractive as compared to other techniques is that each integration step requires solving only two successive tridiagonal systems.
The method resembles alternating direction schemes but its foundation and analysis are different. The main goal of this paper is
to derive this method and to discuss its validity and accuracy.
The theory is then verified by performing two numerical tests, an
azimuthally independent case and azimuthally dependent one. The
first example is for testing the accuracy of the method and for
determining how wide an angle it can accommodate. The second example is for verifying whether angular dependencies are well
handled and for comparing the speed of our method with the speed

of other methods. This last test shows an example where the new method is orders of magnitude faster than existing competing techniques.

2. BACKGROUND

The standard wide angle 3-D wave equation was developed using the classical formulation of the Helmholtz equation in three dimensions, in cylindrical coordinates (r,θ,z):

$$\frac{\partial^2 p}{\partial r^2} + \frac{1}{r}\frac{\partial p}{\partial r} + \frac{1}{r^2}\frac{\partial^2 p}{\partial \theta^2} + \frac{\partial^2 p}{\partial z^2} + k_0^2 n^2 p = 0 \tag{2.1}$$

In the above equation p represents the acoustic pressure, $k_0 = \omega/c_0$, where c_0 is a reference sound speed, $\omega = 2\pi f$, in which f is the frequency of the signal and, finally, $n = n(r,\theta,z) = c_0/c(r,\theta,z)$ is the index of refraction in which $c(r,\theta,z)$ is the sound speed.

We make a standard transformation of the above equation by writing the pressure in the form [7]:

$$p(r,\theta,z) = u(r,\theta,z)v(r)$$

where the factor $v(r)$ represents a rapidly varying portion of the pressure and $u(r,\theta,z)$ is its modulation, a slowly varying function with respect to range. After neglecting small terms, making use of the far-field approximation $(k_0 r \gg 1)$, and rearranging the above equation we obtain:

$$\frac{\partial^2 u}{\partial r^2} + 2ik_0\frac{\partial u}{\partial r} + \frac{1}{r^2}\frac{\partial^2 u}{\partial \theta^2} + \frac{\partial^2 u}{\partial z^2} + (n^2 - 1)k_0^2 u = 0 \tag{2.2}$$

This new equation has been at the origin of the very successful small angle parabolic approximation technique, which consists in simply dropping the second-order derivative with respect to r and integrating the resulting parabolic equation.

It is convenient to define the operators:

$$X = \frac{1}{k_0^2}\frac{\partial^2}{\partial z^2} + (n^2 - 1) \tag{2.3}$$

$$Y = \frac{1}{k_0^2 r^2}\frac{\partial^2}{\partial \theta^2} \tag{2.4}$$

after which the above equation reads as follows:

$$\frac{\partial^2 u}{\partial r^2} + 2ik_0\frac{\partial u}{\partial r} + k_0^2(X + Y)u = 0 \tag{2.5}$$

The standard wide angle PE technique [5] starts by approximately factoring the above operator as the product

$$\left[\frac{\partial}{\partial r} + ik_0 - ik_0 Q\right]\left[\frac{\partial}{\partial r} + ik_0 + ik_0 Q\right] \tag{2.6}$$

in which

$$Q \equiv \sqrt{1 + X + Y} \tag{2.7}$$

Then the operator Q is approximated by a rational function in the form

$$Q \approx \frac{1 + p_1 X + p_2 Y}{1 + q_1 X + q_2 Y} \tag{2.8}$$

which yields the wide angle 'parabolic' equation

$$u_r = \left(-ik_0 + ik_0 \frac{1 + p_1 X + p_2 Y}{1 + q_1 X + q_2 Y}\right) u \tag{2.9}$$

To solve (2.9), Schultz, Lee, and Jackson [4] applied the Crank-Nicolson scheme in conjunction with a preconditioned conjugate gradient method. The application of Crank-Nicolson reduces (2.9) to a sequence of systems of difference equations of the form

$$\left(I - \frac{1}{2} \Delta r L\right) u^{n+1} = \left(I + \frac{1}{2} \Delta r L\right) u^n \tag{2.10}$$

where

$$L \equiv -ik_0 + ik_0 \frac{1 + p_1 X + p_2 Y}{1 + q_1 X + q_2 Y} \tag{2.11}$$

By multiplying both members of (2.10) by the denominator of (2.11), we obtain a marching process, in which a large block-tridiagonal linear system must be solved at each step.

3.   THE NEW APPROACH

Our approach starts with equation (2.5) which is formally considered as an ordinary differential equation with respect to the variable r. For convenience the variables z and θ will therefore be dropped out in the remainder of the paper: u(r) stands for u(r,z,θ). Locally, its formal solution has the form

$$u(r + \Delta r) = e^{-ik_0 \Delta r} e^{ik_0 \Delta r \sqrt{1+X+Y}} u^+(r)$$
$$+ e^{ik_0 \Delta r} e^{-ik_0 \Delta r \sqrt{1+X+Y}} u^-(r)$$

where $u^+(r)$ and $u^-(r)$ are some initial conditions at the range r. The first term in the above solution is the outgoing wave and the second is the incoming wave. In this paper we will neglect back-scattering and therefore the second term will be dropped to yield the local solution

$$u(r + \Delta r) = e^{-\delta} e^{\delta \sqrt{1+X+Y}} u(r) \tag{3.1}$$

in which we have set

$$\delta \equiv i k_0 \Delta r$$

Note that this is also a local solution of the one-way wave equation

$$u_r = (-i k_0 + i k_0 \sqrt{1 + X + Y}) u$$

which is obtained by neglecting the second factor in (2.6).

The approach taken in this paper consists of approximating the term $e^{\delta \sqrt{1+X+Y}}$ in a convenient and accurate manner. An easy way in which this can be done is to use the approximation

$$\sqrt{1 + X + Y} \approx 1 + \frac{1}{2}(X + Y)$$

which yields the standard three-dimensional narrow angle parabolic equation:

$$u_r = \left[\frac{1}{2} i k_0 (n^2 - 1) + \frac{i}{2k_0} \frac{\partial^2}{\partial z^2} + \frac{i}{2k_0 r^2} \frac{\partial^2}{\partial \theta^2} \right] u \equiv Lu$$

This equation was solved by Baer and Perkins [1,3] using a split-step Fourier algorithm. However, this equation accurately represents only narrow angle propagation.

To accommodate wide angle propagation, we consider the higher order approximation

$$\sqrt{1 + X + Y} \approx 1 + \frac{1}{2} X - \frac{1}{8} X^2 + \frac{1}{2} Y \tag{3.2}$$

The corresponding approximation to the wave equation becomes

$$u_r = \left[-i k_0 + i k_0 \left[1 + \frac{1}{2} X - \frac{1}{8} X^2 + \frac{1}{2} Y \right] \right] u \tag{3.3}$$

and formula (3.1) becomes:

$$u(r + \Delta r) = e^{-\delta} e^{\delta (1+\frac{1}{2}X-\frac{1}{8}X^2+\frac{1}{2}Y)} u(r) \tag{3.4}$$

Assuming that $n(r,\theta,z)$ varies slowly with respect to θ, the operators X and Y are nearly commutative, and equation (3.3) yields

$$u(r + \Delta r) = e^{-\delta} e^{\delta(1+\frac{1}{2}X-\frac{1}{8}X^2)} e^{\frac{\delta}{2}Y} u(r) \tag{3.5}$$

In the following we seek an approximation of the term

$$G(\delta,X) = e^{\delta(1+\frac{1}{2}X-\frac{1}{8}X^2)} \tag{3.6}$$

Here, the function G and its approximations should be regarded as functions of the real variable X while δ is an independent parameter. The Taylor series expansion of (3.6) about X = 0 is

$$G(\delta,X) = e^{\delta}\left[1 + \frac{\delta}{2} X + \frac{1}{2!}\left(\frac{\delta^2}{4} - \frac{\delta}{4}\right)X^2\right] + O(X^3) \tag{3.7}$$

We seek an approximation to the function G in the form

$$G(\delta,X) \approx e^{\delta} \frac{1 + pX}{1 + \bar{p}X} \tag{3.8}$$

where p is a complex number to be determined.

Writing that the expansion (3.7) is equal to the right hand side of (3.8) up to $O(X^3)$, we get the equation

$$1 + \frac{\delta}{2} X + \frac{1}{2!}\left(\frac{\delta^2}{4} - \frac{\delta}{4}\right)X^2 = \frac{1 + pX}{1 + \bar{p}X}$$

from which it is easy to obtain p:

$$p = \frac{1}{4} + \frac{\delta}{4}$$

A similar development for the term

$$H(\delta,Y) = e^{\frac{\delta}{2}Y}$$

leads to the approximation

$$H(\delta,Y) \approx \frac{1 + qY}{1 + \bar{q}Y} \tag{3.9}$$

with q = $\delta/4$.

Therefore we get the final expression

$$u(r + \Delta r) = \left[\frac{1 + (\frac{1}{4} + \frac{\delta}{4})X}{1 + (\frac{1}{4} - \frac{\delta}{4})X}\right]\left[\frac{1 + \frac{\delta}{4}Y}{1 - \frac{\delta}{4}Y}\right] u(r) \tag{3.10}$$

This can be rewritten as

$$u(r + \Delta r) = L\ u(r)$$

where

$$L = L_X^{-*} L_X L_Y^{-*} L_Y \tag{3.11}$$

in which

$$L_X = 1 + \left(\frac{1}{4} + \frac{\delta}{4}\right) X \qquad L_Y = 1 + \frac{\delta}{4} Y \tag{3.12}$$

and where X^{-*} stands for the inverse of X^*, the adjoint of X. Equation (3.10) can formally be regarded as an explicit marching scheme. It can easily be seen that the two operators in the denominator of (3.10) are nonsingular because δ is purely imaginary and X and Y are both self-adjoint. In the next section we analyze the accuracy and stability of this scheme, and in Section 5 we will see how to discretize it and use it numerically.

4. THEORETICAL ASPECTS

4.1 Stability

The operators X and Y defined earlier are self-adjoint and therefore their corresponding eigenvalues are real. Since the numerator and denominator of each term between brackets in equation (3.10) are the conjugate of each other, a modal expansion of (3.10) shows that with respect to each mode, the error induced by the marching scheme will not increase exponentially, i.e., the scheme (3.10) is stable. After discretization, the eigenvalues of X and Y will remain real provided boundary conditions are properly handled and the discretizations in the numerators and denominators are calculated at the same range r (see next section). Under these conditions the scheme is stable. Note that the second part of (3.10) is the usual Crank-Nicolson approximation applied here to the term $e^{(\delta/2)Y}$.

4.2 Accuracy

To analyze the local error of the integration scheme (3.10) we must attempt to find an estimate of the difference between the operator

$$e^{-\delta} e^{\delta\sqrt{1+X+Y}} \tag{4.1}$$

and the operator in the right hand side of (3.10). In the following we consider that X and Y are two independent real variables. On the one hand, we find after some calculation that the second order Taylor expansion of the operator (4.1) is

$$e^{-\delta} e^{\delta\sqrt{1+X+Y}} = 1 + \frac{\delta}{2} (X + Y) + \frac{\delta}{8} (\delta - 1) [X^2 + 2XY + Y^2]$$

$$+ O(\| (X,Y) \|^3) \tag{4.2}$$

On the other hand, the second order Taylor expansion of the operator of the right hand side of (3.10) is given by

$$\left[\frac{1 + (\frac{1}{4} + \frac{\delta}{4})X}{1 + (\frac{1}{4} - \frac{\delta}{4})X}\right]\left[\frac{1 + \frac{\delta}{4}Y}{1 - \frac{\delta}{4}Y}\right] = \left[1 + \frac{\delta}{2}X + \frac{\delta}{8}(\delta - 1)X^2 + \cdots\right]$$

$$\times \left[1 + \frac{\delta}{2}Y + \frac{\delta^2}{8}Y^2 + \cdots\right]$$

$$= 1 + \frac{\delta}{2}(X + Y) + \frac{\delta^2}{4}XY$$

$$+ \frac{\delta}{8}(\delta - 1)X^2 + \frac{\delta^2}{8}Y^2 + \cdots \qquad (4.3)$$

The larger terms in the difference between the two (4.2) and (4.3)
are

$$e^{-\delta}e^{\delta\sqrt{1+X+Y}} - \left[\frac{1 + (\frac{1}{4} + \frac{\delta}{4})X}{1 + (\frac{1}{4} - \frac{\delta}{4})X}\right]\left[\frac{1 + \frac{\delta}{4}Y}{1 + \frac{\delta}{4}Y}\right] = -\frac{\delta}{4}XY - \frac{\delta}{8}Y^2$$

$$+ O(\|(X,Y)\|^3)$$

Thus, one can expect good accuracy when Y is very small and X
is small. The above error is better than an error of the form
$O(X^2)$, because the error expression is the product of three small
terms, namely, δ, Y, and X. Moreover, we have assumed that the
term Y is much smaller than the term X.

When using a marching scheme, an upper bound for the global
error at some point can be derived from the above local truncation
error by expressing the error

$$e_h^{j+1} \equiv u_h^{j+1} - u^{j+1}$$

where u_h^{j+1} represents the computed solution at step j and u^{j+1} the
exact solution at the same point. On the one hand, we have

$$u_h^{j+1} = L_h u_h^j$$

where L_h is the discretization of the operator L as defined by
(3.11). On the other hand

$$u^{j+1} = L_h u^j + \varepsilon_j$$

where ε_j is the truncation error incurred at step j. Hence,

$$e_h^{j+1} = L_h e_h^j + \varepsilon_j$$

and since the operator L_h is unitary, we have

$$\|e_h^{j+1}\| \leq \sum_{i=0}^{j} \|\varepsilon_j\|$$

Since the number of steps in range is $O(\delta^{-1})$, the above analysis shows that the norm of the global error $\|e_h^{j+1}\|$ is $O(X^2)$.

5. FINITE DIFFERENCE SOLUTION

To employ formula (3.10) numerically, we must discretize the operators X and Y by central differences and replace the corresponding operators by their discrete analogues. However, we first put equation (3.10) in the form

$$u^{j+1} = \left[1 + \left(\frac{1}{4} - \frac{\delta}{4}\right)X\right]^{-1}\left[1 + \left(\frac{1}{4} + \frac{\delta}{4}\right)X\right]\left[1 - \frac{\delta}{4}Y\right]^{-1}\left[1 + \frac{\delta}{4}Y\right]u^j$$

(5.1)

There are several ways of rewriting (5.1) exploiting the commutativity of the operators L_X and L_X^{-*} and of L_Y and L_Y^{-*}. The near commutativity of X and Y can also be exploited to derive alternative formulae. For example, we may consider the scheme

$$\left[1 + \left(\frac{1}{4} - \frac{\delta}{4}\right)X\right]\left[1 - \frac{\delta}{4}Y\right]u^{j+1} = \left[1 + \left(\frac{1}{4} + \frac{\delta}{4}\right)X\right]\left[1 + \frac{\delta}{4}Y\right]u^j$$

(5.2)

Although not obvious at first, the above scheme is also unconditionally stable. The reason for this is that the corresponding operator $\tilde{L} = L_Y^{-*}L_X^{-*}L_X L_Y$ is also unitary as is readily seen by forming $\tilde{L}\tilde{L}^*$ which is found to be the identity operator. The question as to which of the various schemes is to be preferred is certainly worth further investigation but we will not pursue it in the present paper.

Let us denote by A the finite difference approximation of the operator $L_X = I + (\frac{1}{4} - \frac{\delta}{4})X$ and by B that of $L_Y = I + \frac{\delta}{4}Y$. Both matrices are tridiagonal with the structure indicated in Table 1.

TABLE 1. Coefficients of the Two Tridiagonal Matrices A and B

	Super diagonal	Diagonal	Sub diagonal
A	$\left(\dfrac{1}{4} - \dfrac{\delta}{4}\right)\dfrac{1}{k_0^2 h^2}$	$1 + \left(\dfrac{1}{4} - \dfrac{\delta}{4}\right)(n^2 - 1) - 2\left(\dfrac{1}{4} - \dfrac{\delta}{4}\right)\dfrac{1}{k_0^2 h^2}$	$\left(\dfrac{1}{4} - \dfrac{\delta}{4}\right)\dfrac{1}{k_0^2 h^2}$
B	$-\dfrac{\delta}{4}\dfrac{1}{k_0^2 r^2}\dfrac{1}{(\Delta\theta)^2}$	$1 + \dfrac{\delta}{2}\dfrac{1}{k_0^2 r^2}\dfrac{1}{(\Delta\theta)^2}$	$-\dfrac{\delta}{4}\dfrac{1}{k_0^2 r^2}\dfrac{1}{(\Delta\theta)^2}$

When solving tridiagonal systems with the matrices A and B it is of interest to know whether these matrices are diagonally dominant or not. While the matrix B is always diagonally dominant, the situation is more complicated for A. In the simple case where $n(r,\theta,z) \equiv 1$, the matrix is conditionally, i.e., for $h \geq 1/k_0$, diagonally dominant in the sense that the modulus of the diagonal term is greater than or equal to the sum of the moduli of the off-diagonal terms in the same row. The more general case where n is arbitrary is not easy to analyze.

The scheme (5.1) becomes

$$u^{j+1} = A^{-*}AB^{-*}Bu^j \tag{5.3}$$

Note that we evaluate the matrices A and B at mid distance between u^{j+1} and u^j, i.e., at range $r + \Delta r/2$. This is in order to ensure that the operators A* and A, as well as B* and B, form two pairs of operators that are conjugate of each other. This choice will guarantee stability as was seen in Section 4.1.

To perform one step of (5.3) we must start by computing $w^j = Bu^j$ and solve the tridiagonal system

$$B*v^{j+1} = w^j$$

Then we compute $v^{j+1} = A^{-1}A*u^{j+1}$ and solve the tridiagonal system

$$A*u^{j+1} = Av^{j+1} \tag{5.4}$$

Thus there are two multiplications of a tridiagonal system by a vector and two tridiagonal systems to solve at every step.

6. TEST EXAMPLES

The numerical scheme (5.3) has been implemented into a research computer code which is used to predict wave fields at required ranges. We used the marching scheme (5.2) instead of the original scheme (5.1). The resulting values are compared against a known solution to check the validity as well as the accuracy of our scheme. All computations are made on the VAX-11/780 computer using single precision complex arithmetic. We present two examples. The first one is an azimuthally independent case and the second is an azimuthally dependent one. The input parameters for both examples are shown in Table 2 for convenience.

TABLE 2. Parameters for the Two Test Problems

Input parameters	Problem 1	Problem 2
Source	300 m	10 m
Initial range	100 m	10 m
Source frequency	20 Hz	20 Hz
Bottom depth	400 m	20 m
Sound speed	1500 m/s	1500 m
Reference sound speed	1500 m/s	1500 m/s
Receiver depths	156 m, 312 m	5 m
Propagating sector	-2-1/2, +2-1/2	-5, 5
Depth increment	4 m	0.2 m
Range step size	1 m	0.001 m, 0.25 m
Angular increment	0.5°	1°
Maximum range	1 km	10.5 m
Surface condition	pressure release	Dirichlet
Bottom condition	rigid	Dirichlet
Size of matrices A and B	1000	1000

6.1 An Azimuthally Independent Case

To start the computation, the initial field is taken from the following formula borrowed from [6]:

$$u(r,z) = \frac{i}{2z_h} \sum_{j=0}^{\infty} \sin(kz_0\sqrt{1 - a_j^2})\sin(kz\sqrt{1 - a_j^2})H_0^{(1)}(ka_j r) \tag{6.1}$$

where a_j satisfies

$$a_j = \sqrt{1 - \left[\frac{(j + 1/2)\pi}{kz_h}\right]^2} \tag{6.2}$$

We are concerned with the propagating modes, namely, those for which a_j remains real.

In the computation the sector is divided into 10 portions. The two extreme sectors represent sector boundaries where the solution is supplied from the exact solution. The inner sectors, because of the azimuthal independence have the same initial values. It is expected that at different sectors at the same depth, the computed wave fields should be identical. We examined this hypothesis. Another verification we did was to look at the size of the angle of propagation. To simulate this we took the mode index j to be 9 so that we could obtain an angle of propagation of around 52°. Our method could handle such a wide angle without any major difficulty.

                    D. Lee, Y. Saad and M.H. Schultz

TABLE 3. Wave Field Results at 1 km Range: Accuracy Test

z_r (m) $\diagdown$ θ	$-\frac{1}{2}$ degree		$+\frac{1}{2}$ degree	
156	-0.59261E-04	0.20995E-04	-0.59259E-04	0.20995E-04
	-0.59224E-04	0.20954E-04	-0.59224E-04	0.20954E-04
312	-0.96924E-04	0.34283E-04	-0.96924E-04	0.34285E-04
	-0.96908E-04	0.34287E-04	-0.96908E-04	0.34287E-04

TABLE 4. Wave Field Results at 1 km Range: Angle of Propagation Measurements

Mode j	Angle size (degrees)	Results Complex values	Results dB
6	31.03°	(0.13749E-04, -0.23722E-05)	97.107
		(0.15624E-04, -0.12743E-05)	96.095
7	37.54°	(-0.96649E-05, 0.25102E-05)	100.013
		(-0.68974E-05, 0.23300E-05)	102.757

Our results are summarized in Tables 3 and 4. Table 3 shows the accuracy achieved and Table 4 shows the angle of propagation for the case of 8 modes and 9 modes. In the tables the values appearing in the first row represent the calculated values by our new method; the values in the second row are the exact solutions. The first columns are the real parts and the second colums are the imaginary parts.

6.2 An Azimuthally Dependent Case

This second example deals with a low frequency propagation in shallow water. To construct an azimuthally dependent case, we modified a reference solution tested by Chan, Shen, and Lee [2] and used the same exact input parameters to derive a system of equations with the same dimension in order to compare the computation speed. An exact solution, after the modification, to the wide angle three-dimensional wave equation (3.3) can be expressed by

$$u(r,\theta,z) = e^{-\Omega z} e^{im\theta} e^{im^2/(2k_0 r)} \tag{6.3}$$

The expression is used to generate the initial field. The input parameters used are shown in the second column of Table 2. The surface condition is taken to be

$$u(r,\theta,0) = e^{im\theta} e^{im^2/(2k_0 r)} \tag{6.4}$$

and the bottom condition was

$$u(r,\theta,z_{max}) = e^{-\Omega z_{max}} e^{im\theta} e^{im^2/(2k_0 r)} \tag{6.5}$$

The scalar Ω in equation (6.5) is chosen to be $2k_0$. The angular modal number m is taken to be 3. To obtain an accuracy of 10^{-2}, as in [2], their methods need to take a range step size of 0.001 m.

We tested two different step sizes: 0.001 m and 0.25 m. The experiment with the first range step-size is only done for a comparison with reference [2]. We should point out that such a small step size for the 5-point method of Chan, Shen, and Lee is necessary because the scheme is explicit. In this computation, we found that our method was approximately 1.6 times faster than the 5-point explicit method of [2], and 17 times faster than Crank-Nicolson of [4]. Note that the Crank-Nicolson of [4] uses a stable version of the conjugate gradient method applied to the normal equations, called Craig's method. Using the second step size of 0.25 m, we found that the same accuracy could be achieved by our method as with $\Delta r = 0.001$, but the execution was much faster. Here our method is approximately 160 times faster than the 5-point method and 1600 times faster than Crank-Nicolson. Results are displayed in Table 5.

TABLE 5.   Results for Problem 2

Method	Δr	Relative error	CPU time (h-m-s)
Crank-Nicolson	0.001	(0.18E-01, −0.12E-01)	03-47-10
5-Point explicit	0.001	(0.10E-01, −0.11E-01)	00-21-35
New method	0.001	(0.26E-02, −0.11E-02)	00-13-12
New method	0.25	(0.22E-02, −0.11E-01)	00-00-09

7.   CONCLUSION

Obtaining solutions to ocean acoustic propagation in three dimensions can be very complicated and computationally expensive. Moreover, it is now becoming the general consensus that two-dimensional models are no longer sufficiently representative. Efficient methods and clever implementations for dealing with three-dimensional wave propagation are therefore very important.

The new method proposed in this paper is not only a fast and accurate method, but also has the property of being as representative a model as other well known existing 3-D models. Our numerical results have demonstrated that the method is efficient and have confirmed the theory that it is also stable.

Our new approach is based on considering a form of the 3-D wave equation as an ordinary differential equation with respect to range. Then a formal expression of the solution is written in terms of the exponential of the square-root of some operator. The artifice used in this paper is to approximate this exponential in a clever way by the product of two rational functions of the type (1,1). As a consequence, the resulting ODE-integration process requires only two successive tridiagonal system solutions. The theory shows that our method is unconditionally stable. Moreover, it is so accurate that larger step-sizes can be afforded resulting in substantial savings in computational times. This has been widely confirmed by the numerical tests. Moreover, angles of propagation as wide as 31 degrees have been accurately handled.

ACKNOWLEDGMENTS

This work was supported in part by the Office of Naval Research under contracts N00014-82-K-0184, N00014-85-WR-24068, N00014-85-WR-24268, N00014-86-WR-24233, and in part by Naval Underwater Systeme Center Independent Research Project A65020.

The authors would like to thank George Botseas for his technical assistance in developing the research computer code and for producing the numerical results in this paper.

REFERENCES

1. Baer, R. N., Propagation through a three-dimensional eddy including effects on an array, JOURNAL OF THE ACOUSTICAL SOCIETY OF AMERICA, 69 (1981), 70-75.

2. Chan, T. F., L. Shen, and D. Lee, Difference schemes for parabolic wave approximation in ocean acoustics, JOURNAL OF COMPUTERS AND MATHEMATICS WITH APPLICATIONS, 11 (1985), 747-754.

3. Perkins, J. S., and R. N. Baer, An approximation to the three-dimensional parabolic equation method for acoustic propagation, JOURNAL OF THE ACOUSTICAL SOCIETY OF AMERICA, 72 (1982), 515-522.

4. Schultz, M. H., D. Lee, and K. R. Jackson, Application of the Yale sparse technique to solve the three-dimensional parabolic wave equation, in RECENT PROGRESS IN THE DEVELOPMENT AND APPLICATION OF THE PARABOLIC EQUATION, P. D. Scully-Power and D. Lee, eds., Naval Underwater Systems Center Technical Document 7145, 1984.

5. Siegmann, W. L., G. A. Kriegsmann, and D. Lee, A wide angle three-dimensional parabolic wave equation, JOURNAL OF THE ACOUSTICAL SOCIETY OF AMERICA, 78 (1985), 659-664.

6. St. Mary, D. F., Analysis of an implicit finite difference scheme for very wide angle underwater acoustic propagation, PROCEEDINGS OF THE 11TH IMACS WORLD CONGRESS, IMACS, 1985, pp. 153-156.

7. Tappert, F. D., The parabolic approximation method, in WAVE PROPAGATION AND ACOUSTICS, J. B. Keller and J. Papadakis, eds., Springer-Verlag, New York, 1977.

COMPUTATIONAL ACOUSTICS: Wave Propagation
D. Lee, R.L. Sternberg, M.H. Schultz (Editors)
Elsevier Science Publishers B.V. (North-Holland)
© IMACS, 1988

FINITE DIFFERENCE COMPUTATIONS OF
THREE-DIMENSIONAL SOUND PROPAGATION

William L. Siegmann

Rensselaer Polytechnic Institute
Troy, New York

Ding Lee

Naval Underwater Systems Center
New London, Connecticut

George Botseas

Naval Underwater Systems Center
New London, Connecticut

ABSTRACT

Several parabolic-type approximations have been developed for single frequency, three-dimensional underwater acoustic propagation. One wide angle three-dimensional equation has been solved efficiently by finite difference techniques [5]. Accuracy of this new algorithm is indicated by comparison of computational results with analytic solutions. Sensitivity of results to variations in step sizes is discussed. Computations suggest the types of three-dimensional propagation effects which can be produced. An example is provided of the capabilities of the method for sound propagation through ocean features, such as a frontal zone, which can vary in three dimensions.

1. INTRODUCTION

Two closely linked problems must be resolved in order to develop an effective capability for determining acoustic propagation through ocean environments which vary in three spatial dimensions. The first is the formulation of appropriate model equations, which should describe sound transmission under broad enough conditions for many applications of interest. At the same time, the model equations must lend themselves to approximate numerical solution, for which the construction of an accurate and efficient computational scheme is the second fundamental problem.

The first (or modeling) problem was the focus of a recent paper [1] that presented a framework for wide angle three-dimensional (3-D) parabolic approximations. This work is an evolution

from the 3-D parabolic approximation which was introduced in Ref.
2 for near-axial propagation. The background for, and other features of, parabolic approximations are contained in Ref. 2 and recent modeling advances involving parabolic approximations are summarized in [3]. The second (or computational) problem is complicated by the requirements of efficiency and of varied applications.
Although the first implementation [4] was a novel advance, it was
not computationally fast. A new method of numerical solution has
been developed [5] that requires relatively little computational
effort. Preliminary tests have demonstrated the algorithm's potential for speed and accuracy.

The principal purpose of this paper is to provide new tests
of the algorithm in [5] and to indicate some of its capabilities
for 3-D acoustic propagation. In Section 2 we characterize the
types of physical situations to which the method is expected to be
applicable. Test solutions which generate simple interference
patterns in transmission loss are described in Section 3. These
are used in Section 4 to provide accuracy tests for the algorithm,
to demonstrate variations with computational step sizes, and to
indicate some 3-D versus 2-D influences on propagation modes. A
model propagation problem is considered in Section 5 to further
test the algorithm's accuracy. We remark that all computations
described in this paper were performed on a VAX 11/780. For our
purposes here, the computational power of this machine was adequate. However, its limitations were responsible for some of our
selections of restricted computational step and domain sizes. As
the new algorithm becomes available on larger computers and on
parallel processors, such constraints will be removed.

2. WIDE ANGLE EQUATION

Previous developments of wide angle 3-D acoustic propagation
equations are discussed in detail in [1] and summarized in [5].
The latter paper presents a new and efficient algorithm for solving one of these equations. We indicate here some modeling implications of that algorithm.

One way to view partial differential equations which approximate the Helmholtz equation is by an operator formalism [2]. From
this perspective, the primary operator approximation in [5] [cf.
Eq. (3.2)] is, using notation identical to that paper's,

$$\sqrt{1 + X + Y} \approx 1 + \frac{1}{2} X - \frac{1}{8} X^2 + \frac{1}{2} Y \tag{1}$$

where the operators are

$$X = \frac{1}{k_0^2} \frac{\partial^2}{\partial z^2} + (n^2 - 1) \qquad\qquad Y = \frac{1}{k_0^2 r^2} \frac{\partial^2}{\partial \theta^2} \tag{2}$$

The approximation in Eq. (1) retains the terms shown in an expansion of the square root but neglects X^3, Y^2, and XY terms. In a formal sense, Eq. (1) should be appropriate if the effects of operator Y are smaller than those of operator X but larger than those of operator X^3.

What wide angle 3-D parabolic-type approximation does Eq. (1) generate? We observe that an approximation which is equivalent to Eq. (1) to the powers of the operators retained is

$$\sqrt{1 + X + Y} \approx \frac{1 + \frac{3}{4} X}{1 + \frac{1}{4} X} + \frac{1}{2} Y \tag{3}$$

Applying the formalism in Sec. I of [1], it follows that the propagation equation which corresponds to terms retained in Eq. (3) is

$$\left(1 + \frac{1}{4} X\right)\frac{\partial u}{\partial r} = \frac{ik_0}{2} (X + Y)u \tag{4}$$

Using Eqs. (2) and (4) yields the partial differential equation

$$\left(1 + \frac{1}{4} (n^2 - 1) + \frac{1}{4k_0^2} \frac{\partial^2}{\partial z^2}\right)\frac{\partial u}{\partial r} = \frac{ik_0}{2}\left[(n^2 - 1) + \frac{1}{k_0^2} \frac{\partial^2}{\partial z^2}\right.$$

$$\left. + \frac{1}{(k_0 r)^2} \frac{\partial^2}{\partial \theta^2}\right]u \tag{5}$$

Even though Eq. (5) is third order, it is convenient to refer to it as a 3-D parabolic equation (PE). As the operator approximations Eqs. (1) or (3) show, Eq. (5) accounts for propagation effects in the vertical (z) direction which are higher order than those in the cross-range azimuthal (θ) direction. Because of the correspondence between its terms and those in previous 3-D PEs, Eq. (5) may be characterized as wide angle in the vertical direction and narrow angle in the azimuthal direction. We remark that the right sides of Eqs. (1) and (3) differ from each other, and from a formal Maclaurin expansion of the square root operator, in the X^3, Y^2, and XY terms.

Another way to view approximations to the Helmholtz equation is by asymptotic procedures [2]. In particular, the multiscale formalism of [1, Sec. III] may be carried out to derive an analog of Eq. (5). Principal scaling assumptions are that the variables r, θ, and z can be written in terms of scaled quantities, with asterisks, as

$$r^* = \varepsilon k_0 r \qquad z^* = \varepsilon^{1/2} k_0 z \qquad \theta^* = \theta \qquad (6)$$

One way to define the small parameter ε in Eqs. (6) is as a measure of the deviation of the squared index of refraction from one,

$$n^2(r,z,\theta) = 1 + \varepsilon \eta(r^*,z^*,\theta^*) \qquad (7)$$

With Eqs. (6) and (7) the multiscale development produces a dimensionless equation corresponding to Eq. (5). The result differs only in including a near source term and a term with the range derivative of the refractive index, both of which are explicitly neglected in the operator formalism. It is worth noting that this analog of Eq. (5) is the direct result of Eqs. (6) and (7), which in combination with Eq. (2) produce

$$X = \varepsilon \frac{\partial^2}{\partial z^{*2}} + \varepsilon \eta \qquad Y = \varepsilon^2 \frac{1}{r^{*2}} \frac{\partial^2}{\partial \theta^{*2}} \qquad (8)$$

Consequently, the operator Y is required to have effects comparable to the operator X^2. This result is fully consistent with, and in fact more specific than, the condition mentioned for appropriateness of the approximation in Eq. (1).

The principal conclusion is that both operator and asymptotic formalisms lead to Eq. (5) as the 3-D PE which is, to the order of terms retained, the same as that solved by the algorithm of [5]. Moreover, Eq. (5) is seen to account for relatively wide angle propagation in the channel depth direction. Thus, it should have advantages similar to those of previous wide angle two-dimensional PE models, such as [6] and [7]. It incorporates some coupling between propagation in different azimuthal directions, which distinguishes it from another implementation [8] designed for situations with no exchange of acoustic energy across azimuthal directions. Finally, Eq. (6) indicates conditions on the relative magnitudes of spatial rates of change where Eq. (5) is appropriate. In particular, cross-range variability should be comparable to range variability, since $r^{-1}\partial/\partial\theta$ is scaled like $\partial/\partial r$, and vertical variability should be a factor of $\varepsilon^{-1/2}$ larger. As a numerical illustration, a sound-speed change of 10 m s^{-1} corresponds to a

value of $\varepsilon = 0.0067$. If the 10 m s^{-1} change occurs over a depth
of 1 km, then the same change should occur over a horizontal distance of about 12 km or more. These estimates indicate that sound
propagation through a wide range of ocean volume effects, including mesoscale eddies, rings, and fronts, can be treated by the
algorithm of [5]. Incorporation of other 3-D phenomena, such as
bottom topographical and structural variations, into the algorithm
are under development.

3. TEST SOLUTIONS COMPARISONS

To investigate the accuracy and efficiency of computationally-
intensive algorithms for ocean acoustic propagation problems, the
use of exact solutions as test examples has several advantages [9].
The simplest solutions for this purpose consist of normal modes.
In this section we consider these test examples for Eq. (5) and
related equations, in order to show features of wide angle 3-D
propagation.

Separable modal solutions for PEs are guaranteed to exist
under well-known assumptions, including horizontal bottom and surface boundaries and depth-stratified sound speed. For simplicity
we specialize further to an isospeed channel with sound speed c,
pressure release surface z = 0, rigid bottom z = H, and periodic
behavior in θ. We are particularly interested in the squared magnitude of the complex solutions to Eq. (5), since propagation loss
from a PE solution u is conventionally given by

$$PL = -10 \log_{10} |u|^2 + 10 \log_{10} r \tag{9}$$

Since $|u|^2$ for a single mode of Eq. (5) is always constant at
any depth, we consider a combination of modal solutions. The relative intensity from two interfering modes can have enough range
and azimuth variation to provide significant tests of the accuracy
of numerical computations which attempt to match it. At the same
time, the interference patterns readily show features of wide
angle and narrow angle propagation without the complications of
many interacting modes. Consequently, the initial condition u_0 at
range $r = r_0$ for Eq. (5) is taken as

$$u_0(r_0,\theta,z) = \sum_{j=1}^{2} e^{iM_j\theta} Z_j(z) \tag{10}$$

where

$$Z_j(z) = \alpha_j \sin(\mu_j k_0 z) \tag{11}$$

and

$$\mu_j = (N_j - 1/2)\pi/k_0 H \tag{12}$$

In Eqs. (10)-(12) M_j is an integer azimuthal mode number, α_j is a relative mode amplitude, and N_j is an integer vertical mode number. The constraint for propagating modes is $\mu_j < 1$, i.e., the effective mode number $N_j - 1/2$ should be less than twice the depth in wavelengths.

The magnitude squared of the exact solution u_w to the wide angle Eq. (5) under the aforementioned conditions is

$$|u_w|^2 = Z_1^2(z) + Z_2^2(z) + 2Z_1(z)Z_2(z) \cos \Psi_w(r,\theta) \tag{13}$$

In Eq. (13) the angle Ψ_w is given by

$$\Psi_w(r,\theta) = (M_1 - M_2)\theta - \frac{R_1(r)}{1 - \mu_1^2/4} + \frac{R_2(r)}{1 - \mu_2^2/4} \tag{14}$$

where

$$R_j(r) = \frac{1}{2} \mu_j^2 k_0 (r - r_0) - \frac{M_j^2}{2k_0}\left(\frac{1}{r} - \frac{1}{r_0}\right) \tag{15}$$

The reference sound speed c_0 is chosen as c. At any depth z, the deviation of transmission loss from cylindrical spreading is due to the variation of Ψ_w with r and θ. Note that Ψ_w can be characterized as having three components. The first is periodic in θ with azimuthal wave number $M_1 - M_2$. The second component is the difference in two terms which grow linearly with range. The third component, the difference in two terms which decay algebraically with range, represents a variation due explicitly to three dimensionality in this example.

For comparisons, we consider the "standard" 3-D PE [2],

$$\frac{\partial u}{\partial r} = \frac{ik_0}{2}\left[(n^2 - 1) + \frac{1}{k_0^2}\frac{\partial^2}{\partial z^2} + \frac{1}{(k_0 r)^2}\frac{\partial^2}{\partial \theta^2}\right]u \tag{16}$$

This equation is produced by a linear approximation of the square root operator in Eq. (1) and hence is restricted to a relatively narrow aperture of effective propagation angles. The magnitude squared of the exact solution u_N to Eq. (16) is the same as Eq. (13) but with Ψ_w replaced by

$$\Psi_N = (M_1 - M_2) - R_1(r) + R_2(r) \tag{17}$$

Also, the standard wide angle two-dimensional PE (or narrow-angle 2-D PE) is Eq. (5) [or Eq. (16)] with the last term on the right absent. The squared magnitudes of the exact solutions for the same conditions again follow from Eq. (13) with either Eq. (14) or Eq. (17), respectively. However, the last term in Eq. (15) is dropped for the 2-D PEs. We shall see in the next section how effects of this term can influence propagation.

4. NUMERICAL COMPUTATIONS

Propagation loss plots for the test solutions of Section 3 are presented in this section. Parameter values selected were $c = 1500$ m s^{-1}, $H = 100$ m, $f = 50$ Hz, $M_1 = 24$, $M_2 = 36$, $N_1 = 3$, $N_2 = 5$, and $\alpha_1 = \alpha_2 = 1$. Source depth is 10 m, and curves are shown for receiver depth of 60 m. Unless stated otherwise, initial range r_0 was taken as 500 m. Solutions are calculated within an angular sector $|\theta| \le \theta_0 = 2$ deg in order to limit run times.

We first describe some of the computations performed to determine the sensitivity of results to computational step sizes. The exact propagation loss at $\theta = 0$ deg, from evaluating Eqs. (9) and (13)-(15), is shown in Fig. 1(a) for ranges between 500 m and 3 km. To generate the other figures using the 3-D algorithm, the exact initial condition, Eq. (10), was applied. Also, the exact values of the solution u_w were used at the sector boundaries, which were taken at $\theta = \pm(\theta_0 + \Delta\theta)$ where $\Delta\theta$ is the azimuthal step size. The upper and lower boundary conditions are $u = 0$ and $u_z = 0$, respectively, as noted in Section 3. Figure 1(b) shows the computed result with range and vertical step sizes $\Delta r = \Delta z = 0.5$ m and with $\Delta\theta = 0.25$ deg. Note the resolution of the fade and peak locations and the overall level behavior, compared to the exact solution. We increased Δr and Δz individually and found only a slow overall deterioration of the prop loss pattern. For example, Fig. 1(c) has Δr as a factor of 8 larger than in Fig. 1(b). When only Δz is increased by a factor of 8, the computed result is also very close to the exact solution. In Fig. 1(d), both Δz and Δr are increased to values of 4 m, which in this example is about 13% of the acoustic wavelength. Only relatively large step sizes showed a significant degradation of the pattern. We continued calculations as far out in range as 20 km, with the same results.

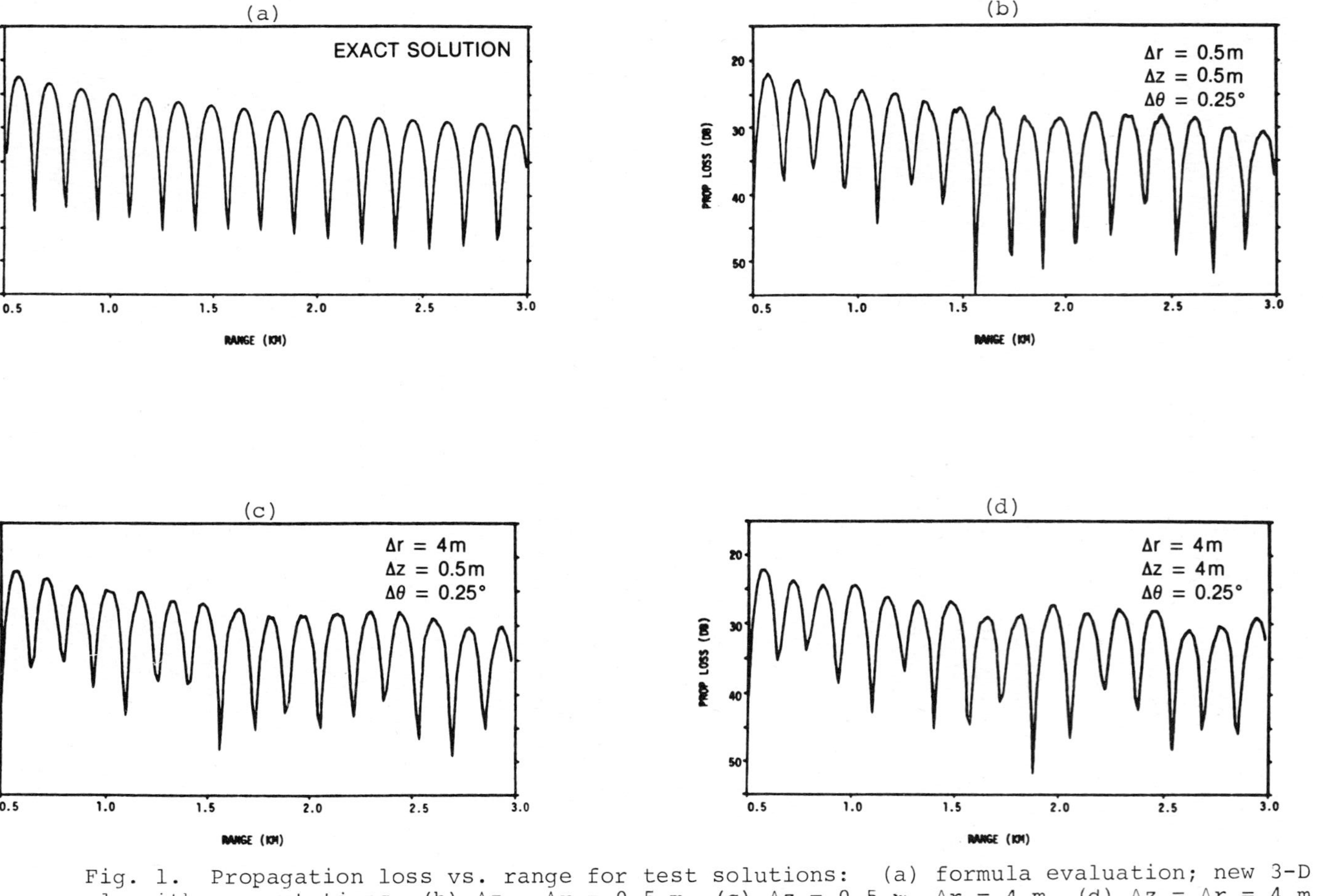

Fig. 1. Propagation loss vs. range for test solutions: (a) formula evaluation; new 3-D algorithm computations, (b) $\Delta z = \Delta r = 0.5$ m, (c) $\Delta z = 0.5$ m, $\Delta r = 4$ m, (d) $\Delta z = \Delta r = 4$ m.

From these and other computations, we conclude that the algorithm is robust with respect to variations in range and vertical step sizes.

Reducing the values of Δr and Δz did not noticeably change the agreement between Figs. 1(b) and 1(a). However, reducing the value of $\Delta\theta$ to 0.125 deg did produce changes, such as smoothing near the peaks and making more uniform the fade depths. Note from Eqs. (13) and (14) that prop loss has an azimuthal periodicity of 30 deg for our example. This was confirmed by noting the phase shift between the computed patterns at $\theta = \pm 2$ deg. Thus, it might be questioned why such small $\Delta\theta$ values should be needed to resolve the pattern or to improve agreement with the exact solution.

To investigate the behavior with $\Delta\theta$, we looked more closely at the pattern between 500 m and 1 km at $\theta = 0$ deg. The curve in Fig. 2(a), with $\Delta\theta = 0.1$ deg and $\Delta r = \Delta z = 1$ m, is very close to the exact solution (not shown). In Fig. 2(b), $\Delta\theta$ is increased to 0.25 deg, with negligible change in the overall levels of the prop loss. The level starts to deteriorate in Fig. 2(c), where $\Delta\theta = 0.5$ deg, and it mismatches by several dB in Fig. 2(d), where $\Delta\theta = 1$ deg. The reason is that the solution for u_w contains a component with azimuthal periodicity of 6 deg in our example. Even though this component does not appear in the prop loss, it is generated as the algorithm solves Eq. (5). Evidently this component requires a relatively finer grid than the 30 deg periodicity of the prop loss would suggest. We remark that the plot corresponding to Fig. 2(d) with $\Delta r = \Delta z = 0.5$ m differs very little. The conclusion from these and other computations is that the algorithm is insensitive to computational step sizes which are not large fractions (i.e., less than about 10%) of the solution wavelengths. We also remark that computations confirm theoretical predictions that doubling any step size nearly halves execution time [5].

We next illustrate the differences between the wide angle and narrow angle 3-D exact solutions and the corresponding 2-D solutions. Figure 3(a) shows the solution from Eqs. (9) and (13)-(15) for ranges between 1 m and 2.5 km at $\theta = 0$ deg. The peaks of the curve show cylindrical-spreading decay, and the effects of the second term in Eq. (15) appear as closer spacing of the arches for smaller ranges. The same characteristics are visible in Fig. 3(b), which follows from the exact 3-D narrow angle solution u_N. The principal difference between Figs. 3(a) and (b), the more rapid near-periodicity of the arches for larger ranges, corresponds to

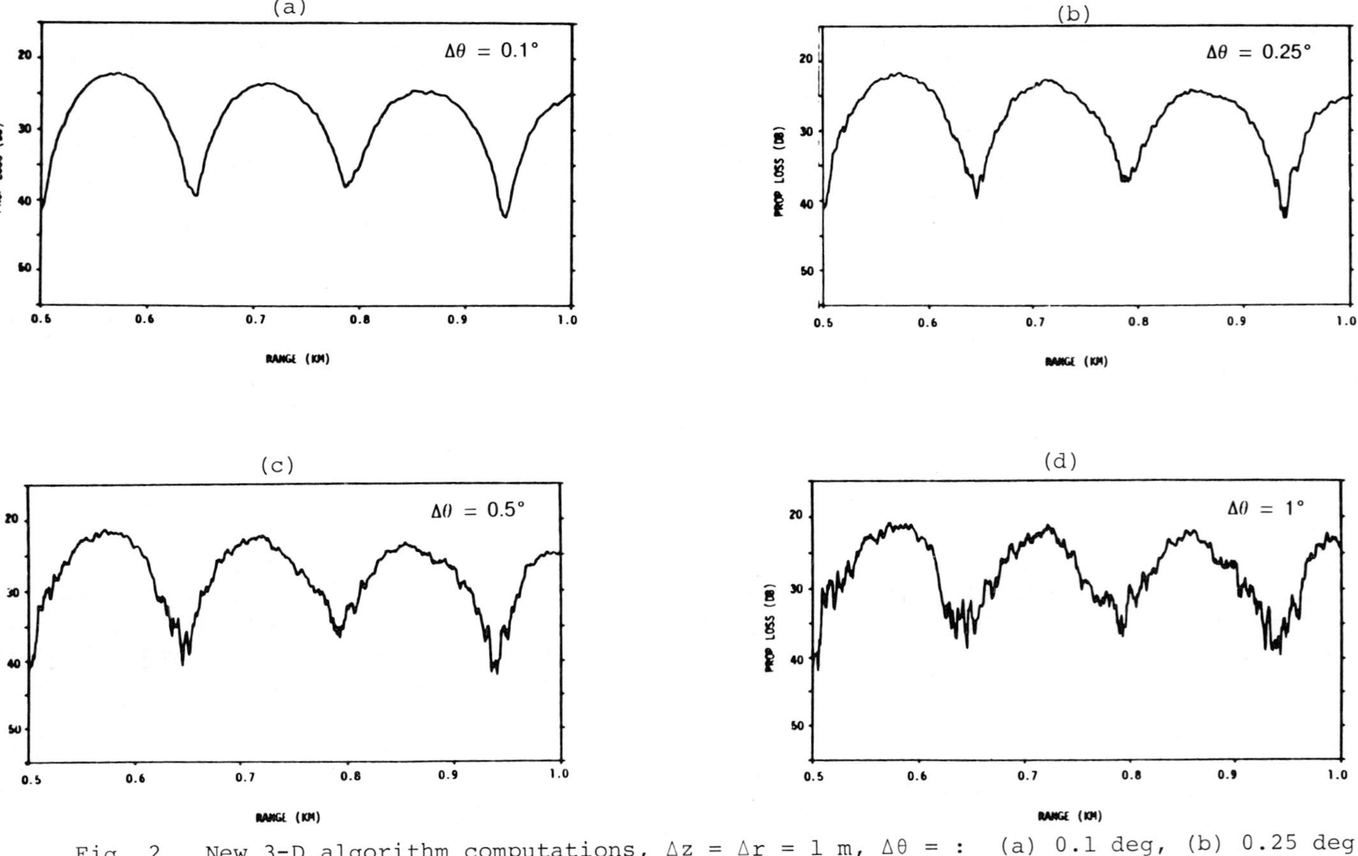

Fig. 2. New 3-D algorithm computations, $\Delta z = \Delta r = 1$ m, $\Delta\theta = :$ (a) 0.1 deg, (b) 0.25 deg, (c) 0.5 deg, (d) 1 deg.

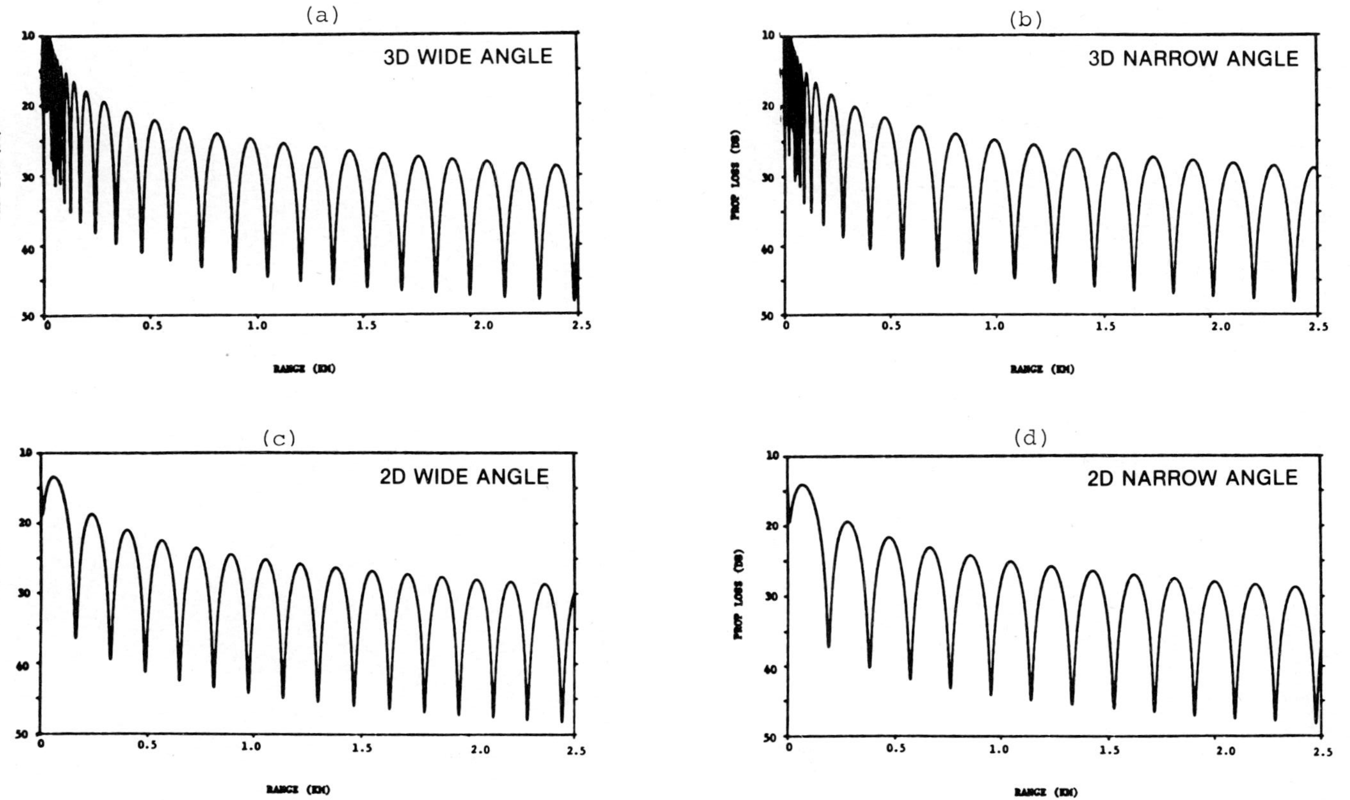

Fig. 3. Formula evaluations for test solutions: (a) wide angle 3-D, (b) narrow angle 3-D, (c) wide angle 2-D, (d) narrow angle 2-D.

the differences between Eqs. (14) and (17). The exact solutions
for wide and narrow angle 2-D equations are shown in Figs. 3(c)
and (d). Note that the arch peaks and fades are periodic for all
ranges for the 2-D solutions. Also, the periods for the 2-D solutions are close to the arch spacing at the right side of the corresponding 3-D figures. The corresponding 2-D and 3-D solutions
do agree for longer ranges, where the last term in Eq. (15) is
negligible. Figure 3 illustrates for our simple exact solutions
that wide angle effects persist for all ranges, while 3-D effects
are confined to smaller ranges. In view of this, can 3-D effects
exert any influence on prop loss beyond smaller ranges?

To answer this question using the test solutions, we considered different side boundary conditions for the sector while keeping the initial, top, and bottom conditions exact. For the 3-D
computations in Figs. 1 and 2, the side boundary values are those
of the exact solution u_w. However, in typical 3-D propagation
problems, such values are not available. In practice, side boundary conditions would be either some assumed physical model, such
as an absorbing layer, or values obtained from a 2-D PE calculation. The latter assumes that 3-D effects can be ignored at the
side boundaries. An example of this procedure is to use a narrow
angle 2-D PE to obtain values for the pressure u at the side
boundaries, and then use the new 3-D algorithm to compute the solution throughout the sector. For our problem, we obtain the side
boundary values from the 2-D narrow angle exact solution. The re-
sult in Fig. 4(a), with $\Delta\theta = 0.125$ deg and $\Delta r = \Delta z = 1$ m at the
sector center, fails to match either level or pattern of the exact
solution Fig. 1(a).

Since the 3-D algorithm is wide angle in the vertical, a seemingly consistent approach would be to use the wide angle 2-D solution values at the side boundaries. Figure 4(b) indicates that
the 3-D algorithm can reproduce only a little more than one arch
of the exact solution before the computed solution starts to
seriously disagree. However, Fig. 3 shows that 3-D effects in
our exact solutions, and specifically in the side boundary values,
decay for sufficiently long ranges. Is it possible that the 3-D
solution computed as in Fig. 4(b) eventually matches the exact
solution far enough out in range? Figure 4(c), showing the continuation of the 3-D calculation out to 20 km, demonstrates that
the answer is no. Even though 3-D effects do not occur in the
side boundary condition at longer ranges, the computed 3-D

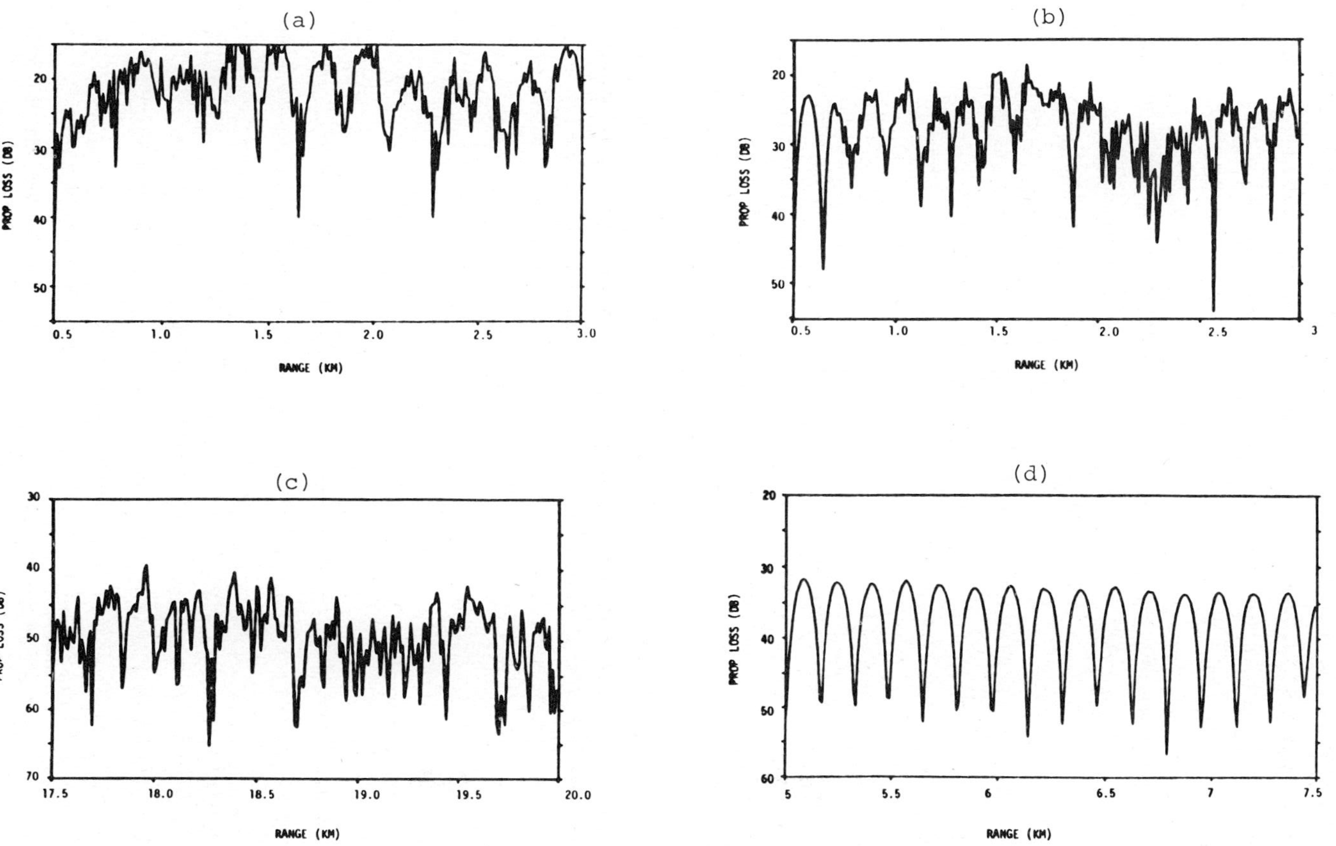

Fig. 4. New 3-D algorithm computations, illustrating influence of 2-D side conditions on solutions at sector center.

solution at longer ranges is seriously in error from the exact
solution. To confirm the source of the discrepancies in Figs.
4(a)-(c) from the exact solution values, we performed another 3-D
calculation, with side boundary values obtained from the 2-D wide
angle exact solution, as before. However, the initial range was
chosen large enough (r_0 = 5 km) so that 3-D effects in the side
boundary values are truly negligible. The result, in Fig. 4(d),
shows that the solution from the 3-D algorithm now well approximates the exact solution. We conclude that even though in the
simple test examples 3-D effects occur only for relatively small
ranges, failure to treat them correctly at the side boundary can
influence the solution at much longer ranges. This represents a
type of long-distance influence of localized 3-D effects.

5. EXAMPLE: OBLIQUE FRONT

In the previous section, exact test solutions illustrated the
accuracy and efficiency of the computational algorithm of Ref. 5.
Another type of example for this purpose, which also may be constructed to indicate capabilities of the algorithm, is a model
propagation problem. In this section we consider a simple representation of a front in a shallow ocean channel, as might occur
on a continental shelf or in a strait. The example is not meant
to be a realistic model of these often complex environmental situations, but only to suggest one plausible situation where 3-D
propagation can arise. In the absence of the front, the example
reduces to one which has proven useful in previous testing of 2-D
PE implementations.

The environment for this example with the front absent is
nearly the same as Case 3B of Ref. 10. The range-independent
basic state is a model of Pekeris type, with environmental parameters as follows: channel depth H = 100 m, water sound speed c =
1500 m s^{-1}, water density 1.0 gm cm^{-3}, bottom sound speed 1590 m
s^{-1}, and bottom density 1.0 gm cm^{-3} (in Ref. 10 this value is 1.2
gm cm^{-3}). There is no attenuation in the water, while in the bottom attenuation is 0.5 dB per wavelength. The upper surface is
pressure release. Other parameters are f = 250 Hz, c_0 = 1500 m
s^{-1}, source depth 99.5 m, and receiver depth 50 m. In a calculation to maximum range R = 10 km, all eleven propagating modes are
excited. Comparisons with benchmark normal mode and Hankel transform solutions verified that narrow angle 2-D PE solutions could
not match the range pattern of propagation loss, but the wide

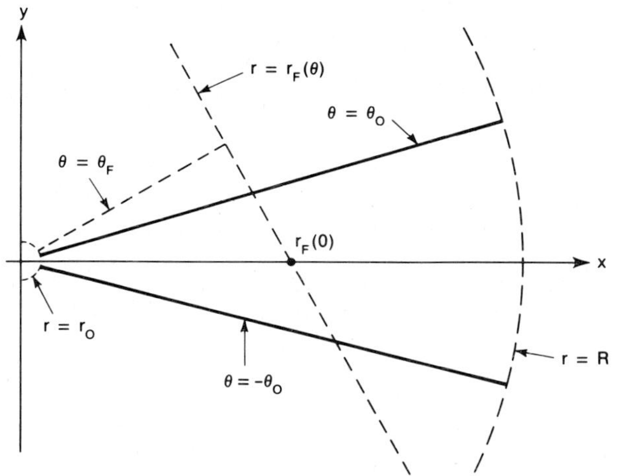

Fig. 5.   Geometry for oblique frontal transition zone.

angle IFD 2-D PE could do so [6,10].   This example demonstrated
the capability of the wide angle IFD model.

To represent a frontal transition zone, we retain all parameter values above but replace the constant sound speed by

$$c(r,\theta) = c_0 + \frac{1}{2} (\Delta c)\{1 + \tanh([r - r_F(\theta)]/w)\} \tag{18}$$

for $r_0 \le r \le R$, $-\theta_0 \le \theta \le \theta_0$.   In Eq. (18) the quantity Δc denotes
the jump in sound speed across the zone, which has an effective
width w.   The range $r_F(\theta)$ is the location of the middle of the
front, where $c = c_0 + \Delta c/2$.   Choices of these quantities are made
so that the minimum over θ of $(r_F(\theta)/w)$ and $([R - r_F(\theta)]/w)$ is
larger than about two.   It follows from properties of tanh that
for any θ, c will be very close to c_0 for r small and to $c_0 + \Delta c$
for r near R.   Thus, Eq. (18) represents a smooth transition between these two sound speed values for any θ in the sector.   The
value of 1500 m s^{-1} is retained for the reference sound speed,
since this is very nearly the sound speed near the source.   The
frontal midpoint is assumed to lie along a straight line.   This
line is a distance $r_F(0)$ along the x-axis ($\theta = 0$) from the origin,
and the angle between $\theta = 0$ and the perpendicular from the origin
to the line is θ_F (see Fig. 5).   It follows that

$$r_F(\theta) = \frac{r_F(0) \cos \theta_F}{\cos(\theta - \theta_F)} \tag{19}$$

From Eqs. (18) and (19), lines parallel to $r_F(\theta)$ are isospeed

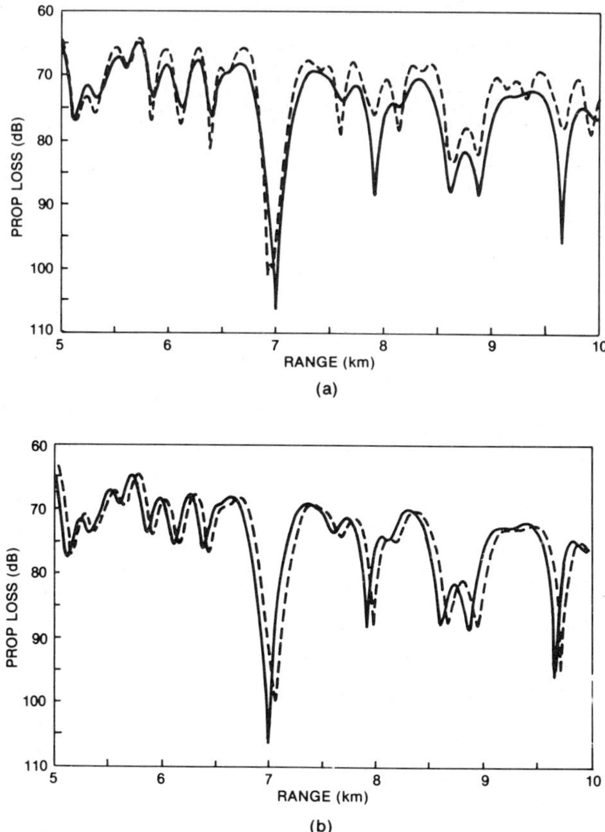

Fig. 6. Propagation loss vs. range from: (a) wide angle 2-D IFD, with (solid) and without (dashed) front; (b) wide angle 2-D IFD (solid) and new 3-D algorithm (dashed), with front.

contours in horizontal planes. A similar model has been constructed for a "radial front," for which the frontal midpoint lies along a ray θ = constant.

Computations using the new algorithm of Ref. 5 were performed. We shall discuss these with parameters chosen as w = R/10, $r_F(0)$ = R/2, and θ_F = 60 deg. A relatively large value of Δc = 40 m s^{-1} compensates for the relatively small value of θ_0 = 2 deg used in the VAX computations. The side boundary conditions are obtained from wide angle 2-D IFD computations [6], using sound-speed profiles from Eqs. (18) and (19). As mentioned in Section 4, this presupposes sufficiently weak azimuthal coupling that the side boundary values can be accurately specified by 2-D calculations. The initial condition is taken from the wide angle 2-D IFD solution at r_0 = 500 m. The computational step sizes used were Δz = 0.5 m, Δr = 1 m, and $\Delta\theta$ = 1 deg and 0.5 deg.

The dashed curve in Fig. 6(a) is the result of a 2-D wide angle IFD computation with Δc = 0. The solid curve is the corresponding calculation with the profile of Eqs. (18) and (19) at θ = 0 deg. The curves are identical at ranges less than 5 km. The relatively large differences in level and pattern of the prop loss curves arise from the relatively large sound-speed jump. The solid curve in Fig. 6(b) is the same as in Fig. 6(a), while the dashed curve is the result from the new 3-D algorithm at the sector center. The close correspondence of the two curves verifies that no significant azimuthal coupling has occurred for these propagation ranges and angles. Furthermore, the wide angle 2-D IFD code has produced accurate results when Δc = 0, by comparison with benchmark FFP calculations. Assuming this accuracy is maintained when Δc is nonzero, then the close comparison of the curves in Fig. 6(b) verifies the accuracy of the new 3-D algorithm. The small shift in the pattern (exaggerated for display) between the two curves in Fig. 6(b) most probably arises from the small difference between the right sides of Eqs. (1) and (3). We note that the weak azimuthal coupling demonstrated here need not persist for other frontal parameters, orientations, source-receiver ranges, etc. In such cases, results from the new 3-D algorithm would not agree with 2-D wide angle calculations. This has occurred, for example, in computations for cases with a sharp sound-speed jump along a radial line, as will be discussed elsewhere.

6. SUMMARY

This paper focuses on characteristics of, and numerical experiments with, the new 3-D algorithm developed in Ref. 5. Operator and asymptotic formalisms show that the new algorithm solves a 3-D PE which is wide angle in the channel depth direction and narrow angle in the cross range direction. Layered bottoms, general side boundary conditions, and 3-D sound-speed variations can be handled by the computational algorithm. Simple modal test solutions have propagation loss patterns with sufficient variation to test the new algorithm. Results are reasonably insensitive to variations in computational increments whenever solution components are adequately sampled. Two-dimensional results can be efficiently and accurately matched by the new 3-D algorithm. Some 3-D effects are indicated by computations with test and model problems. Example computations show how incorrect treatment of 3-D variations can influence propagation loss at ranges well beyond the location of those variations. Therefore, a correct description of the propagation loss over the entire region of interest needs a proper treatment of 3-D variations where they occur.

REFERENCES

1. Siegmann, W. L., G. A. Kriegsmann, and D. Lee, A wide-angle three-dimensional wave equation, JOURNAL OF THE ACOUSTICAL SOCIETY OF AMERICA 78 (1985), 659-664.

2. Tappert, F. D., The parabolic approximation method, in WAVE PROPAGATION AND UNDERWATER ACOUSTICS, J. B. Keller and J. S. Papadakis, eds., Springer-Verlag, New York, 1977.

3. Lee, D., The state-of-the-art parabolic equation approximation as applied to underwater acoustic propagation, JOURNAL OF THE ACOUSTICAL SOCIETY OF AMERICA 76 (1984), S9.

4. Perkins, J. S., R. N. Baer, L. F. Roche, and L. B. Palmer, Three-dimensional parabolic-equation-based estimation of the ocean acoustic field, Report 8685, Naval Research Laboratory, Washington, D.C., 1983.

5. Lee, D., Y. Saad, and M. H. Schultz, An efficient method for solving the three-dimensional wide angle wave equation, PROCEEDINGS OF THE FIRST IMACS SYMPOSIUM ON COMPUTATIONAL ACOUSTICS, North Holland, New York, 1986.

6. Botseas, G., D. Lee, and K. E. Gilbert, IFD: Wide angle capability, TR 6905, Naval Underwater Systems Center, New London, CT, 1983.

7. Greene, R. R., The rational approximation to the acoustic wave equation with bottom interaction, JOURNAL OF THE ACOUSTICAL SOCIETY OF AMERICA 76 (1984), 1764-1773.

8. Perkins, J. S., and R. N. Baer, An approximation to the three dimensional parabolic equation method for acoustic propagation, JOURNAL OF THE ACOUSTICAL SOCIETY OF AMERICA 72 (1982), 515-522.

9. Siegmann, W. L., and D. Lee, Aspects of three-dimensional parabolic equation computations, COMPUTING AND MATHEMATICS WITH APPLICATIONS 11 (1985), 853-862.

10. Davis, J. A., D. White, and R. C. Cavanagh, eds., NORDA PARABOLIC EQUATION WORKSHOP, Naval Ocean Research and Development Activity, Bay St. Louis, MS, TN 143, September 1982.

COMPUTATIONAL ACOUSTICS: Wave Propagation
D. Lee, R.L. Sternberg, M.H. Schultz (Editors)
Elsevier Science Publishers B.V. (North-Holland)
© IMACS, 1988

THE SOLUTION OF WIDE-ANGLE THREE-DIMENSIONAL
PARABOLIC APPROXIMATIONS ON HYPERCUBES

Martin H. Schultz

Department of Computer Science
Yale University
New Haven, Connecticut

1. INTRODUCTION

The wide-angle three-dimensional parabolic approximation as
already described by Ding Lee [5] has very important application
to a number of difficult underwater acoustic problems including,
for example, acoustic propagation through an eddy. Since this
parabolic approximation is in reality a pseudo-differential equation rather than a partial differential equation, we need to develop some special numerical algorithms for it. Fortunately for
us, this has already been done. We refer to the paper of Don St.
Mary [6] which contains an implicit finite difference approximation which is expensive (cf. [4] for a discussion of the application of iterative methods and to the paper of Ding Lee, Youcef
Saad, and the author [2], which contains an implicit alternating
direction method which is significantly less expensive per step
in range). These two algorithms have already been discussed at
this workshop by Ding Lee.

Unfortunately, even our fastest algorithm takes a long time
for realistic problems and thus we are motivated to consider the
impact of the new (massively) parallel supercomputer architectures
for speeding up these algorithms. In particular, we discuss the
mapping of the alternating direction method [1] onto the hypercube
architectures which are a product of microelectronics. We will
show that hypercube architectures have great potential for significantly speeding up these types of computations.

2. MATHEMATICAL MODEL AND ALGORITHM

Siegmann et al. [5] have developed the wide-angle three-dimensional parabolic approximation

$$u_r = \left(-ik_0 + ik_0 \frac{1 + p_1 x + p_2 y}{1 + q_1 x + q_2 y}\right) u \tag{2.1}$$

where

$$x \equiv \frac{1}{k_0^2} \frac{\partial^2}{\partial z^2} + (n^2 - 1)$$

$$y \equiv \frac{1}{k_0^2 r^2} \frac{\partial^2}{\partial \theta^2}$$

for acoustic wave propagation. The domain of integration of (2.1) is a "pie-shaped" region of the ocean with an initial condition specified at the center or vertex of the pie and boundary conditions specified on the top and bottom. The complex-valued function u, the acoustic pressure, is a function of the three independent variables z, θ, and r.

The alternating direction method [2] for solving (2.1) reduces to

$$u(r + \Delta r) = \left[\frac{1 + \left(\frac{1}{4} + \frac{\delta}{4}\right) x}{1 + \left(\frac{1}{4} - \frac{\delta}{4}\right) x} \right] \left[\frac{1 + \frac{\delta}{4} y}{1 - \frac{\delta}{4} y} \right] u(r) \qquad \delta \equiv ik_0 \Delta r \qquad (2.2)$$

for advancing the solution from range r to range r + Δr. When we replace x and y by three point difference approximations we see that each range step requires two complex, tridiagonal matrix vector products and the solution of two complex, tridiagonal linear systems. Thus the total work is proportional to the product of the number of range steps and the number of grid points in the r - θ plane. Unfortunately, for realistic problems this amounts to a huge amount of work, although the algorithm is perfectly satisfactory otherwise. The main thrust of this paper is the rapid solution of (2.2) on hypercube processors. Our goal is to speed up the computation of (2.2) as much as possible by the use of modern massively parallel supercomputers.

3. HYPERCUBES

Modern microelectronics has led to the development of several different multiprocessor architectures. For example, there are shared memory bus-oriented multiprocessors such as the ENCORE and SEQUENT computers; there are ring-based architectures such as the LCAP system of Dr. Enrico Clementi of IBM/Kingston; and there are hypercubes. In this paper we consider only the application of hypercubes. The reader is referred to [1] for a discussion of the application of other architectures to this problem.

The main technical difficulty in applying large hypercubes to (2.2) involves the solution of the tridiagonal systems in (2.2). First, we must transpose the data in the $z - \theta$ plane for every range step, i.e., we first process the tridiagonal systems in the z-direction and then, after a transpose, in the θ-direction. As we increase the number of processors in the hypercube, it eventually exceeds the number of tridiagonal systems to be solved in each direction and therefore it no longer suffices to solve each tridiagonal on a single processor. Since we are interested in exploring how far we can push hypercube technology for this problem, it becomes important to study how to solve n tridiagonal systems of size n on hypercubes with greater than n nodes.

There are now several hypercubes commercially available which are loosely coupled multiprocessors based on the binary n-cube networks. An n-dimensional hypercube, or n-cube, consists of 2^n nodes that are numbered by n-bit binary numbers from 0 to $2^n - 1$ and interconnected so that there is a link between two processors if and only if their binary representations differ by one bit.

The commercially available hypercubes span a large range in parameter design space. Three examples are the Intel Scientific Computing IPSC-D7 which has 2^7 nodes each of which is in essence a PC-AT, the NCUBE/10 which has 2^{10} nodes, each of which is roughly equivalent to a VAX 11/780, and the Connection Machine which has up to 2^{16} bit-serial nodes and 2^{11} floating point accelerators which can deliver several gigaflops on compute intensive kernels. Clearly, these machines already have immense power with much more to come. The fundamental question is: Can this power be used effectively for these problems?

4. HYPERCUBE ALGORITHMS

In this section we summarize a few results of Saied et al. [3] on hypercube implementations of the alternating direction methods. The reader is referred to [2] for complete details.

Suppose we have P tridiagonal systems, each of order P, i.e., a P^2 grid in the $z - \theta$ plane, and a hypercube with N processors. If $N \leq P$, we can use one-dimensional domain decomposition and solve P/N tridiagonal systems of size P per processor. This requires $O(P^2/N)$ work. We then do a transpose which requires $O(P^2 \log N/N)$ work (cf. [3]) and then solve again P/N tridiagonal systems of size P per processor. Clearly, the work is $O(P^2/N + P^2 \log N/N)$ and we see that we get almost linear speedup as we let N → P. This is very satisfactory.

   If N > P, that is, we have a massive hypercube, the above approach does not quite work because there are not enough systems to go around. However, in this regime we can use two-dimensional domain decomposition techniques. This involves separating the tridiagonal systems into groups that are confined to disjoint sub-cubes and the problem on each subcube corresponds to one-dimensional domain decomposition, with more processors than systems. This reduces the problem to one already solved. We refer to [3] for details of how to handle this problem efficiently.

   We have run some simple benchmarks on the Intel IPSC to measure the time required for one complete step of the alternating direction method for a 128 × 128 grid. The timings are as follows:

N	Time (sec)
2	16.1
4	8.5
8	4.5
16	2.4
32	1.4

The time on a VAX 8600 for this problem was estimated to be 1.5 sec and on a CRAY-XMP 0.15 sec. These timings would lead one to conjecture that the NCUBE/10 would be faster than a CRAY-XMP for this problem yet cost one-tenth as much. We hope to conduct experiments on massively parallel hypercubes to verify this conjecture.

ACKNOWLEDGMENT

   This work was supported in part by the following: ONR grant N00014-86-K-0564, NSF/CER grant DCR 8106181, and ARO FSP DAAL03-38-6-0029. Approved for public release; distribution unlimited.

REFERENCES

1. Johnsson, S. L., Y. Saad, and M. H. Schultz, Alternating direction methods on multiprocessors, Yale Computer Science Department Research Report YALEU/DCS/RR-382, October 1985.

2. Lee, D., Y. Saad, and M. H. Schultz, An efficient method for solving the three-dimensional wide-angle wave equation, Yale Computer Science Department Research Report YALEU/DCS/RR-463, October 1986.

3. Saied, F., C.-T. Ho, S. L. Johnsson, and M. H. Schultz, Solving Schrodinger's equation on the Intel iPS6 by the alternating direction method, Yale Computer Science Department Research Report YALEU/DCS/TR-502, January 1987.

4. Schultz, M. H., D. Lee, and K. Jackson, Application of the Yale sparse technique to solve the three-dimensional parabolic wave equation, NUSC Technical Report, Technical Document 7145, Naval Underwater Systems Center, New London, CT, 1984.

5. Siegmann, W. L., G. A. Kriegsmann, and D. Lee, A wide angle three-dimensional parabolic wave equation, JOURNAL OF THE ACOUSTICAL SOCIETY OF AMERICA 78 (1985), 659-664.

6. St. Mary, D. F., Analysis of an implicit finite difference scheme for very wide angle underwater acoustic propagations, PROCEEDINGS OF THE 11TH IMACS WORLD CONGRESS, IMACS, 1985, pp. 153-156.

COMPUTATIONAL ACOUSTICS: Wave Propagation
D. Lee, R.L. Sternberg, M.H. Schultz (Editors)
Elsevier Science Publishers B.V. (North-Holland)
© IMACS, 1988

THE CONSISTENCY OF MODELS OF
SOUND WAVES IN FLUIDS

Garrett Birkhoff

Department of Mathematics
Harvard University
Cambridge, Massachusetts

ABSTRACT

Many analytical and numerical models have been proposed for predicting the behavior of sound waves in fluids. Among analytical models, ray models, models based on Euler's equations for an elastic fluid, and 'parabolic' models have been those most intensively studied and tested. Models defined by the Navier-Stokes equations (with bulk viscosity and heat conduction) have also been studied. Moreover, many numerical schemes have been proposed for implementing each of these analytical models on a digital computer. A report will be given on various questions concerning the consistency of these three models with each other, with their discretizations, and with experimental observations of sound waves in real fluids.

1. INTRODUCTION

Many analytical and numerical models have been proposed for predicting and analyzing the behavior of sound waves in fluids. Among analytical models, 'ray' models and models based on Euler's equations for an elastic fluid have been the most intensively studied and tested. Over the past decade, 'parabolic' models for the propagation of simply harmonic ('monochromatic') sound waves in layered 'oceans' have also been intensively studied. The main concern of this report will be with the consistency of these three models with each other and with available experimental data.

However, to conclude from the preceding summary remarks that only three models are in need of comparison with each other (and with experiment) would grossly oversimplify the facts. The first two "models" named above really refer to classes of models. Besides homogeneous fluids with and without free surfaces in various domains, a wide variety of inhomogeneous fluids have been studied, with hypothetical 'layered' fluids as the most important subclass

                    G. Birkhoff

of inhomogeneous elastic fluids, but also with a variety of hypothetical ocean bottoms.

In addition to this proliferation of analytical models, which
extends to parabolic approximations (for which several different
differential equations have been proposed), each specific analytical model must be discretized to be treated on a digital computer.
Even if we ignore the vagaries of computer arithmetic (which is
probably reasonable if 14-digit accuracy is maintained), a great
variety of difference, finite element, and 'boundary element' approximations are clamoring for acceptance.

Originally, I hoped to make a systematic review of the consistency of the analytical models and numerical approximations briefly alluded to above. Instead, I have become increasingly concerned
with the many inconsistencies involved in practical acoustic modeling. For example, the "physical" acoustics defined by Euler's
equations neglects various real physical effects, including viscosity, heat conduction, air bubbles, turbulence, and molecular relaxation, to mention only five. Attempts to incorporate these into
a more broadly conceived "physical acoustics" will be reviewed in
Sections 3-5 and Appendix A.

I will next recall, in Sections 6 and 7, the usual derivatives from 'ray theory'of classical formulas for the refraction of
sound. The consistency of Huygens' principle with Fermat's principle will then be analyzed in Section 8. (Their partial consistency with Euler's equations will be taken up in Section 12.)

In Section 9, I call attention to the fact that the usual
mathematical formulations of 'physical acoustics' are actually
inconsistent with Euler's equations in inhomogeneous fluids, because sound waves generate vorticity in them. In Sections 10 to
14, I try to clarify the implications of this inconsistency, exploiting the fact that linear acoustics (with viscosity and molecular relaxation effects neglected) can be consistently formulated
in 'Chaplygin' fluids without loss of generality. Reasons are
given in Section 14 for doubting the validity of 'Huygens' second
premise' even for plane waves, in inhomogeneous fluids. In Section 15, the basis for the eikonal approximation is examined,
while the likely interference of sound waves of specified frequency emitted by multiple sources having independent random
phases is discussed in Section 16.

<u>The integrity of science</u>. It is, of course, not necessary
for all mathematical models of sound wave propagation to be

consistent to be useful: In practice, one often uses different models to describe different phenomena. For example, one can use 'ray theory' to predict the refraction of sound waves in homogeneous fluids, while using Euler's equations to predict diffraction, and molecular models to predict absorption effects.

However, not only is such consistency desirable philosophically for the integrity of science, but the clear recognition of inconsistencies (especially between theories and observations), sometimes suggests new and more satisfactory models. For these reasons, I hope that the results presented below will be of some interest.

To limit possible confusion and to make it easier for readers to investigate more thoroughly the questions involved, I have tried to base my discussion on the (mostly standard) references listed in the Bibliography. Moreover, to provide a simple yet reliable perspective on these questions, I will base my exposition mainly on four landmark documents: Rayleigh's THEORY OF SOUND [17] (1877, 1895); Lamb's classic [11] (1932), and his scholarly DYNAMICAL THEORY OF SOUND (1925); a careful scientific review [14] by Markham, Beyer, and Lindsay of experimental research stimulated by World War II findings (for which see [21]); and the third edition [25] of A. B. Wood's classic text (1955). For the current status of the questions discussed, I will rely primarily on Allan Pierce's admirable and comprehensive survey [16].

2. HISTORICAL REMARKS

In assessing models for sound wave propagation, one should keep in mind the fact that Euler's concept of a homogeneous elastic fluid, with 'equation of state' $p = p(\rho)$, has stood the test of time extremely well. In particular, by neglecting suitably chosen second-order infinitesimals [16, pp. 14-17], one can derive from Euler's equations the wave equation $\phi_{tt} = c^2 \nabla^2 \phi$ and from this the Helmholtz equation $\nabla^2 \phi + k^2 \phi = 0$. Supplemented by suitable initial and boundary conditions, these equations define the most widely used model of analytical acoustics, and I have little to add to the enormous literature concerned with solving them.

My main concern will be rather with inhomogeneous elastic fluids, with absorption and dispersion phenomena not described by Euler's model, and with possible future developments in numerical acoustics. In this connection, I remind you that the first effective numerical methods for treating inhomogeneous fluids were

based on ray theory (see [5]). Moreover, the basic concepts of
wave front (Huygens) and ray theory (Fermat), finally reconciled
around 1850 by Hamilton-Jacobi theory ([18] and [2a, Sec. 4]),
seem also to follow from Euler's equations in the limiting case of
very short wave length. I will discuss this last connection in
Section 15 below; for the present, I simply remind you that it is
the classic argument for the consistency of ray theory (geometrical acoustics) with 'physical' acoustics based on Euler's equations.

   <u>Thermodynamics</u>. One reason for the continuing usefulness of
Euler's mathematical model is its flexibility. Euler's unspecified
"equation of state" $p = p(\rho)$ is flexible enough to accommodate both
the isothermal and the adiabatic compression of air. As Newton had
shown in his PRINCIPIA (Prop. 48), the speed of sound in any elastic fluid is related to this equation by the formula $c^2 = dp/d\rho$.
As was also known in Euler's time, if one assumes Boyle's law $p \propto \rho$
of isothermal compression of air in Newton's formula, the predicted
speed of sound is nearly 20% too low. It was Laplace (1823) who
first resolved this discrepancy, by deriving for adiabatic compression the formula $c^2 = \gamma dp/d\rho$, where $\gamma = C_p/C_V$ is the ratio of the
specific heat at constant pressure to that at constant volume.
(In liquids, Laplace's adiabatic correction is ordinarily negligible [16, p. 30]; thus in water at 10°C it is less than 0.1%.)

   <u>Classical kinetic theory</u>. By the "classical" kinetic theory
of gases, I mean the arguments used by Maxwell and others to deduce the adiabatic equation $p = k\rho^{1.4}$ for sound waves in air, and
the 'ideal gas' equation $pv = RT$, with R an absolute constant.
This theory and its limitations are clearly set forth in Jeans'
DYNAMICAL THEORY OF GASES (4th ed., 1925).

   According to this theory, the pressure in a pure ideal gas is
related to the mean squared velocity of its molecules by the formula $p = \rho\overline{v^2}/3$ [Jeans, (312)], a formula already conjectured by
Daniel Bernoulli before 1738. From this one can derive the formula
$c^2 = \gamma RT/m$, where m is the molecular weight [Jeans (313), (316)].

   Using ingenious arguments, Maxwell also derived a general
'principle of equipartition of energy.' This asserts in particular that, in static thermal equilibrium, molecules of all gases
have the same expected mean squared kinetic energy $E_m = m\overline{v^2}/2$,
where m is the molecular mass, and that this E_m is proportional to
the absolute temperature $T = E_m/k_1$.

In a pure gas, clearly $\rho = Nm$, where N is the expected number of molecules per unit volume. Hence, $p = \rho\sqrt{\overline{v^2}}/3 = Nm\sqrt{\overline{v^2}}/3 = 2k_1NT/3$, so that $N = 3p/2k_1T$, showing that all (pure) gases have the same number of molecules per unit volume at any given pressure and temperature (Avogadro's Law).

Using his principle of equipartition of energy, Maxwell also suggested that the adiabatic constant γ of any pure gas is related to the number n of (vibrational + rotational) internal degrees of freedom of molecular motion by the formula $\gamma = 1 + 2/(3 + n)$ [Jeans (490)], where for diatomic gases n = 2 so that $\gamma = 1.4$.* (Air is mostly diatomic.) It is a corollary that the squared speed of sound $c^2 = \gamma p/\rho$ is inversely proportional to the atomic weight for diatomic gases.

Analogies with optics. I have referred to "geometrical" and "physical" acoustics to emphasize their analogies with the well-established distinction between "geometrical" and "physical" optics. Thus "ray theory" obviously stems from the Principles of Fermat and Huygens. These men had optics and not acoustics in mind when they applied their principles to deduce Snell's law of refraction, but their arguments apply equally well to sound wave propagation. Conversely, the wave equation $\phi_{tt} = c^2\nabla^2\phi$ can be derived (in a homogeneous medium) from Maxwell's as well as from Euler's equations. These common features are instances of a deep-seated optical analogy, which led Rayleigh to refer to acoustics and optics as "sister sciences." This analogy will be discussed in Appendix B below.

3. CLASSICAL VISCOSITY EFFECTS

By replacing the formula $c^2 = p/\rho$ for isothermal pressure waves in a perfect gas with the formula $c^2 = \gamma p/\rho$, the idea of adiabatic compression superficially reconciled Euler's model of a homogeneous elastic fluid with observations of sound wave propagation in air. Indeed, until recently, practical gas dynamics** was usually based on the concept of a perfect gas with $pv = RT$

---

*Indeed, Maxwell predicted that C_V should be "independent of temperature and the same for all diatomic gases ($\simeq 7$ calories per mole per °C)." See R. D. Present, KINETIC THEORY OF GASES, McGraw-Hill, 1958, p. 85.

**See the treatises of Oswatitch, Courant-Friedrichs, and von Mises (the last completed by Geiringer and Ludford).

and constant C_p, C_v. For adiabatic compression, these imply the
(Eulerian) 'polytropic' equation of state $p = k\rho^\gamma$. (In water, the
adiabatic correction is negligible.)

On the other hand, already by 1825, Navier and Poisson had
pointed out the deeper inconsistency of Euler's model with the
then nascent kinetic theory of gases. They recognized that the
irreversible phenomena of viscosity and heat conduction should
occur, whereas the models of Newton and Lagrange for pressure
waves in homogeneous fluids are reversible (they preserve mechanical energy).

In 1845, Stokes showed how to derive concepts of shear and
bulk viscosity in a homogeneous isotropic fluid from the stress
tensor concept of continuum physics. His derivation is applicable
to liquids* as well as to gases, and is accepted as standard today.
To obtain a well-posed boundary value problem, Stokes also replaced
Euler's scalar boundary condition $u_n = 0$ by the condition $\underline{u} = \underline{0}$ of
'sticking to the wall.'

Until 1920, the resulting Navier-Stokes equations (with bulk
viscosity $2\lambda + 3\mu = 0$), supplemented by Fourier's heat conduction
equation, were generally accepted as the main causes of acoustic
and energy dissipation in fluids. Their implications will be reviewed in the present section; more recent models based on concepts
of molecular relaxation will be taken up in the next two.

All the models to be reviewed in this paper predict an exponential decay in the energy of sound waves of given frequency f
with distance and time. Most classical models also agree dimen-
sionally in predicting that the rate of exponential fall-off in
amplitude, α, should be proportional to the squared frequency, f^2,
so that α/f^2 should be a material constant (mildly temperature-
dependent). Thus, Stokes' considerat-on of shear viscosity led
him to conclude that

$$\alpha = 8\pi^2 \nu f^2/3c^3 = 8\pi^2 \nu/3\lambda^3 c = 2k^2\nu/3c \qquad (3.1)$$

where $\nu = \mu/\rho$ is the kinematic viscosity, λ is the wave length,
and k the corresponding spatial wave number.

More suggestive than α is the related decay rate per wave
length $\lambda = c/f$,

---

*Note that, whereas $\mu = \mu_0\sqrt{T}$ in gases is nearly independent of the
pressure, as predicted by Maxwell from kinetic theory, the shear
viscosity of liquids usually decreases with increasing temperature.

$$\lambda\alpha = 8\pi^2 \nu f / 3c^2 \tag{3.1'}$$

In a dilute gas, by kinetic theory (cf. Appendix A), the viscosity $\mu = \mu(T)$ depends only on the temperature T [Jeans, p. 773], while (cf. Sec. 2)

$$\rho = p/RT = \gamma p/mc^2$$

Hence $\nu = mc^2\mu(T)/\gamma p$; substituting into (3.1'), we get

$$\lambda\alpha = 8\pi^2 m\mu(T) f / 3\gamma p \tag{3.1''}$$

Thus we conclude that, in a dilute gas, $\lambda\alpha$ should be a single-valued function of f/p at fixed T.

The Navier-Stokes equations also predict a small boundary layer correction for acoustic measurements in cylindrical Kundt tubes which, until 1920, provided the most accurate laboratory data on sound wave velocity. Today, observations in Kundt tubes are still corrected by the Kirchhoff-Rayleigh formula

$$c(f)/c_0(f) = 1 - (\nu'/4\pi f)^{1/2}/r \tag{3.2}$$

where ν' is a combined "kinematic viscosity and heat diffusion coefficient" [25, p. 254]. Note that this boundary layer correction decreases with increasing frequency f (in proportion to $1/r\sqrt{f}$), and that the predicted absorption effects

$$\alpha = (\pi\nu' f)^{1/2}/c_0 r \tag{3.2'}$$

increase with f much more slowly than that predicted by (3.1).

The decay predicted by (3.1) is not easy to measure. To quote Lamb [11, p. 655]: "for distances of very many wave lengths, the attenuation due to viscosity is altogether negligible in comparison with that due to spherical divergence." To incorporate dissipation effects into "physical" acoustics, it is natural (and usual) to replace the inverse square law of sound intensity (amplitude squared) decrease with

$$A(r) \propto e^{-2\alpha r}/r^2 \tag{3.2''}$$

where α is associated with plane waves. Since differentiation multiplies this expression by twice $-\alpha - 1/r$, the $e^{-2\alpha r}$ factor does not become dominant until $r > 1/\alpha$. Since $1/\alpha = 4.1 \times 10^{10}/f^2$ meters in air and $3.84 \times 10^{19}/f^2$ meters in water [26, p. 352], this effect is negligible in the audible range. But it can be measured in the ultrasonic range, especially in air. Indeed, setting rs = exp(iωt − ikr) in (3.2'), where k is complex, we get $\omega^2 = c^2 k^2 - 4i\nu k^2 \omega/3$, whence (cf. [11, p. 670])

$$k = [\omega^2/(c^2 - 4i\nu\omega/3)]^{1/2} = \omega/c[1 - 2i\nu\omega/3c^2 + \cdots]$$

and $\alpha = -\text{Im}\{k\}$ is given by (3.1).

<u>Thermal conductivity</u>. One can interpolate mathematically between adiabatic and isothermal sound propagation by considering a thermoelastic fluid, in which the heating by compression is diffused by conduction. As is shown in [16, pp. 34-36], the dispersion (nonlinear dependence of the (complex) spatial wave number k on the temporal wave number ω) can be determined from the biquadratic equation

$$k^2 - (\omega/c)^2 = \frac{\kappa}{\rho C_p}\frac{k^2}{i\omega}[k^2 - (\omega/C_T)^2] \tag{3.3}$$

where C_T is the isothermal sound speed. As Herzfeld and Rice first emphasized in 1928 [8, p. 45], noting that "Stokes's view of the matter appears to be incorrect" [25, p. 351], it is very high frequency waves and not low frequency waves that propagate isothermally.

A more refined classical model assumes what may be called a thermoviscous elastic fluid; cf. [22] and [26, pp. 3-45 ff].

As Kirchhoff pointed out in 1868, thermal conductivity should dissipate sound intensity $I \propto \overline{s^2}$ in the same way as viscosity. Indeed, assuming kinetic theory, Maxwell concluded that the effect of conductivity should be to replace the kinematic viscosity ν of a viscoelastic gas by an effective viscous plus thermal diffusivity $\nu' = 2.5\nu$ [25, p. 351], and the estimates of $1/\alpha$ given above were based on $\nu' = 2.5\nu$. (Lamb [11, p. 649 (30)], probably by mistake, interprets ν'/ν as $k/\rho C_p \nu$ instead.)

A careful reconsideration of ideal gases by Rayleigh [17] led him to replace α in (3.1) with

$$\alpha_{c\ell} = (2\nu\omega^2/3c^3)[1 + 3(\gamma - 1)/r\text{Pr}] \tag{3.4}$$

instead of 2.5α. In (3.4), the "Prandtl number" $\text{Pr} = \mu C_p/\kappa$ is the ratio of the kinematic viscosity $\nu = \mu/\rho$ to the adiabatic thermal diffusivity $\kappa/\rho C_p$, and is about 0.7 in air.

The $\alpha_{c\ell}$ defined by (3.4) is called the "classical" rate of sound amplitude dissipation in gases [16, (10.10-2.13)]. Note that, like almost all theoretical estimates of the rate 2α of acoustic energy dissipation made before 1940, formula (3.4) ignores possible effects of bulk viscosity.

Bulk viscosity. "Stokes had already recognized the possibility of a volume viscosity," but "it seemed hopeless to measure effects connected with it, and therefore he omitted it from his equations" [8, p. 38]. Lamb [11, p. 358] is one of the few later authors to even define a classical "bulk viscosity" μ_B. As he observes, in a perfect gas, we should have

$$p = \rho RT + \mu \; \mathrm{div}(D\underline{u}/Dt) \qquad (3.5)$$

but even Lamb soon sets $\mu_B = 0$, although this is believed to be true only in monatomic gases, and not in air.

Indeed, astonishing though it may seem, most books on fluid mechanics even today ignore the fact that "the usual (metaphysical) derivation of the Navier-Stokes equations... involves two coefficients of viscosity, λ and μ,"* and to assume that the "bulk viscosity" $\mu_B = \mu + (2\lambda/3) = 0$. This assumption is reasonable in aerodynamics, because there (especially in boundary layers), rates of shear deformation are orders of magnitude greater than rates of expansion and compression.

In acoustic near fields, where boundary layers and flow separation may strongly affect the generation of sound, this approximation seems reasonable. But in far fields, it seems preferable to view observed rates of sound attenuation as the best means for measuring an effective bulk viscosity μ_B.

Truesdell's analysis. Unfortunately, as Truesdell was the first to show in a landmark 1953 paper [22], this empirical bulk viscosity μ_B cannot be assumed to be independent of the sound frequency $f = \omega/2\pi$. In his paper, which contains many interesting critical and historical remarks, Truesdell determined the rates of absorption and dispersion of plane sound waves in a general idealized "Kirchhoff-Langevin fluid" having arbitrary bulk and shear viscosities as well as thermal conductivity. After trying to fit empirical data with these rates, he concluded that "The result has been negative."

As will be explained in Section 4, it had previously been discovered empirically (see Section 4 and Appendix A) that the ratio α/f^2 typically decreases with f in the ultrasonic range,

---

*Cf. G. Birkhoff, HYDRODYNAMICS: A STUDY OF LOGIC, FACT AND SIMILITUDE, 2nd ed., Princeton University Press, 1960, p. 47.

contrary to the predictions of (3.1) and (3.3). However, by showing that this variability also occurs in general thermoviscous
(alias "Kirchhoff-Langevin") fluids, but not in ways that are
physically realistic, Truesdell's result is much sharper.

4. WORLD WAR II FINDINGS

Although generally adequate for the Natural Philosophy of the
19th century, the classical models discussed in Section 3 failed
to explain the ultrasonic absorption experiments of the more
sophisticated "high-tech" era which began to take shape after 1920.
It may be of interest to recall a few milestones in the path of
subsequent discovery.

In the mid-1920s, systematic experiments by G. W. Pierce going up to megacycle frequencies showed that: "At frequencies of
10^5 and 2×10^5 the absorption in CO_2 was, respectively, four
times and more than eighty times the absorption in air" [25, p.
354]. Likewise, in the 1930s, V. O. Knudsen found that "The observed absorption in moist air was found to be from 10 to 100
times that calculated" (by the formulas of Section 3). These effects were first rationalized by H. O. Kneser in terms of the concept of molecular relaxation to be discussed in Appendix A. (See
[25, pp. 275 and 353] for references.)

During World War II, the importance of submarine warfare
focused attention on the total real 'transmission loss' of acoustic energy, especially in the ocean. On this question, the National Defense Research Council Report [21] of 1946 is the definitive
reference for the state of knowledge about underwater acoustics
as of that time.

As I have stated, Euler's model attributes all acoustic energy
intensity dissipation to the spreading of a fixed total energy over
a larger area. The report [21] essentially contrasts empirical
evidence with the deviation from this 'elastic fluid' model attrib-
utable to viscosity with the deviations actually observed.

This report begins by observing (p. 27) that, for spherical
waves, the sound "intensity loss tends to be much greater than the
value predicted" by the inverse square law deducible from Euler's
equations. It then quotes Stokes' formula (3.1), as [21,(97)].

Somewhat later [21, p. 39] comes a thoughtful section provoc-
atively entitled: 2.9 INADEQUACY OF WAVE ACOUSTICS. This section summarizes the shortcomings of existing theories, on the
basis that: (i) the boundary conditions p = 0 (on the ocean

surface) are inexact, (ii) the ocean surface is "usually disturbed and uneven," (iii) "Many inhomogeneities... such as bubbles, floating plant and animal life, fish...may produce a very important part of the observed transmission loss" [21, p. 40].

The transmission loss (or absorption) of sound energy is taken up again later in [21], and the facts summarized in the statement that "the observed absorption of sound in the ocean is far greater than can be accounted for by viscosity." (As I stated above, shear viscosity μ is intended.) For each fixed sound frequency or 'pitch' f, the data can thus be fitted by replacing the theoretical $\nu - \mu/\rho$ in (3.2) by an empirical value of $\tilde{\nu}$ given by

$$\beta = 8\pi^2\tilde{\nu}/3\lambda^3 c = 2k^2\tilde{\nu}/3c \qquad \tilde{\nu} = \tilde{\nu}(f) \tag{4.1}$$

where k is the wave number, and $\tilde{\nu}$ is an empirical, kinematic, bulk viscosity.

From a scientific standpoint, it is Fig. 17 of [21, p. 105] that is most informative. At the lowest range (16-30 kHz) for which substantial data are available, the observed attenuation of sound in sea water is about 100 times that given by (3.1). The ratio decreases to about 10 in the range 500-5000 kHz.

Specifically, around 20 kHz ($\lambda \doteq 9.5$ cm in water), Stokes' formula (3.2) would predict an attenuation of about 1/40 db in 1000 yds = 5000λ in water, or 1 db in 200,000 wave lengths! The observed attenuation is about 1 db in a distance of 2000λ instead.

Around 1000 kHz ($\lambda \doteq 2$ mm in water), the same formula would predict an attenuation in sea water of about 60 db in 1000 yds, or 1 db in 10,000λ. The observed attenuation is about 6 or 7 times as much, all according to [21, Fig. 17].

These astonishing World War II findings stimulated intensive postwar laboratory research, which was carefully reviewed in 1951 by Markham, Beyer, and Lindsay [14]. Following a brief review of a number of physical phenomena ignored by Kirchhoff and Rayleigh, they wisely remarked that [14, Sec. 11]:

> "In any fluid, the problem of sound propagation is far more complicated than the treatment given so far, since one must correct for all the effects acting at the same time. We therefore desire an equation which includes heat radiation, heat conduction, viscosity, and the various relaxational phenomena. One should even include effects of diffusion of various gases in gas mixtures, such as air."

Having made this wise cautionary observation, [14] then concentrated on the many fascinating "relaxation phenomena" that had

been so intensively explored in so many laboratories in the preceding decade.

Naturally, the unexpectedly high rate of absorption of sound
in sea water was the subject of special attention, and laboratory
measurements of α in the range 20 kHz to 2 mHz are displayed [14,
Fig. VI-1]. Nearby [14, p. 404], one reads that "The relaxational
effect in sea water was at first attributed...to a shift in the
chemical equilibrium of ionized sodium chloride....Recent experiments indicate that the excess of absorption in sea water can be
explained by the presence of small amounts of $MgSO_4$." Cf. also
[16, p. 552] and [26, pp. 3-70] for equally cautious statements
on this topic.

More generally, as regards "relaxation processes and sound
absorption," I fear that it is still true that, as F. V. Hunt
wrote in 1957 [26, pp. 3-53]: "Much more remains to be done....
before this basic acoustic problem can be said to be understood."

5. PRACTICAL IMPLICATIONS

However, to avoid undue emphasis on what is ordinarily a very
secondary effect, I will postpone a systematic review of molecular
relaxation phenomena until Appendix A. In this section, I will
concentrate on a few obvious and important practical implications
of molecular relaxation phenomena for sound absorption.

First, the studies reported in [14] were performed on a laboratory scale, measurements being made of plane wave propagation
in cylindrical tubes having lengths presumably 10 cm to 10 m.
Since audible sound waves have negligible attenuation over such
short ranges, at least in air and water, statements about the
attenuation of audible sound energy are primarily based on indirect evidence.

An interesting case in point concerns the graph shown in [16,
p. 560] of the sound absorption coefficient in air at 20°C and
atmospheric pressure, with relative humidity of 20%. This graph
indicates an amplitude decay for f = 25 Hz of 0.1% in 110 meters,
or one "neper" (= factor of e) in 110 Km, and a decay of one neper
in about 10 Km for f = 200 Hz. This last rate is about ten times
that quoted in [25, p. 352], where the influence of humidity is
ignored.

This contrast illustrates the extent to which ideas about the
attenuation of acoustic energy have changed since the "classical"
era, but it should be repeated that both estimates are presumably

based on indirect experimental evidence.

A closer inspection of the graph referred to above brings out
another interesting contrast. The predicted rate of decrease in
sound wave amplitude specified there, of about one neper in 110
Km at 25 Hz, increases to about one neper in 0.9 meters at 250
kHz. Although dramatic, the increase by a factor of 10^5 in distance is more nearly proportional to $f^{5/4}$ than to the f^2 of (3.1).
Variations in the absorption per wave length, $\alpha\lambda$, are much more
gradual than those in the rate of absorption per unit distance, α.
Reinterpreted in this way, the graph of [16, p. 560] indicates
that the distance in which 0.1% of the acoustic energy is absorbed,
in air having the specified characteristics, decreases (monotonic-
ally) from about 4λ at 25 Hz to 0.35λ at 250 kHz.

Surprisingly, deviations from the Stokes-Kirchhoff prediction
seems to be smallest in the high ultrasonic range above 250 kHz,
and largest in the audible range, contrary to the statement made
in [24, p. 350] and quoted earlier. This is because, as was
first shown by Tisza (cf. [16, p. 561]): "the apparent bulk viscosity within a given frequency range is composed of contributions
from all relaxation processes whose relaxation frequencies are
higher than the upper limit of that frequency range."

Before deferring until Appendix A the further discussion of
molecular relaxation effects, three more important facts should
be mentioned. First, the "bulk viscosity" effects of molecular
relaxation are really associated with heat transfer (like those
of heat conduction), and not with the momentum transfer postulated
by Stokes. Second, as the absence of a simple kinetic theory of
liquids would suggest, molecular relaxation effects in liquids
(unlike the "classical" effects discussed in Section 3) are much
more complicated than they are in gases. And finally, all of the
experimental evidence about ultrasonic absorption and dispersion
is concerned with homogeneous (if not necessarily chemically pure)
fluids. I will now take up the very different subject of refraction in inhomogeneous fluids, and especially in those which are
layered (i.e., stratified).

---

*For a scholarly analysis of this subject, see Calvin Wilcox,
SOUND PROPAGATION IN STRATIFIED FLUIDS, Springer-Verlag, 1984.)

6. REFRACTION OF SOUND

Having much more effect usually on sound intensity then the dissipative mechanisms discussed in Sections 3-5, is the refraction of sound due to variations in the sound speed c and wind velocity U. To a first approximation, these variations depend primarily on altitude (in the atmosphere) and depth (in water). In this section and the next, we will assume that these are the only variations, so that $c = c(y)$ and $U = U(y)$, where y is a vertical coordinate. Fluids with this property are said to be stratified.

Already in 1877 [17, Art. 288ff], Rayleigh applied the geometrical principles of Fermat and Huygens to predict the refraction of sound in stratified fluids; their consistency with the "physical" acoustics defined by Euler's elastic fluid will be taken up in Part B. In the present section, I will restrict attention to effects due to variations in c; the effects of variations of U will be taken up in Section 7. Note the contrast between the geometrical acoustics of heterogeneous fluids to be treated next, and the sophisticated physical acoustics of homogeneous fluids discussed in Sections 3-5. Also, that energy conservation must be assumed in Sections 6 and 7 to correlate theoretical predictions with observations of sound intensity, such as are mentioned by Woods [25, pp. 322-5].

In air, variations in the sound speed c are associated almost exclusively with variations in the temperature T. This is because, as Maxwell emphasized, isothermal variations in the pressure p have a negligible effect on c (and μ) in gases. The same is often true in lakes and seas, but in oceans, c varies appreciably with the pressure p as well as with the temperature.

Snell's Law. For any vertical sound velocity profile $c = c(y)$, one easily deduces from Fermat's Principle the fact that the ratio $(\sin \theta)/c$ is preserved under refraction across any horizontal interface separating two homogeneous fluids. Passing through an intermediate series of horizontal slices of homogeneous fluid between which c changes by smaller and smaller increments, one is easily led to the limiting case of a static stratified fluid with continuously varying $c = c(y)$. In such a fluid, therefore,

$$c(y)/\sin \theta = \text{const.} \qquad \text{or} \qquad \sin \theta = kc(y) \tag{6.1}$$

To determine the acoustic paths of phonons, we set $\sin \theta = dx/ds$ in (6.1), and note that $ds^2 = dx^2 + dy^2$. If we square both sides and then take their reciprocals, we obtain $y'^2 + 1 = 1/k^2c^2(y)$.

Solving for y', we get

$$dy/dx = [1/k^2 c^2(y) - 1]^{1/2} = [1 - k^2 c^2(y)]^{1/2}/kc(y)$$

Separating variables, we get finally

$$x - x_0 = \int \frac{kc(y)\,dy}{[1 - k^2 c^2(y)]^{1/2}} \tag{6.2}$$

where x_0 is an arbitrary constant of integration.

   <u>Sound speed linear in altitude</u>. In the simplest nontrivial case, $c(y) = Cy$ is linear in the altitude, and (6.2) reduces to

$$-[1 - k^2 C^2 y^2]^{1/2}/kC = x - x_0$$

Squaring both sides and transposing, we get

$$(x - x_0)^2 + y^2 = 1/k^2 C^2 \tag{6.3}$$

which shows that the acoustic paths are arcs of circles.

   The case of the half-plane $y > 0$ is especially interesting mathematically since, in this case, the acoustic paths of shortest time are precisely the geodesics of Poincaré's model for the non-Euclidean geometry of Bolyai and Lobachewsky.

   <u>Wave fronts</u>. In a stratified fluid, the wave fronts emanating from any acoustic source are surfaces of revolution having a common axis of symmetry containing the source and perpendicular to the surfaces $y = $ const. Setting the source at $(0,y,0)$, the acoustic paths and wave fronts can thus be found by first computing them as orthogonal families of curves in the (x,y)-plane, and then rotating about the y-axis. When $c(y) = Cy$, they are respectively semicircles through the source and the orthogonal family of spherical wave fronts.

   <u>Curvature</u>. We next determine the curvature $\kappa = d\theta/ds$ of acoustic paths. Differentiating the formula $(\sin \theta)/c = k$ with respect to y, we get

$$0 = \frac{\cos \theta}{c} \frac{d\theta}{dy} - \frac{\sin \theta}{c^{2\alpha}} \frac{dc}{dy}$$

Simplifying, we get $d\theta/dy = \tan \theta\, d(\ln c)/dy$. Since $dy/ds = \cos \theta$, this implies

$$K \equiv d\theta/ds = (\cos \theta)d\theta/dy = \sin \theta\, d(\ln c)/dy$$

This is however the normal gradient of $\ln c$ in the direction of

decreasing velocity. We conclude, as in Wood [25, p. 322], that
the vector curvature of an acoustic path is in the direction of
the negative gradient of ln c:

$$\kappa = \partial(\ln\ c)/\partial n \tag{6.4}$$

in the plane normal to the path.

7. VARIABLE WIND SPEEDS; FOCUSING

As Rayleigh remarked [17, p. 289], "Another cause of atmospheric refraction [of sound waves] is the action of wind." He
then used simple physical (actually, kinematic!) considerations
to estimate the magnitude of the atmospheric refraction of sound
by a horizontal wind in a stratified fluid. His discussion was
not quite complete (see Section 8), because he considered the angle
ϕ with the horizontal made by a sound ray relative to the moving
fluid around it, rather than that made by its path as seen by a stationary observer. Rayleigh's choice simplifies the analysis by
making the wave fronts perpendicular to the acoustic paths relative
to the fluid. Denoting by ϕ the first of these two angles with the
vertical, he reasoned essentially as follows.

Let a horizontal interface separate two homogeneous elastic
fluids, having sound speeds c and c' and wind speeds U and U',
respectively. Relative to a stationary observer, the point where
a wave front impinges on the interface from the lower fluid moves
along the latter with total speed c csc ϕ + U. The corresponding
point on the refracted wave front moves with speed c' csc ϕ' + U'.
Since the two speeds must be the same, we have

$$c\ \csc\ \phi + U = c'\ \csc\ \phi' + U' \tag{7.1}$$

Note that when U = U' = 0, $\phi = \theta$ and formula (7.1) reduces to
Snell's law c/sin θ = c'/sin θ'.

This 'conservation law' applies to any stratified fluid composed of homogeneous elastic horizontal layers separated by interfaces. Passing to the limit as the jumps in c and U get smaller
and smaller, to the limiting case of a stratified fluid with continuously varying sound speed c(y) and horizontal wind speed U(y),
we obtain the following generalized Snell's Law

$$c(y)\ \csc\ \phi + U(y) = \text{const.} \qquad \text{(Rayleigh)} \tag{7.2}$$

To complete the discussion, we now derive the connection between ϕ and the angle θ with the vertical made by the acoustic
path, as seen by a stationary observer. By considering the

horizontal and vertical components, we obtain

$$\frac{\cos\,\theta}{\sin\,\theta} = \frac{c\,\cos\,\phi}{c\,\sin\,\phi + U} = \frac{\cos\,\phi}{\sin\,\phi + M}$$

where $M = U/c$ is the Mach number.

Note that $\theta = \phi$ in two important cases: when $M = 0$ (no wind) and when $\phi = 90°$ (so that $\cos\,\phi = 0$), i.e., in the case of a horizontal path. If we assume that $U = 0$ at ground level, therefore, Rayleigh's limiting angle below which all rays "traveling upward at an inclination $((1/2)\pi - \theta)$ to the horizon" less than β are totally reflected, is given by

$$\sec\,\beta = 1 + U'/c' = 1 + M' \tag{7.4}$$

precisely as stated in [17, Sec. 289 (3)]. In particular, "all rays whose inclination is less than 11° are totally reflected by a wind moving at...15 miles per hour." The admirable simplicity of Rayleigh's argument glosses over the fact that "reflection" of a sound wave from a vortex sheet in an otherwise homogeneous fluid seems unlikely, especially since such vortex sheets are unstable (Helmholtz). In applications, it would seem more realistic to assume that $U(y)$ increases continuously with y. This assumption would make each acoustic path making sufficiently small initial angle ϕ_0' with the horizontal approach asymptotically a corresponding level line $y = y_0$ from below, where $\phi' = \pi/2 - \phi$. Hence, in the notation of (7.1),

$$M(y_0) = \csc\,\phi_0 - 1 \qquad (M = U/c) \tag{7.1'}$$

Next, Rayleigh [17, Sec. 290] reviews many experimental observations confirming qualitatively the atmospheric refraction of sound in air, while cautioning readers that "the auxiliary assumption on which [formulas (7.1)-(7.3)] are founded is by no means strictly or generally true." From a quantitative standpoint, it is relevant that, whereas winds with Mach number $M = U/c$ of 0.03 are not uncommon in the atmosphere, currents in lakes, seas, and oceans seldom exceed 10 f/s, so that $M \leqq 0.002$. This shows that in nature the refraction of sound by currents in water is ordinarily much less than the refraction of sound by brisk winds.

<u>Focusing</u>. In some cases, refraction can lead to local concentrations of sound intensity greater than those produced by the trapping of spherical waves inside cylindrical slabs. Specifically, in layered fluids, all the nearby rays emitted by a source on

a level where the speed of sound is a minimum will return to that
level at nearly the same distance. This 'focusing' in nature is,
however, much less common in acoustics than in optics, and only
occurs in very limited localities.

8. FERMAT'S PRINCIPLE

Fermat's Principle of geometrical acoustics asserts that the
acoustic path from a stationary emitter (speaker) at P to a stationary receiver (hearer) at Q is the quickest path. The determination of this path is mathematically equivalent to the solution
of Zermelo's navigation problem: given the speed $c = c(y)$ of a
boat and the current $U = U(y)$, how should the boat be steered
across a river to reach a given point Q on the opposite bank in
minimum time, from a given starting point P?

Whereas quickest paths are geodesics in an associated Riemannian manifold in any stationary (homogeneous or inhomogeneous) fluid
or crystal, they are geodesics of an associated <u>Finsler space</u> in any
moving fluid. In particular, the quickest path from Q to P does not,
in general, coincide with that from P to Q: when P is upstream
of Q, one should head to the center of the river in going from P
to Q, but hug the banks in returning.

Although general methods for determining such paths have been
discussed thoroughly in Caratheodory's VARIATIONSSTRECHNUNG, to
solve specific problems is difficult. So is the theoretical verification of the equivalence of Fermat's Principle to Huygens'
Principle. If a wavefront is defined as the set of all points Q
attainable from P in time T, but no shorter time, then one can
establish this equivalence under weak differentiability hypoth-
eses, even in Finsler spaces.

Rayleigh's ingenious derivation of formula (7.1) assumed
Huygens' Principle for plane wave fronts. We now show by direct
calculation that Fermat's Principle leads to the same formula,
because (i) the method of calculation is of intrinsic interest,
and (ii) it establishes the equivalence of Huygens' with Fermat's
Principle in the case considered. The problem is to show that if
the net downstream travel $\Delta x + \Delta x'$ is fixed, then the condition
$\delta(\Delta t + \Delta t') = 0$ of stationary time is equivalent to (7.1).

Consider a parallel flow of constant total width $a + b$, flowing between two parallel walls. Let the flow speed be U in a
strip of fluid of width a, adjacent to the nearer wall, and let
it be U' adjacent to the further wall. The problem is to find the

quickest path from a point P on the nearer wall to a point Q on the further wall, given also that the sound speeds in the two strips are c and c', respectively. Letting Δt and $\Delta t'$ be the times spent in the strips, clearly (in the notation of Sec. 7):

$$\Delta t + \Delta t' = (a \sec \phi)/c + (b \sec \phi')/c' \qquad (8.1)$$

Likewise, the total distance traveled downstream will be $\Delta x + \Delta x'$, where

$$\Delta x = (U + c \sin \phi)\Delta t = (M + \sin \phi)a \sec \phi$$
$$\Delta x' = (U' + c' \sin \phi)\Delta t = (M' + \sin \phi')b \sec \phi' \qquad (8.2)$$

with $M = U/c$ and $M' = U'/c'$ the local Mach numbers.

Since the downstream distance $\Delta x + \Delta x'$ from P to Q is fixed, $d(\Delta x) = -d(\Delta x')$, where

$$d(\Delta x) = a \sec^2 \phi \, d\phi [M \sin \phi + 1]$$
$$d(\Delta x') = b \sec^2 \phi' \, d\phi'[M' \sin \phi' + 1] \qquad (8.3)$$

Likewise, for $\Delta t + \Delta t'$ to be an extremum, we must have $d(\Delta t) = -d(\Delta t')$, where

$$d(\Delta t) = a \sec^2 \phi \, d\phi [(\sin \phi)/c]$$
$$d(\Delta t') = b \sec^2 \phi' \, d\phi'[(\sin \phi')/c'] \qquad (8.3')$$

For the conditions $d(\Delta t + \Delta t') = 0$ and $d(\Delta x + \Delta x') = 0$ to be equivalent, the condition is

$$[M \sin \phi + 1][(\sin \phi')/c'] = [M' \sin \phi' + 1][(\sin \phi)/c]$$

Multiplying through by $cc' \csc \phi \csc \phi'$ and expanding, we get (setting $Mc = U$, $Mc' = U'$):

$$U + c \csc \phi = U' + c' \csc \phi' \qquad (8.4)$$

which agrees with (7.1).

Note that this is also the condition that the ray leaving the near side at an angle ϕ (relative to the fluid) should reach the other side at a maximum or minimum distance downstream in the fixed total time $\Delta t + \Delta t'$. For related analytical applications to ray tracing techniques in layered fluids, cf. [16, Ch. 8]; for numerical applications, see [5].

9. GEOMETRICAL VS. PHYSICAL ACOUSTICS

I now come to the most interesting question: To what extent is geometrical acoustics, based on suitably extended and interpreted principles of Fermat and Huygens, consistent with the physical acoustics stemming from Euler's concept of what I have called an 'elastic fluid?' In trying to answer this question, one should distinguish between homogeneous fluids, layered fluids, and general inhomogeneous elastic fluids.

As regards homogeneous fluids, I have little to add to the vast literature concerned with solutions of the constant-coefficient wave equation $\phi_{tt} = c^2 \nabla^2 \phi$. It is well known that Snell's Law follows from Euler's equations and boundary conditions, which (cf. Sec. 6) suggests that geometrical and physical acoustics are consistent as regards refraction phenomena in layered media. However, it is also well known (and obvious) that, as regards diffraction phenomena, geometrical acoustics is at best asymptotically consistent with physical acoustics, in the limiting case of very high frequencies. Thus Rayleigh commented in 1878 [11, Sec. 286] (cf. [4, Sec. 3]) that:

> The fundamental assumption of the smallness of the
> wave-length, on which the doctrine of rays is built,
> having a far wider application to the phenomena of
> light than to those of sound, may properly be left to
> the cultivators of that sister science.

A few years later, in 1882, Kirchhoff showed that Euler's physical model rigorously implies Huygens' Principle in a three-dimensional homogeneous fluid (cf. [11, Sec. 273]). On the other hand, as Volterra showed in 1894, Huygens' Principle fails for cylindrical waves, i.e., in two dimensions.[*] The accepted contemporary view is that Euler's model implies Huygens' Principle in n space dimensions if and only if n is odd.

<u>Wave equations</u>. Most derivations of the wave equation for sound waves depend on Kelvin's theorem that, in a homogeneous elastic fluid, the circulation $\Gamma = \oint \underline{u} \cdot d\underline{x}$ around any closed curve γ moving with the fluid is a constant. More precisely, if the integral $\int dp/\rho(p) = \tilde{\omega}(p)$ is the same for all fluid particles, and if the external gravity field $g(x) = \nabla G$ is conservative [i.e., the gradient of a gravitational potential $G(x)$], then Γ is constant in time for any closed curve (of fluid particles) moving with the fluid. It is a corollary that if a train of sound waves passes through a homogeneous elastic fluid initially at rest, so that $\Gamma = 0$ for all γ at $t = 0$, the velocity field will be

---

[*]Cf. [11, Sec. 275]. For a historical discussion, see [1] or the author's SOURCE BOOK IN CLASSICAL ANALYSIS, p. 405.

irrotational (i.e., the gradient $\underline{u}(\underline{x};t) = \nabla\phi(\underline{x};t)$ of a velocity potential ϕ) at all later times.

Rayleigh [17, Chap. XI], and following him Lamb [11, Secs. 285-88], Morse and Ingard [15a, p. 244], Pierce [16, Secs. 1-6] and many others derive the (linear) constant-coefficient wave equation for ϕ,

$$\phi_{tt} = c^2\nabla^2\phi \qquad c^2 = dp/d\rho \tag{9.1}$$

in essentially this way. It may be shown that in the linearized 'small amplitude' approximation, variations $\delta p = p - p_0$ in the pressure, $\delta\rho$ in the density ρ, and the scaled density $s = (\rho - \rho_0)/\rho_0$ also satisfy (1.1) if ϕ does.

Inhomogeneous elastic fluids. The mathematical theory of inhomogeneous elastic fluids is much more difficult. In particular, the standard wave model for refraction across interfaces predicts nonzero vorticity, even in static layered fluids: when the sound wave changes discontinuously in direction, continuity in $\partial\phi/\partial n$ is inconsistent with continuity in $\partial\phi/\partial s$. For this reason, I shall defer its general discussion to Appendix C. After discussing earlier formulations in the rest of this section, I shall take up a few relatively simple examples.

In the context of optics, Sommerfeld [20a, Sec. 35] related 'rays' to 'wave functions' through the concept at an eikonal function. However, he restricted his attention to waves of constant frequency $f = k/2\pi$, and simply wrote down the wave equation

$$\nabla^2 u + k^2 u = 0 \qquad k = \sqrt{\varepsilon\mu\omega} = 2\pi/\lambda \tag{9.2}$$

Moreover, he never specified which physical variable u was intended to signify.

Pierce [16, Sec. 8-1] has applied this concept to the refraction of sound in moving fluids, giving many references. However, he does not discuss the question of consistency, which is my main concern here.

In an ingenious and original treatment of sound waves in an elastic fluid as a Lagrangian dynamical system, Morse and Ingard [15, pp. 248-50] expressed $p = \rho\phi_t$ and $\underline{u} = -\nabla\phi$ in terms of an assumed velocity potential $\phi(x;t)$, to derive a Lagrangian L of the form

$$-L = \rho[\nabla\phi \cdot \nabla\phi + \rho\kappa\sigma_t^2]/2 \tag{9.3}$$

where κ is a compressibility coefficient. Koopman [10, p. 19] assumed a 'wave equation' of the form

$$\phi_{tt} = (c^2/\rho)\nabla \cdot (\rho\nabla\phi) \tag{9.4}$$

where again ϕ is a velocity potential.

After an especially rigorous analysis which allows for a vector source density $\underline{G}_1$ satisfying $\rho_0(\underline{x})\underline{G}_1 = \nabla G$, Wilcox[*] derived the wave equation

$$\phi_{tt} = c^2(\underline{x})\rho_0(\underline{x})\nabla \cdot [\nabla\phi/\rho_0(\underline{x})] \tag{9.5}$$

for the case $\underline{G}_1 \equiv 0$. Here $\phi(\underline{x};t)$ is an 'acoustic potential.'

Unfortunately, vorticity is generated by sound waves in inhomogeneous elastic fluids, even if gravity is ignored and only first-order effects are considered. This general fact is illustrated by the following almost trivial example.

Consider a layered elastic fluid with equation of state $p = p_0 + c^2[\rho - \rho_0(y)]$, where the equilibrium density $\rho_0(y)$ is variable but the speed of sound is constant. The linearized equations $\underline{u}_t = -\nabla p/\rho_0(y)$ and

$$p_t = c^2\gamma\rho_t = -\rho_0(y)c^2\nabla \cdot \underline{u}$$

are then satisfied by the velocity field

$$u = A(y)\cos(kx - \omega t) \qquad v \equiv 0 \tag{9.6}$$

and associated pressure distribution

$$p = P_0\cos(kx - \omega t) \tag{9.7}$$

provided that $\omega = ck$ and $A(y) = P_0/I(y)$, where $I(y) = c_{0}(y)$ is the one equilibrium impedance distribution.

A more sophisticated example of vorticity generation (by pressure waves traveling in circles) will be described in Section 11.

10. CHAPLYGIN FLUIDS

From a mathematical standpoint, homogeneous 'elastic fluids' having an equation of state of the linearized form

$$p = p_o - I^2/\rho \qquad\qquad I = \text{constant} \tag{10.1}$$

---

[*]Calvin Wilcox, SOUND PROPAGATION IN STRATIFIED FLUIDS, Springer-Verlag, 1984, p. 5. (His u is our ϕ.)

are the easiest to treat. Since they were first treated by
Chaplygin, I shall call them Chaplygin fluids. (They were also
used by von Kármán and Tsien to treat linearized compressible
flows.)

Differentiating, we obtain

$$c^2 = dp/d\rho = I^2/\rho^2$$

which shows their defining property as being that their impedance
$\rho c = I$ is unchanged by compression and expansion.

<u>Plane waves</u>. In the case of plane waves, Chaplygin fluids
have another theoretical advantage. Namely, they make it possible to include convection effects. To include these in plane
waves, it suffices to replace the independent position variable x
with the Lagrangian cumulative mass variable $a = \int \rho(x)\,dx$, and to
consider the location $x = x(a,t)$ of a particle initially at $x_0 = x(a,0)$ as a function of a and t. The particle velocity is then
evidently $u = x_t$, where subscripts signify derivatives with re-
spect to the variable indicated, while $x_a = 1/\rho = \sigma$ is the specific volume. Hence the equation of motion is

$$x_{tt} = u_t = -p_a \tag{10.2}$$

while the equation of 'continuity' (conservation of mass) is ful-
filled automatically.

Using these Lagrangian coordinates, one easily derives from
Eqs. (10.1) and (10.2) the linear wave equation

$$x_{tt} = I^2 x_{aa} \tag{10.3}$$

having as its general solution

$$x = f(a + It) + g(a - It) \tag{10.3'}$$

<u>Inhomogeneous Chaplygin fluids</u>. The preceding well known
results suggest the study of plane waves in inhomogeneous Chaplygin fluids. These are defined as fluids having, in Lagrangian
coordinates, an equation of state of the special form

$$p(a,t) = P(a) - I^2(a) x_a \tag{10.4}$$

In static equilibrium, such a fluid must be at some constant pressure p_0. Hence the function P(a) in (10.4) must be of the form

$$P(a) = p_0 + I^2(a)/\rho_0(a) \tag{10.4'}$$

where $\rho_0(a)$ is the equilibrium density of the fluid with Lagrangian coordinate a.

The analog of (10.3) in an inhomogeneous Chaplygin fluid is therefore the (inhomogeneous) second-order linear differential equation (DE)

$$x_{tt} = P'(a) + [I^2(a)x_a]_a \qquad (10.5)$$

in which the last term is self-adjoint.

Huygens' minor premise. The literature concerned with Huygens' minor premise (for which see [1] and my comments in [19, p. 653]) has emphasized distinctions between even and odd dimensions in homogeneous fluids. These correspond to the effects of the parity of n on solutions of the second-order differential equation

$$\phi_{tt} = c^2\left[\phi_{rr} + \frac{n-1}{r}\phi_r\right] \qquad (10.6)$$

satisfied by the velocity potential of spherical sound waves in homogeneous n-dimensional fluid. (Equivalently,

$$r^{n-1}\phi_{tt} = c^2[r^{n-1}\phi_r]_r.)$$

It would be interesting to discover those functions $I(a)$ for which Eq. (10.5) is consistent with Huygens' minor premise. This would be especially interesting since, for reasons to be explained in Appendix C, the theory of sound waves of 'infinitesimal amplitude' in inhomogeneous Chaplygin fluids should apply to inhomogeneous fluids in general!

11. PRESSURE-GENERATED VORTICITY

We will now give a second example of sound waves in an inhomogeneous elastic fluid which generate vorticity; in it the 'rays' are circular paths. This example is most easily motivated by thinking of it as an example of self-similar circular refraction of sound waves.

Consider an elastic fluid whose equilibrium density $\rho = \rho_1/r^2$ is inversely proportional to the square of the distance from the z-axis. Around the cylindrical shell $x^2 + y^2 = r^2$, let the elasticity be governed by the Chaplygin equation of state $p = 1 - 1/\rho r^2$. Its equilibrium pressure will then be $p = 1 - \rho_1$, where ρ_1 is a constant independent of r.

We look for sound waves which progress around circular paths $r = $ const., so that, if α is the (Lagrangian) angular coordinate of a particle in equilibrium, $\theta = \alpha + \varepsilon h(\alpha,t)$, where ε is a small positive parameter chosen to make $|h|_{max} = 1$. Then the specific volume $\sigma = 1/\rho$ is

$$\sigma = r^2 \sigma_1 [1 + \varepsilon(\partial h/\partial \alpha)]$$

The circumferential equation of motion is

$$rh_{tt} = \varepsilon r\ddot{\theta} = -\varepsilon r \, \partial p/\partial \alpha = \varepsilon r\sigma_1 h_{\alpha\alpha}$$

i.e., the linear, constant-coefficient wave equation. This is exactly satisfied if and only if

$$h(\alpha,t) = f(\alpha - ct) + g(\alpha + ct) \tag{11.1}$$

where $c = \sqrt{\varepsilon c_1}$, while f, g are smooth.

The radial component of the equation of motion is

$$\ddot{r} = -u_\theta^2/r = \dot{\theta}^2 r = \varepsilon^2 h_t^2 r \tag{11.2}$$

Changes in r, being quadratic in ε, are asymptotically negligible as $\varepsilon \downarrow 0$. It follows that all superpositions of two circumferential waves having arbitrary profiles, and proceeding with constant angular velocity clockwise and counterclockwise, respectively, are solutions of the linearized equations.

The essential properties making this possible are that (i) the equilibrium fluid density $\rho_0 = \rho_1/r^2$ is inversely proportional to r^2, and (ii) the local sound speed $c = (dp/d\rho)^{1/2} = (1/\rho_0^2 r^2)^{1/2} = r/\rho_1$ is directly proportional to r.

Because of these properties, energy is propagated around circles (acoustic paths) which do satisfy Fermat's Principle of Least Time.

The vorticity of these sound waves is

$$\partial(ru_\theta)/\partial r - \partial u_r/\partial \theta = \varepsilon \partial(r^2 y_t)/\partial r \tag{11.3}$$

or $2\varepsilon rh_t = 2u_\theta$. The variability of u_θ in time for fixed (r,α) clearly violates the assumption of the existence of a velocity potential.

12. SNELL'S LAW; LAYERED FLUIDS

It is not hard to extend Snell's Law of refraction to inhomogeneous elastic fluids. The extension is based on Green's derivation of Snell's Law, which does not assume Fermat's Principle.

Let $y = 0$ be an interface separating two homogeneous elastic fluids. Consider the reflection and refraction of simply periodic plane waves traveling at an angle α with the horizontal in the lower fluid (i.e., where $y < 0$). If the coefficient of reflection is r, we will then have

$$
\begin{aligned}
u &= A \cos \alpha \cos(kx + \ell y - \omega t) \\
&\quad + rA \cos \alpha \cos(kx - \ell y - \omega t) \\[4pt]
v &= A \sin \alpha \cos(kx + \ell y - \omega t) \\
&\quad - rA \sin \alpha \cos(kx - \ell y - \omega t) \\[4pt]
p &= I_0 A\{\cos(kx + \ell y - \omega t) \\
&\quad + r \cos(kx - \ell y - \omega t)
\end{aligned}
\tag{12.1}
$$

where $\tan \alpha = \ell/k$. Likewise, if the coefficient of refraction is r', we will have in the upper fluid (where $y > 0$):

$$
\begin{aligned}
u &= r'A \cos \beta \cos(kx + my - \omega t) \\
v &= r'A \sin \beta \cos(kx + my - \omega t) \\
p &= I_1 r'A \cos(kx + my - \omega t)
\end{aligned}
\tag{12.2}
$$

Here β is the angle made with the horizontal by the refracted waves, so that $m/k = \tan \beta$; $I_0 = \rho_0 c_0$ and $I_1 = \rho_1 c_1$ are the impedances of the two fluids.

The interface conditions are the continuity of p and of v. Cancelling the common factor A, these give (when $my = 0$)

$$
I_0(1 + r) = I_1 r'
\tag{12.3}
$$

and

$$
(1 - r) \sin \alpha = r' \sin \beta
\tag{12.3'}
$$

The fact that we have the same k in (12.2) as in (12.1) corresponds to Green's derivation of Snell's Law: $k = c_0 \cos \alpha = c_1 \cos \beta$, where the c_i are the speeds of sound in the two media.

The condition for zero vorticity is the continuity of u across the interface $y = 0$, since the velocity fields on both sides are irrotational. This amounts to $(L + r)\cos \alpha = r' \cos \beta$ or, by Snell's Law and (12.3),

$$
\frac{r'}{1 + r} = \frac{\cos \alpha}{\cos \beta} = \frac{c_1}{c_0} = \frac{I_1}{I_0}
$$

which is evidently equivalent to $\rho_1 = \rho_0$: the condition that the two fluids have the same density.

By induction and passage to the limit as the number of plane interfaces increases, we are thus led to the following conclusion.

THEOREM. Geometrical and physical acoustics are consistent in layered fluids; physical acoustics has the advantage of permitting one to predict the intensities of the reflected and refracted waves. Sound waves are irrotational in layered fluids if and only if these have constant density.

13. COLLIDING PULSES

The concept of a simple pulse extends naturally to homogeneous Chaplygin fluids. Physically, it is a wave behind which the fluid velocity U is a constant. Since the wave front traverses a mass $I_0 = \rho_0 c_0$ per unit time, such a pulse is described in Lagrangian coordinates by the formula

$$x(a,t) = \begin{cases} a/\rho_0 + U\tau & \text{if } \tau \geq 0 \\ a/\rho_0 & \text{if } \tau \leq 0 \end{cases} \tag{13.1}$$

where $\tau = t - a/I_0$ is the lapse of time since the front passed the particle a. If $U > 0$, the wave is a compression wave; if $U < 0$, it is a rarefaction wave. As noted before, if $\rho_0 U > I_0$ (i.e., in a supersonic compressive pulse), matter is unrealistically 'turned inside out.' We can rewrite (13.1) in the form

$$x(a,t) = a/\rho_0 + f(\tau) \qquad \tau = t - a/I_0$$

where the displacement $f(\tau) = a - x$ is $U\tau$ if $\tau \geq 0$ and zero if $\tau \leq 0$.

We next consider the effect of two colliding pulses (in an otherwise static homogeneous fluid), timed to reach the origin from opposite sides at time $t = 0$. For $t < 0$, let the first pulse be defined by (13.1) for $a < 0$, and the second (for $a > 0$) by

$$x(a,t) = a/\rho_0 + g(\tau') \qquad \tau' = t + a/I_0 \tag{13.1'}$$

where $g(\tau')$ is $U'\tau'$ if $\tau' \geq 0$ and zero if $\tau' \leq 0$ (i.e., if $a < -I_0 t$, where $t < 0$). Since the Chaplygin model is linear, the motion induced by the colliding pulses will be given simply by

$$x = a/\rho_0 + f(\tau) + g(\tau') \tag{13.2}$$

In the interval $(-I_0 t, I_0 t)$, the resultant velocity will be $U + U'$, and the displacement $U\tau + U''\tau'$.

<u>Transmission and reflection</u>. Consider next the transmission and reflection of the pulse (13.1) by an interface at $x = a = 0$, to the right of which the density is ρ_1, the speed of sound c_1, and the impedance $I_1 = \rho_1 c_1$. We wish to determine the transmitted velocity $U' = r'U$ and the reflected velocity $U'' = r''U$, as well as the pressures in the three regions $a < -I_0 t$, $-I_0 t < a < I_1 t$, and $a > I_1 t$. If $a > I_1 t$, clearly $p = p_0$ is the equilibrium pressure of the undisturbed fluid. Since a velocity U' is being imparted to a mass I_1 of fluid per unit time, $p = p_0 + I_1 U'$ in the interval $(-I_0 t, I_1 t)$, while $p = p_0 + I_0 U$ if $a < -I_0 t$.

In the subinterval $(-I_0 t, 0)$, the rate $I_0 U''$ of momentum change is therefore $\Delta p - \Delta p' = I_0 U - I_1 U''$. Dividing through by U, we get the equation $I_0 = r'I_1 + r''I_0$. Finally, by continuity, the velocity must be the same on the two sides of the interface, whence $U + U'' = U'$, and $1 = r' - r''$. Solving, we get

$$r' = \frac{2I_0}{I_0 + I_1} \qquad r'' = \frac{I_1 - I_0}{I_0 + I_1} \qquad\qquad (13.3)$$

as in [2a, Appendix A].

Finally, for reasons to be explained in the next section, we consider the collision of two pulses meeting at an interface separating two homogeneous fluids. By the linearity of the equations for Chaplygin fluids, if the pulse velocities are U and V, respectively, the resulting velocity in the interval $(-I_0 t, I_1 t)$ will be $U' + V'$ for $t > 0$, where $V' = [2I_1/(I_0 + I_1)]V$.

14. PLANE WAVE DISCRETIZATIONS

It would be straightforward to solve the initial value problem for plane waves in a Chaplygin fluid numerically, using (10.5). However, one gets a simpler formula by taking the pressure p instead of the position x as the dependent variable. By (10.4),

$$p_t = c^2 \rho_t = [I^2(a)/\rho^2]\rho_t$$

where, since $-(\ln \rho)_t = u_x$, $\rho_t = -\rho^2 u_a$. Substituting back, we have in a one-dimensional Chaplygin fluid,

$$p_t = -I^2(a) u_a \qquad\qquad (14.1)$$

Combining with the equation of motion $u_t = -p_a$, valid in any inviscid fluid, we obtain

$$p_{tt} = -I^2(a)u_{ta} = -I^2(a)u_{at} = I^2(a)p_{aa} \qquad (14.2)$$

which has many well-known discretizations.

I think it would be especially interesting to use the formulas of Section 13, because these permit one to compute exactly (up to roundoff error) the progress of plane waves through any piecewise homogeneous Chaplygin fluid, in which all homogeneous slabs have the same traverse time $\Delta t = \Delta a_i / I_i = \Delta x_i / c_i)$ in the jth slab). This can be done by the method of characteristics, and should make possible a clearer understanding of the extent to which Huygens' second premise holds in a one-dimensional inhomogeneous elastic fluid (i.e., for plane waves). (As I have recalled, every elastic fluid is equivalent to some Chaplygin fluid in the linear acoustic approximation.) The two are obviously consistent if $I_0(a) = \rho_0(a)c_0(a)$ is constant, but I suspect that they are inconsistent otherwise.

Let the a_i be the slab interfaces, so that $\Delta a_i = a_i - a_{i-1}$, and let $t = n\Delta t$. Then the characteristics through the initial mesh points $(a_i, 0)$ on the initial line $t = 0$ will constitute two families of piecewise straight broken lines, passing through the mesh points (a_i, t_j) with $r = i + j$ and $s = i - j$ constant, respectively. These broken line characteristics divide the (a,t)-plane into rhombi in the interior of each of which p and u are constants, provided that the initial $u(x,0)$ and $p(x,0)$ are constant in each slab.

<u>Initial conditions</u>. To define a well posed initial value problem, one must specify not only the piecewise constant initial $u(x,0)$ and $p(x,0)$, but also related values u_i and p_i at mesh points $(a_i, 0)$. These can be, but need not be, the arithmetic means of the values in the adjacent slabs, and should be interpreted as the (constant) previous values in the rhombus whose top vertex is at $(a_i, 0)$.

From these data, interpreted as arising from pulses in the rhombi having opposite vertices (a_{i-1}, a_i) and spreading for $t < 0$ into the intermediate rhombi with top vertices at $(a_{i-1}, 0)$ and $(a_i, 0)$, one can compute by use of (13.3), analogous formulas for transmitted and reflected pressures, and linearity (the superposition principle), a unique new pressure and velocity resulting (for $t > 0$) from the collision of these pulses at the ith interface at time $t = 0$. By Newtonian relativity (using moving axes), one can reduce to the case that the fluid velocity U_i in the rhombus whose top vertex a_i is zero: one considers only the velocity jumps Δu_i across the 'shock points' (characteristics) separating adjacent rhombi, and transmitting the values of p and U in one rhombus above the front to that below it.

At the center of the mesh rectangles $[a_{i-1}, a_i] \times [t_{j-1}, t_j]$, we have the simpler case of the collision of two pulses in a homogeneous fluid, tractable using (13.2).

To develop a computer program based on the methods and formulas outlined above seems to me a very worthwhile, if substantial, project.

15. EIKONAL APPROXIMATION

In acoustics, the eikonal function may be interpreted as an amalgamation of ray theory (Fermat-Huygens) and the concept of an inhomogeneous elastic fluid. To the extent that these two are consistent, the eikonal function is consistent with both (cf. [2, p. 650] and [2a, Sec. 5 ff]).

The preceding interpretation is clearest in the case of an acoustic point-source. Hamilton-Jacobi theory, formulated in the Riemannian manifold with travel time $\int ds/c(\underline{x})$ as arc-length, then defines spherical coordinates in which the 'radii' are the geodesics emanating from the source, while latitude and longitude are established at the source. (These comprise the 'normal coordinates' of Riemann.)

If we now introduce 'workless constraints' by inserting frictionless walls bounding stream tubes consisting of bundles of rays, a new conservative Lagrangian dynamical system will be defined. Simply harmonic radial pulsations of the source will produce "simple" [2, p. 650] waves with pressure variations of the form

$$\delta p = A(x,y,z)\cos(\omega[t - \tau(x,y,z)]) \tag{15.1}$$

relative to a suitably chosen time origin.

If the stream tube has constant cross-section, then $\delta p_{tt} = I^2(a)\delta p_{aa}$ as in (14.2), where a designates the cumulative mass along the tube from the source. Substituting $\delta p = P(a)\cos \omega t$ into this differential equation, we obtain the Sturm-Liouville equation

$$p'' + \omega^2 I^2(a)P = 0 \tag{15.2}$$

Note that although the local half wave-lengths (defined as the distances between successive nodes) are asymptotically equal to

$2\pi c/\omega$, the nodes cannot be computed by setting $P = 0$ in the formula

$$P = A(\tau)\cos \omega\tau \qquad\qquad (15.3)$$

where $\tau = \int ds/c(\underline{x})$ is the travel time.[*] In general, doubling ω does not exactly halve the wave-length.

It would be interesting to work out the analog of the formula $\delta p_{tt} = I^2(a)\delta p_{aa}$ for stream tubes having variable cross-section area.

<u>Consistency condition</u>. For oscillations of infinitesimal amplitude in three-dimensional space, constraining particles to oscillate along acoustic 'rays' is evidently equivalent to the condition that the 'wave fronts' (i.e., the concentric spheres with center at the source in the Riemannian manifold in which 'distance' is the travel time τ) be isobars [2, p. 650]. Note that since $ds = c\, d\tau$, the orthogonality of rays with spheres in this Riemannian manifold is equivalent to geometric orthogonality in ordinary Euclidean space.

For this to be the case for all ω, we must have $A = A(\tau)$, so that the surfaces of constant phase must be isobars. By (15.2), it seems probable that $I^2(\underline{a})$ must be independent of latitude and longitude as well. Finally, for this independence to hold for latitude and longitude relative to every acoustic source, since vorticity is generated when ρ varies even when $I = \rho c$ is constant, it seems probable that a three-dimensional elastic fluid must be homogeneous in order for the eikonal model to be exact.[**]

16. INTERFERENCE EFFECTS

In addition to refraction, diffraction, absorption, and dispersion effects, sound propagation can produce important interference effects. Thus the anomalous absorption of ultrasonic sound in CO_2 was detected by C. W. Pierce (see Appendix A) using an acoustic interferometer. Likewise, 'shadow zones' in underwater sound are attributed to interference by reflected sound.

---

[*]For the relevant asymptotic formulas, see Chapter 10 of G. Birkhoff and G.-C. Rota, ORDINARY DIFFERENTIAL EQUATIONS, 4th ed., Wiley, 1988; cf. [16, Sec. 8-1] and [21, Sec. 3.6].

[**]Cf. J. Hadamard, Annals of Math. 43 (1942), 510-222, and its review by Bourgin in Math. Revs. 4 (1943), p. 45. In his influential VARIATIONAL PRINCIPLES OF MECHANICS, p. 277, C. Lanczos calls wave surfaces "surfaces of equal phase."

<u>Violin ensembles</u>. An interesting question concerns the intensity of the musical note produced by a violin ensemble playing in unison. It seems reasonable to assume that each violin produces the same note with about the same intensity, but with uncorrelated random phases. The question naturally arises: Why do there not exist zones in which these sound waves cancel by interference?

For many violins in the same direction, the superposition will produce oscillations $\delta x = A \cos \omega t + B \sin \omega t$, where A and B are independent random variables having the same, roughly Gaussian distribution. The random sound wave amplitude will therefore have a probability density Kre^{-r^2} in the (A,B)-plane.

In space, the resulting amplitude will have a probability density $Kr^5 e^{-r^2}$; the chance of this being inaudible is thus extremely small indeed!

APPENDIX A: MOLECULAR RELAXATION EFFECTS

The "classical" kinetic theory of gases, briefly recalled in Section 2, was just the first (and simplest) of many molecular models of fluids. Equally influential was a more sophisticated model proposed in 1873 by van de Waals. This model, originally intended to be applicable to liquids as well as to imperfect gases, is associated with equations of state of the form*

$$\left(p + \frac{a}{v^2} \right) (v - b) = RNT \tag{A.1}$$

(The limiting case $a = b = 0$ evidently corresponds to a perfect gas.)

The still more sophisticated concept of molecular relaxation can be traced back to 1875, when** "Stefan and Boltzmann made the assumption that the translational energy of the molecules was passed on from place to place with greater speed than the remaining energy.... Meyer seems to have accepted these ideas in the first edition of his KINETIC THEORY OF GASES (1877), but repudiated them in the second edition (1899)."

----

*James Jeans, KINETIC THEORY OF GASES (4th ed., 1935), pp. 125-154. Here, in Sec. 149, it is observed that the perfect gas law predicts correctly the "osmotic pressure of weak solutions," and the pressure exerted by electrons surrounding a hot solid.

**Jeans, op. cit., p. 297. Here, in Secs. 395-405, "classical" kinetic theory is used to estimate the Prandtl numbers of gases.

This concept was applied by Herzfeld and Rice in 1928 to explain the anomalous absorption and dispersion effects observed by G. W. Pierce in 1925. They suggested that "at these very high frequencies (10 kHz-1 mHz) the interchange between the translational and vibrational energy of the molecules lags behind the rapid oscillations of pressure, so that the molecular motions have insufficient time to adjust themselves to the adiabatic temperature variations" [25, p. 275]. Their idea finds qualitative confirmation in the fact that "anomalous absorption takes place in the frequency range of strong dispersion" [25, p. 357]. As developed by H. O. Kneser and others in the 1930s, it is now generally accepted. To quote [25, p. 356]:

> It has become increasingly evident that the theories
> of Stokes and Kirchhoff for the sound attenuation in
> gases and liquids due to viscosity and heat conduction
> are insufficient to explain...experimental observations
> at higher frequencies.

Superficially, this behavior might seem analogous to that of electromagnetic waves (cf. Appendix B). Thus visible light, like audible sound, is restricted to a limited frequency band. Moreover, in both, dispersion becomes appreciable only at very high frequencies (exceeding 10^5 kHz; cf. J. A. Stratton, ELECTROMAGNETIC THEORY, McGraw-Hill, 1941, p. 321). However, the suggested analogy is unsound for two reasons. First, electromagnetic wave dispersion is generally attributed to resonance, and not to relaxation [8, Sec. 9]. And second, although acoustic energy dissipation by molecular relaxation does occur primarily in the ultrasonic range, at the highest frequencies viscosity and heat conduction are currently believed to be responsible for it [16, p. 561].

<u>Gases</u>. In dilute gases, the rate $2\alpha\lambda$ of energy absorption per wave length is predicted by the molecular relaxation model to have maxima in "frequency ranges of strong dispersion," each associated with a particular relaxation time τ_m. Moreover, the maximum value $(2\alpha\lambda)_m$ of $2\alpha\lambda$ is predicted to depend for each gas only on the ratio f/p of the frequency to the pressure, at a given temperature. This ratio corresponds to the number of collisions per period (in time) and hence (for given c) to the ratio λ/λ_f of the wave length of sound to the molecular mean free path.

In most fluids, there are usually only one or two observable bands of strong anomalous dispersion (and relevant "relaxation times"). Indeed, in pure monatomic gases there are none; moreover,

Stokes' relation $3\lambda + 2\mu = 0$ is believed to hold, so that the classical Stokes-Kirchhoff theory seems to be fully consistent with physical reality in monatomic gases.

However, even for dilute gases, the concept of "molecular relaxation" is not clearly defined. The "molecules" of the classical kinetic theory of gases, which inspired the original concept of "molecular relaxation" associated with "vibrational and rotational" energy, are evidently inconsistent with modern (Rutherford) molecular models consisting of electrically charged nuclei and electrons. In addition, as Jeans (op. cit., p. 402) already realized by 1908, quantum theory makes "the fundamental bases of kinetic theory lose all claim to general validity." In particular, "classical kinetic theory and statistical mechanics fail to account for the energy associated with internal degrees of freedom." Today, one should ideally deduce this energy from the appropriate Schrödinger equation, a still different model!

<u>Liquids</u>. Molecular relaxation effects seem to be quite different in liquids than in gases. In the first place, since the total adiabatic correction for sound wave velocity is so much less (see Section 2), so are dispersion effects due to molecular relaxation. Moreover, "relaxation times" in a given liquid are not inversely proportional to the pressure; hence correlations of "frequency ranges of strong dispersion" with ranges of anomalous absorption must be quite different in liquids than in gases.

On the other hand, absorption effects seem to be even more complex in liquids. In pure water, the observed rate α of sound wave attenuation seems to be about three times that predicted by the Stokes-Kirchhoff model: according to [25, p. 358], $3 < \alpha_{cl} < 3.33$ between 0°C and 70°C. This may be a classical 'bulk viscosity' effect.

Sound absorption in other liquids may deviate much more from α_{cl}. Thus, in CS_2, it is believed that $\alpha/\alpha_{cl} > 1000$ in the megacycle frequency range. For a careful analysis of the data available in 1959, see [8], which distinguishes between three qualitatively different kinds of chemically pure liquids: "Kneser," "associated," and "highly viscous," each with its own molecular model.

Although the additional absorption which may occur in impure liquids is not analyzed in [8], it has been intensively investigated, especially in water, ever since the startling discoveries of World War II (see Section 4). In water, where much is known

about the behavior of ions in solution, the latest models attribute
acoustic attenuation with (perhaps two-stage) chemical relaxation
times. These are qualitatively different from the physical delay
times in gases associated with 'rotational' and 'vibrational' internal energy.*

APPENDIX B: OPTICAL ANALOGIES

At the end of Section 2, it was recalled that the basic formulas of 'geometrical' acoustics and of 'physical' acoustics arise
also in optics. This appendix will be concerned with other analogies between these two "sister sciences."

An attractive feature of the basic models of both is their
theoretical applicability to all frequencies, scales, and amplitudes as well as to all homogeneous media,** each medium being
characterized by a few empirical material constants. This flexibility has made their common formulas useful for treating ultrasonic waves, and for treating electromagnetic waves of all sizes,
from radio waves a kilometer long down to electron size (Davisson-
Germer experiment).

Obviously, there are major differences of scale between acoustics and optics in the historical sense. Thus the speed of sound
(340 meters/sec in air, 1500 meters/sec in water) is 5-6 orders of
magnitude less than that of light, while audible wave lengths λ of
sound (of the order of a meter) are 5-6 orders of magnitude greater
than visible wave lengths of light (around 0.35-0.7 microns).
Correspondingly, the frequencies $f = c/\lambda$ of audible sound waves
(25-2500 Hz) are more than 10^{10} less than those of visible light.

These differences in scale have little theoretical significance, for the reasons stated in the preceding paragraph. However,
differences in scale and physical differences force acoustics and
optics to use very different laboratory techniques. Moreover, the
mathematical analogies between acoustics and optics are also ob-
scured by related basic psychological differences. Such psychological differences can actually be very stimulating; our intuitive

---

*Cf. R. H. Mellen, D. G. Browning, and V. P. Simmons, J. Acoust.
Soc. Am. 74 (1983), 987-993, and references given there and in
Ibid. 68 (1980), 248-57.

**This statement obscures the physical fact that light waves can
be transmitted through a vacuum (e.g., from distant stars), whereas sound waves cannot. Likewise, optical indices of refraction
are all less than two, whereas sound speeds can differ by a factor
of ten.

recognition of the concepts of Fourier analysis provides a case
in point.

Thus, as Newton showed in his OPTICKS, one can make a spectral resolution (partial Fourier analysis) of a beam of light into pure colors by refracting it through a glass prism. This is possible because there is a dispersion (variation in c with λ) of about 2% when light passes through glass. In contrast, the dispersion of sound in the audible range is negligible!

The most familiar acoustic illustrations of Fourier analysis, also in contrast, are associated with overtones of stringed instruments and harmony (musical chords). These, and transpositions of melodies from one scale to another, are possible because the range of audible pitches extends over 10 octaves. Their intuitive appreciation can have no analog in optics, since the visible range of light extends over barely one octave!

Modern mathematical theories of Fourier analysis have been enriched by both sets of intuitions. Analogies between phonons and photons, and between sonar and radar, have proved equally suggestive. Finally (see Section 15), the eikonal functions of optics applies just as well to acoustics.

APPENDIX C: INHOMOGENEOUS ELASTIC FLUIDS

An inhomogeneous elastic fluid differs from a homogeneous elastic fluid in that its equation of state varies from one part of the fluid to another. Its defining equations are easily written down in Lagrangian coordinates. They include, besides the usual equations of continuity,

$$D(\ln \rho)/Dt = -\nabla \cdot \underline{u} \qquad \underline{u} = D\underline{x}/Dt \tag{C.1}$$

and motion

$$D\underline{u}/Dt = -\nabla p/\rho \tag{C.2}$$

a new equation of state of the form

$$p = p(a,\rho) \tag{C.3}$$

Here only (C.3) differs from Euler's original formulas, which are reproduced in [16 (15-2)] in a slightly different form. Notice that, in Lagrangian coordinates, D/Dt is the partial time derivative.

Unfortunately, these equations are much more difficult to integrate analytically[*] than Euler's original equations, especially for solving "steady flow" problems in which one wishes to determine $\underline{u} = \underline{u}(\underline{x})$ and not $\underline{x} = \underline{x}(\underline{a}, t)$. Moreover, the $\partial p/\partial x_i$ are not easily calculated in Lagrangian coordinates.

<u>Chaplygin fluids</u>. A "Chaplygin fluid" is a fluid in which (C.3) assumes the special forms

$$p(\underline{a};t) = P(\underline{a}) - I^2(\underline{a})|\partial x_i/\partial a_j| \qquad\qquad (C.4)$$

This reduces to (10.4) for plane waves and, as in that case, the initial value problem ceases to be physically well-posed when the Jacobian $|\partial x_i/\partial a_j|$ changes sign. Loosely speaking, this failure of well-posedness is associated with local 'densities' $\rho_0/|\partial_x i/\partial a_j|$ which become negative after becoming infinite, and is analogous to the formation of shock waves (discontinuities of $\underline{u}(\underline{x};t)$) in real fluids. Mathematically, both singularities arise in pressure waves of sufficiently large finite amplitude.

<u>Linear acoustics.</u> The usual approximation of linear acoustics (cf. [16 (15-3)]) nominally assumes that $\delta p = c^2 \delta\rho$; i.e., it assumes an equation of state of the form $p = \kappa\rho^\gamma$ with $\gamma = 1$. However, this is just intended as a tangent approximation, and (C.4) (the nominal assumption $\gamma = -1$, will serve about as well. This has the advantages of yielding linear partial differential equations of motion in Lagrangian coordinates, as we verified for plane waves in Sections 10 and 14.[**]

A more common linearization in Eulerian coordinates is obtained by making the assumption of small displacements, i.e., by replacing $\underline{x}(\underline{a};t)$ with its initial value $\underline{x} = \underline{a}$, as defined above. One then writes $\rho_0 = \rho_0(\underline{x})$ and $c^2 = c^2(\underline{x})$ in [16, (1-5.3)]. It would be interesting to know whether or not, with these substitutions, one obtains the same partial differential equations!

Conceptually, sound waves in a general (inhomogeneous) elastic fluid are most simply interpreted as "small oscillations" in the vicinity of some static equilibrium configuration. This configuration is necessarily at some constant pressure p_0. In

---

[*]The ALE codes developed at Los Alamos by Francis Harlow, C. A. Hirt, and others, have solved some problems numerically.

[**]For the case of two-dimensional steady flow, see Sec. 12.5 of R. von Mises, MATHEMATICAL THEORY OF COMPRESSIBLE FLUID FLOW, Academic Press, 1958.

                         G. Birkhoff

Lagrangian coordinates, we can let $\underline{x}(\underline{a},0) = \underline{a}$ be the equilibrium position ('cumulative mass' is meaningless). The kinetic energy of the fluid is clearly (for $\dot{x}_k = \partial x_k/\partial t$)

$$T = \frac{1}{2} \iiint (\Sigma\, x_k^2)\, \rho_0(a)\, da_1 da_2 da_3 \tag{C.5}$$

To treat the fluid as a Lagrangian dynamical system, we must also derive an expression for the strain energy V; this will extend the Morse-Ingard formula (9.3) to inhomogeneous elastic fluids.

Relative to its equilibrium state, the potential energy per unit mass of the fluid at $\underline{a}$ is

$$E(\sigma) - \int_{\sigma_0}^{\sigma} p(\sigma)\, d\sigma = \frac{1}{2}\, I^2(\underline{a})\, (\sigma - \sigma_0)^2 + O(\sigma - \sigma_0)^3$$

where the subscript 0 signifies its equilibrium state, and $I = \rho_0 c_0$ is the local specific impedance. The function $E(\sigma)$ is quadratic for the Chaplygin equation of state. Also, $I^2(\underline{a})\sigma_0^2 = c_0^2(\underline{a})$, and $\sigma/\sigma_0 = J(\underline{a};t) = |\partial x_i/\partial a_j|$. Therefore, in the quadratic small amplitude approximation,

$$V(t) = \frac{1}{2} \iiint \rho_0(\underline{a})\, c_0^2(\underline{a})\, [J(\underline{a};t) - 1]^2 da_1 da_2 da_3 \tag{C.6}$$

It thus seems that (C.5) and (C.6) should replace the Morse-Ingard formula (9.3) in an inhomogeneous elastic fluid.

In more than one dimension, the potential energy is only non-negative definite, and not positive definite, because there are many possible volume-conserving vortex flows with zero strain energy. In two dimensions, the vorticity is transported by convection, but in three dimensions, its evolution becomes much more complicated.

ACKNOWLEDGMENT

Research supported by the U. S. Office of Naval Research.

REFERENCES

1. Baker, B., and E. T. Copson, THE MATHEMATICAL THEORY OF HUYGENS' PRINCIPLE, 2nd ed., Clarendon Press, Oxford, 1950.

2. Birkhoff, Garrett, Some mathematical problems of numerical ocean acoustics. Published in [15, pp. 643-54].

2a. Birkhoff, Garrett, Sound waves in fluids. Published in [23, pp. 3-24].

3. Brekhovskikh, L., and Yu. Lysanov, FUNDAMENTAL OCEAN ACOUS-
    TICS, Springer-Verlag, New York, 1982.

4. Caratheodory, Constantin, VARIATIONSRECHNUNG, Teubner, 1935.
    English translation entitled CALCULUS OF VARIATIONS, Holden-
    Day, 1967. References are to volume 2 of this translation.

5. Comyn, John J., GRASS: A digital computer ray-tracing and
    transmission-loss-prediction system, NRL Reports 7621 and
    7642, U.S. Naval Research Laboratory, 1973.

6. De Santo, J. A. (ed.), OCEAN ACOUSTICS, Springer-Verlag, 1979.

7. DiNapoli, Frederick R., Fast field program for multilayered
    media, NUSC Report 4103, U.S. Naval Underwater Systems Center,
    1971.

8. Herzfeld, K. F., and T. A. Litovitz, ABSORPTION AND DISPERSION
    OF ULTRASONIC WAVES, Academic Press, 1959.

9. Keller, Joseph B., Rays, waves, and asymptotics, BULLETIN OF
    THE AMERICAN MATHEMATICS SOCIETY 14 (1978), 727-52.

9a. Keller, Joseph B., Progress and prospects in the theory of
    linear wave propagation, SIAM REVIEW 21 (1979), 229-46.

9b. Keller, J. B., and J. S. Papadakis (eds.), WAVE PROPAGATION
    AND UNDERWATER ACOUSTICS, Lecture Notes in Physics, No. 70,
    Springer-Verlag, 1977.

10. Koopman, B. O., New ray methods in propagation (report pre-
    pared in 1975 for the Office of Naval Research by Arthur D.
    Little, Inc.).

11. Lamb, Horace, HYDRODYNAMICS, 6th ed., Cambridge University
    Press, 1933 (Esp. Ch. X).

12. Lee, Ding, George Botseas, and J. S. Papadakis, Finite-dif-
    ference solution to the parabolic wave equation, JOURNAL OF
    THL ACOUSTICAL SOCIETY OF AMERICA 70 (1981), 795-800.

13. Lindsay, Ronald B., Sound, ENCYCLOPEDIA BRITANNICA, 14th ed.,
    Vol. 20, pp. 925-52.

14. Markham, J. J., R. T. Beyer, and R. B. Lindsay, Absorption
    of sound in fluids, REVIEWS OF MODERN PHYSICS 23 (1951),
    353-411.

15. Morse, Philip M., VIBRATION AND SOUND, 1st ed., Mc-Graw-Hill,
    1936.

15a. Morse, P. M., and K. U. Ingard, THEORETICAL ACOUSTICS, McGraw
     Hill, New York, 1968.

16. Pierce, Allan D., ACOUSTICS: AN INTRODUCTION TO ITS PHYSICAL
     PRINCIPLES AND APPLICATIONS, McGraw Hill, New York, 1981.

17. Rayleigh, Lord, THEORY OF SOUND, two vols, 1st ed., Macmillan
     1877; 2d ed., 1894, 1896. Dover reprint with a historical
     introduction by R. B. Lindsay, 1945.

18. Rund, Hanno, THE HAMILTON-JACOBI THEORY, Van Nostrand, 1966.
     Geometrical optics discussed in Chap. 2, Sec. 11.

19. Schultz, Martin H., and Ding Lee (eds.), COMPUTATIONAL OCEAN
     ACOUSTICS, Pergamon Press, New York, 1985.

20. Sommerfeld, A., and I. Runge, Anwendung der Vektorrechnung...
     geometrische Optik, ANNALEN DER PHYSIK 35 (1911), 277-98.

21. Tate, John T., and Lyman Spitzer (eds.), THE PHYSICS OF SOUND
     IN THE SEA, Summary Report of Division 6, NDRC, Washington,
     1946.

22. Truesdell, Clifford, Hydrodynamic theory of ultrasonic waves,
     J. RAT. MECHANICAL ANALYSIS 2 (1953), 643-741.

23. Vichnevetsky, R. (ed.), PROCEEDINGS of a Workshop on NUMERICAL
     FLUID DYNAMICS, North-Holland, 1987.

24. Whitman, G. B., LINEAR AND NONLINEAR WAVES, Wiley, New York,
     1974.

25. Wood, A. B., A TEXTBOOK OF SOUND, 3rd ed., G. Bell and Sons,
     London, 1955.

26. AMERICAN INSTITUTE OF PHYSICS HANDBOOK, McGraw-Hill, New York,
     1957.

COMPUTATIONAL ACOUSTICS: Wave Propagation
D. Lee, R.L. Sternberg, M.H. Schultz (Editors)
Elsevier Science Publishers B.V. (North-Holland)
© IMACS, 1988

BOUNDARY INTEGRAL EQUATION/ELEMENT METHODS
IN COMPUTATIONAL ACOUSTICS

R. P. Shaw, T. Fukui, and H. C. Wang

Department of Civil Engineering
State University of New York at Buffalo
Buffalo, New York

ABSTRACT

Boundary integral equation/element methods are the almost perfect technique for solving problems of acoustics computationally.
Problems such as acoustic radiation or scattering frequently involve domains which are so large that reflections from some boundaries need not be taken into account, i.e., 'infinite' domains.
Boundary methods would then involve solution only over the (usually) finite radiator or scatterer. Furthermore, these problems are
often linear and involve homogeneous media; these are two properties for which the boundary integral equation/element approach is
most appropriate. The real difficulty lies not in whether boundary
methods should be applied but in how to apply them. A distinction
should be made early on between the boundary integral equation
which is a formulation of acoustics problems as valid as the wave
equation and the boundary element method which is merely one way
to solve this formulation numerically. A number of researchers
prefer to think of this method as inherently numerical, e.g., as a
variation on the finite element theme. Actually, the boundary integral equation may be thought of in very physical terms as a
representation of Huygen's principle. Thus in addition to the ad-
mittedly very powerful boundary element approach for solution,
there are many alternatives including asymptotic expansions, Galerkin methods using a basis of continuous functions, etc. This paper will examine the standard boundary element approach and some
alternatives as applied to acoustic problems.

INTRODUCTION

The application of boundary integral equation methods to problems of computational acoustics is well established, e.g., Shaw
[1970, 1979]. Such methods have definite advantages in dealing
with infinite domains and arbitrary scattering or radiative geometries. What is less well established are the relationships

between the various solution techniques available for this approach. Such techniques fall into two general categories—approximate methods which try to obtain exact solutions to approximated equations, usually in the form of truncated infinite series or iterations which lead to functional form solutions and numerical which directly approximate the original equation by some numerical scheme leading to discrete or nodal values for the dependent variables. In each class, there appears to be a dominant or "preferred" scheme, the "T matrix" for the first class and the "boundary element method" or BEM for the second, but there are alternatives. Neumann iteration methods, Born series, low frequency asymptotic expansions, geometrical asymptotic expansions, the Galerkin method, etc., may all be considered as approximate solution methods. The preferred BEM numerical scheme is so dominant that no real alternatives appear to be in use, but even here there are a number of choices to be made within the method itself regarding the types of elements to be used, the orders of numerical integration schemes, etc. The purpose of this review is to discuss some of these methods and their relationships to one another and to illustrate by simple example some of their similarities and differences.

BASIC BOUNDARY INTEGRAL EQUATION FORMULATION

Consider the time harmonic (using $\exp[-i\omega t]$) wave equation for linear acoustics,

$$\nabla^2 \phi(\bar{r}) + K^2 \phi(\bar{r}) = -Q(\bar{r})$$

with appropriate boundary conditions, e.g., Dirichlet, Neumann, Robin, or some combination of these on a boundary surface, e.g.,

$$\phi = f_1(\bar{r}) \qquad \text{on } S_1 \tag{2a}$$

$$d\phi/dn = f_2(\bar{r}) \qquad \text{on } S_2 \tag{2b}$$

$$\phi + K \, d\phi/dn = f_3(\bar{r}) \qquad \text{on } S_3 \tag{2c}$$

where $S_1 + S_2 + S_3$ sum to the total surface S. For problems involving an infinite domain, a Sommerfeld radiation condition would also be required:

$$\lim_{r\to\infty} r[d\phi/dn - iK\phi] = 0 \tag{3}$$

Classically, for problems where 'THE' Green's function was known, the use of Green's theorem would reduce the solution of this problem to a quadrature of known values. Unfortunately, THE

Green's function must satisfy appropriate boundary conditions on
the given boundary as well as the governing partial differential
equation with an appropriate singular forcing function and such
solutions are unavailable in all but the simplest cases. The significance of the boundary integral equation method is that it allows use of 'A' Green's function, i.e., one which satisfies the
appropriate equation but not necessarily the appropriate boundary
conditions; such Green's functions are readily available for a variety of problems including those governed by the wave equation.
To illustrate this, consider a Neumann boundary condition problem.

$$\nabla^2 \phi(\bar{r}) + K^2 \phi(\bar{r}) = -Q(\bar{r}) \qquad \text{in } V \qquad (4a)$$

$$d\phi(r)/dn = f(\bar{r}) \qquad \text{on } S \qquad (4b)$$

Define 'A' Green's function as the solution to

$$\nabla^2 G(\bar{r},\bar{r}_0) + K^2 G(\bar{r},\bar{r}_0) = -\delta(r - r_0) \qquad \text{in } V \qquad (5)$$

Application of Green's theorem amounts to multiplying Eq. (4a) by
G, multiplying Eq. (5) by ϕ, subtracting and integrating over $V(r)$
using the divergence theorem to replace part of the volume integrals by equivalent surface integrals.

$$\int_V [G\nabla^2\phi - \phi\nabla^2 G] \ dV(\bar{r}) = \int_V [-GQ + \phi\delta(\bar{r} - \bar{r}_0)] \ dV(\bar{r})$$

$$= \int_S [G \ d\phi/dn - \phi \ dG/dn] \ dS(\bar{r})$$

$$= -\oint_V GQ \ dV(\bar{r}) + \alpha\phi(\bar{r}_0) \qquad (6)$$

where n is the outward normal from the volume of integration. The
coordinate $\bar{r}_0$ will be referred to as the field point and the coordinate $\bar{r}$ as the integration or source point. The value of α depends on the integration of the Dirac delta function and thus on
the location of the field point. For field points within V, clearly α will be one while for field points outside of V, α will clearly be zero since the delta function does not operate. If G were
'THE' Green's function, dG/dn would vanish on S, and since dϕ/dn
is specified there, placing $\bar{r}_0$ in V gives a direct solution for
$\phi(\bar{r}_0)$ as

$$\phi(\bar{r}_0) = \int_V G(\bar{r},\bar{r}_0)Q(\bar{r}) \ dV(r) + \oint_S G(\bar{r},\bar{r}_0)f(\bar{r}) \ dS(\bar{r}) \qquad (7)$$

If, on the other hand, 'THE' Green's function is not known, placing

$\bar{r}_0$ at some smooth boundary point on the surface S leads to a value of α of 1/2 and to a boundary integral equation

$$(1/2)\phi(\bar{r}_0) = \int_V G(\bar{r},\bar{r}_0)Q(\bar{r}) \; dV(\bar{r})$$
$$+ \oint_S [G(\bar{r},\bar{r}_0)f(\bar{r}) - \phi(\bar{r}) \; dG(\bar{r},\bar{r}_0)/dn] \; dS(\bar{r}) \qquad (8)$$

This is called a boundary integral equation since it only involves values of the dependent variable at points on the boundary S and not in the entire field V which in turn means a reduction of one order in the dimensionality of the problem. The price paid for this is the solution of an integral equation which involves Cauchy principal values due to the strong singularity at the point $\bar{r} = \bar{r}_0$. Once the boundary values have been found, the field values may be obtained by quadrature of known values by using Eq. (6) with α equal to one.

For scattering problems, it is clear from the Sommerfeld radiation condition that the solution must be written in terms of the scattered field, $\phi_s = \phi - \phi_w$, rather than the total field since frequently an incident wave field, ϕ_w, is assumed that does not vanish at infinity. While it is possible to work with this scattered field directly, it is clearly more convenient in many cases to retain the total field, e.g., if there are homogeneous boundary conditions on the total field. This may be done simply by using Eq. (6) on the scattered field potential as the variable within the acoustic domain, V, and recognizing that the integral terms represent 'jumps' in strengths of the scattered field which, since the incident field is continuous, must also represent the same jump in the total field. Alternatively, an integral equation for points exterior to V may be written in terms of the scattered field and added to the equation for the field equation within V. In either case, the final equation on the total field may be written as

$$(1/2)\phi = \phi_w + \int_S [G(\bar{r},\bar{r}_0)d\phi(\bar{r})/dn - \phi(\bar{r}) \; dG(\bar{r},\bar{r}_0)/dn] \; dS(\bar{r})$$
$$(9)$$

where the source Q has been taken as zero for convenience since most scattering problems do not also contain volume sources, e.g., Friedman and Shaw [1962]. For the rigid boundary (Neumann) problem, $d\phi/dn$ would be replaced by zero, thereby eliminating one of the integral terms.

A general difficulty with this overall approach appears in
exterior scattering and radiation problems; the boundary integral
equation formulation fails at 'fictitious interior eigenvalues,'
i.e., at eigenvalues for the interior Dirichlet problem for the
above exterior Neumann problem. This difficulty is well known and
many attempts have been made to circumvent it, e.g., Schenck [1968],
etc. This difficulty is not surprising when one considers that
the same boundary and thus boundary integral equation holds for
interior as for exterior problems with only a sign change in the
direction of the 'outward' normal. This difficulty is described
in a simple illustration in the appendix where the same problem,
plane time harmonic wave scattering by a rigid circular cylinder,
is solved by a number of techniques.

APPROXIMATE METHODS

The basic approach of 'approximate' as opposed to 'numerical'
solution techniques is that these attempt to solve the original
problem in a functional form, usually as some infinite series of
specified functions whose coefficients are to be determined to
complete the solution process. A large number of classical solution techniques for integral equations fall into this category,
e.g., Neumann series, asymptotic expansions both in frequency and
in geometrical parameters, Born series (which in the present context are frequently taken to be very similar to the Neumann series
methods), Galerkin methods, etc. Delves and Walsh [1974] and Atkinson [1976] give overviews of solution methods for integral equations in general. The most widely used, however, is basically an
eigenfunction expansion method called the "T matrix" or extended
boundary condition method. Although this is, in the author's view,
a boundary integral equation method, it is not a boundary element
method and its roots in integral equation theory are not usually
discussed.

The T matrix approach was developed for time harmonic wave
scattering problems in acoustics, electromagnetics, and elastody-
namics. There are actually several forms of the T matrix approach,
at least one of which is really akin to a Galerkin method, but they
are all basically eigenfunction expansion solutions to the boundary integral equation with different expansions and thus expansion
coefficients inside of and outside of the scattering obstacle.
The original concept was developed by Waterman [1965] for electromagnetic wave scattering and extended by him to acoustic waves

[1968]. Varatharajulu (Varadan) and Pao [1976] and Waterman [1978] extended the method further to elastodynamic wave problems and a review is given in the proceedings of a conference, i.e., Varadan and Varadan [1979].

Consider a time harmonic acoustic scattering problem as an illustration. Although an acoustic case will be given for simplicity, the general electromagnetic and elastodynamic cases follow directly, e.g., Varadan and Varadan [1979], but are, of course, more complicated algebraically. The scattering body is assumed to be of finite size and arbitrary shape. An interior circle (in two dimensions) or sphere (in three dimensions) is inscribed such that all points on the circle or sphere lie within the scatterer boundary. Similarly, an exterior circle or sphere may be circumscribed such that all points on it lie outside of the boundary. The dependent variables and the forcing terms of the problem may be expanded in an infinite series of eigenfunctions which are defined by solutions to the Helmholtz wave equation on these known simple surfaces; these expansions are then valid everywhere inside and outside of the scattering boundary, respectively, by analytic continuation since there are no 'sources' or singularities between these circles or spheres and the scattering surface. The key to the approach is that the expansion of the fundamental Green's function inside the scatterer is different from that outside of the scatterer as is the Helmholtz integral equation. In essence, the scattering surface may be thought of as having been replaced by an equivalent surface of sources and doublets whose strength is such that the boundary conditions are satisfied and the incident wave is extinguished inside of the scattering surface, i.e., a 'saltus' problem as described by Baker and Copson [1950] and used, for example, by Friedman and Shaw [1962] in rewriting the acoustic boundary integral equation scattering problem in terms of total rather than scattered fields. Consider the Helmholtz integral equation on $\phi(\bar{r})$,

$$\alpha\phi(\bar{r}_0) = \phi_w(\bar{r}_0) - \oint_S [\phi(\bar{r}) \, dG(\bar{r},\bar{r}_0)/dn$$
$$- G(\bar{r},\bar{r}_0) \, d\phi(\bar{r})/dn] \, dS(\bar{r}) \qquad (10)$$

The boundary condition will be taken to be a homogeneous Neumann condition, i.e., $d\phi/dn$ equal to zero and thus the second term under the integral vanishes. The incident wave is known to have an expansion in terms of circular or spherical wave functions, i.e.,

$$\phi_w(\bar{r}) = \sum_{m=0}^{\infty} a_m \; \mathrm{Re}\,[\psi_m(\bar{r})] \qquad (11)$$

where the real part of these wave functions is required in order that $\phi_w(\bar{r})$ be finite at the origin, within the scatterer. The scattered wave has a similar expansion,

$$\phi_s(\bar{r}) = \sum_{m=0}^{\infty} f_m \psi_m(\bar{r}) \qquad (12)$$

where the entire wave function is used to have outgoing waves, i.e., to satisfy the radiation condition. Finally, the fundamental Green's function has an expansion in the same eigenfunctions,

$$G(\bar{r},\bar{r}_0) = (iK/4\pi) \sum_{m=0}^{\infty} \psi_m(\bar{r}_>) \; \mathrm{Re}\,[\psi_m(\bar{r}_<)] \qquad (13)$$

where $\bar{r}_>$ is $\bar{r}$ and $\bar{r}_<$ is $\bar{r}_0$ when $\bar{r}_0 < \bar{r}$ and $\bar{r}_>$ is $\bar{r}_0$ and $\bar{r}_<$ is $\bar{r}$ when $\bar{r}_0 > \bar{r}$. Since the original expansions have been developed on inscribed and circumscribed circles or spheres, the question of which r is greater is easily answered. Apply the Helmholtz equation to a point exterior to the infinite medium in which the incident wave is traveling; clearly α is zero, and

$$0 = a_m - (iK/4\pi) \oint_S \phi(\bar{r}) \; d\psi_m(\bar{r})/dn \; dS(\bar{r}) \qquad (14)$$

using the orthogonality of the basis functions. Similarly, application of the Helmholtz integral equation to a point within the infinite medium of interest leads to

$$f_m = -(iK/4\pi) \oint_S \phi(\bar{r}) \; \mathrm{Re}\,[d\psi_m(\bar{r})/dn] \; dS(\bar{r}) \qquad (15)$$

The boundary value of ϕ or the 'surface source' also has an eigenfunction expansion; it may be taken in terms of any convenient set of complete functions, e.g., $\mathrm{Re}\,[\psi_p(\bar{r})]$ or even $\mathrm{Re}\,[d\psi_p(\bar{r})/dn]$ would be appropriate. Take the former for convenience. Then

$$\phi(\bar{r}) = \sum_{p=0}^{\infty} \alpha_p \; \mathrm{Re}\,[\psi_p(\bar{r})] \qquad (16)$$

Defining

$$Q_{pm} = \oint_S \mathrm{Re}\,[\psi_p(\bar{r})] \; d\psi_m(\bar{r})/dn \; dS(\bar{r}) \qquad (17)$$

leads to a relationship between the scattered wave as defined by

f_m and the incident wave as defined by a_m,

$$f_q = T_{qm} a_m \tag{18}$$

where the T or transition matrix is defined by

$$T_{qm} = -\text{Re}\,[Q_{qp}]\,[Q_{pm}^{-1}] \tag{19}$$

Using the second set of basis functions for ϕ would give the same
form but a modified definition of Q, i.e.,

$$Q_{pm} = \oint_S \text{Re}\,[d\psi_p(\bar{r})/dn]\,d\psi_m(\bar{r})/dn \; dS(\bar{r}) \tag{20}$$

Various eigenfunction expansions are used depending on the problem
at hand, but the essence of the T matrix approach is the same
throughout. A simple example is given in the appendix for the
case of time harmonic plane wave scattering by a rigid circular
cylinder in which the T matrix is found, as expected, to be diagonal. This example illustrates one of the pitfalls of this approach not usually mentioned, i.e., the 'fictitious' interior
eigenvalue problem.

The most obvious of the remaining approximate methods is an
iteration method which is actually used in classical integral equation theory, leading to Neumann series. Simply stated for integral
equations of the second kind as applied to acoustic Neumann boundary value problems, e.g.,

$$\phi^{(n+1)}(\bar{r}_0) = 2f(\bar{r}_0) - 2 \oint_S [\phi^{(n)}(\bar{r}) \; dG(\bar{r},\bar{r}_0)/dn] \; dS(\bar{r}) \tag{21}$$

with $\phi^{(0)}(\bar{r})$ equal to zero. The function $f(\bar{r})$ is either an incident wave field (with $d\phi/dn$ zero) or some boundary integral of
known form (with $d\phi/dn$ given). For the present case, this series
will not in general converge. However, Chertock [1968] gives a
modified scheme which does converge, but which requires knowledge
of the eigenvalues of the corresponding homogeneous problem to determine the modification. He proposed to rewrite the iterative
scheme as

$$\phi^{(n+1)}(\bar{r}_0) = H\phi^{(n)}(\bar{r}_0) + 2(1 - H)\,[f(\bar{r}_0)$$
$$- \oint_S [\phi^{(n)}(\bar{r})\,dG(\bar{r},\bar{r}_0)/dn]\;dS(\bar{r})\,] \tag{22}$$

where H is a complex constant. If the corresponding eigenvalue
problem is defined by

$$-\lambda_j \oint \psi_j(\bar{r}) \; dG(\bar{r},\bar{r}_0)/dn \; dS(r) = \psi_j(\bar{r}_0) \tag{23}$$

then $f(\bar{r})$ and $\phi(\bar{r})$ may be expanded in terms of ψ_j. This leads to

$$\phi^{(n+1)}(\bar{r}_0) = \sum_{j=0}^{\infty} \psi_j [2<f,\psi_j>((1 - H)(1 - A_j^n)/(1 - A_j))$$

$$+ <\phi^{(n)},\psi_j>A_j] \tag{24}$$

where $A_j = H + (1 - H)/\lambda_j$ and $<a,b>$ represents an inner product, i.e., the integral of a b over the surface, $S(r)$. This series clearly converges when abs[A] < 1. Then H must be chosen such that all of the reciprocal eigenvalues, $1/\lambda_j$, lie within a circle which passes through (1,0), has a radius of $1/(1 - H)$ and has a center at $H/(1 - H)$. It is of interest to note that failure of the unmodified scheme to converge has implications for the convergence of iterative schemes such as Gauss-Seidel for solution of the system of linear algebraic equations resulting from the boundary element method described above.

Another approximate method is that of asymptotic expansion either in frequency (i.e., low frequency approximations) and in geometry (i.e., in 'almost' separable geometries). Consider the low frequency case. Clearly, both the dependent variables and the Green's functions may be expanded in an asymptotic series in orders of the frequency or, equivalently, the wave number K for time harmonic problems. The lowest order solution corresponds to the static solution which may already be available in some cases. Higher order terms may then be developed using the static solution procedure with additional forcing terms resulting from the lower order solutions. This is certainly a classical approach to such problems, but is still finding application, e.g., Everstine, Henderson, Schroeder, and Lipman [1986]. The same concept may be applied to geometries which are 'almost' separable, i.e., a shell composed of two circular cylinders whose centers are slightly offset, as solved by Shaw [1978]. Results here indicate how neighboring eigenfunctions are introduced into the solution.

The Galerkin approach may also be considered as an approximate method. The standard Galerkin procedure, Kantorovich and Krylov [1956], applies to integro-differential equations as well as differential equations. One of the advantages of this approach is the 'blurring' of the significance of the singularity since it is integrated over twice, e.g., Kobayashi and Nishimura [1979].

Finally, an alternative point of view may be taken which is somewhat classical in nature. Rather than approximating the exact equation based on 'A' Green's function, 'THE' Green's function may be constructed approximately, e.g., as suggested by Boley [1956]. This then would provide an approximate solution as a direct quadrature of (approximately) known surface integrals.

While these approximate methods, other than the T matrix, do not appear to be widely used in present calculations, some interest in these approaches, especially the Neumann or Born series, has been recently aroused through the study of multiple scatterers, e.g., Schuster [1985b], Achenbach and Kitahara [1986], etc. Here essentially the iterative approach is based on the direct scattering as a first iterate, first reflections as the next, etc. Such an approach clearly has advantages over a direct solution attempt involving all surfaces simultaneously.

NUMERICAL METHODS

While there are numerical techniques other than the boundary element method, due to its widespread popularity there is little point in even mentioning them. The boundary element method rather than the boundary integral equation is in fact the starting point for many researchers. Many texts begin with the development of the boundary integral equation formulation by means of a weighted residual approach, e.g., Brebbia and Walker [1980], which leads directly into the boundary element method. While such a view overlooks some alternative solution procedures, it does focus attention on the approach which has dominated the field and brought it to the standing of a general purpose method comparable to the finite element method. The first application to time harmonic acoustic wave scattering appears to be that of Banaugh and Goldsmith [1963] which includes scattering of plane waves by rigid circular cylinders, which is the example chosen for the appendix here. The approach is both simple and complicated; simple in that the concept is quite straightforward and yet complicated in its implementation. The basic concept is that the requisite integral is to be written as a sum of integrals over subsections of the boundary surface referred to as boundary elements in much the same manner as the volume integral in the variational formulation for finite elements is written as a sum of integrals over volume elements. The dependent variable is then approximated by an algebraic expression such that its variation over the boundary element is determined by its value

at various node points. The boundary element itself may be represented by an approximate geometry, frequently in terms of the same
'shape' functions as the dependent variable, i.e., an isoparametric
formulation. This leads to compatibility with finite element formulations that might occur in the same problem, e.g., if there are
regions of nonlinearity or inhomogeneity or even in fluid-structure
interaction problems. Once the integrations have been performed
to obtain 'influence coefficients' relating the various nodal values, the resulting system of linear algebraic equations may be
solved by standard methods. The major difficulty with such formulations is that integrations over boundary elements where the field
point and the integration variable coincide require integration of
a singular function. When the Cauchy principal value is to be
taken, such integrations become quite difficult and a 'trick' is
frequently used to avoid the worst of these. The trick, introduced
by Cruse [1974] and extended by Watson [1979], is based on the concept in elasticity that a rigid body displacement should not introduce a stress or strain field. The corresponding trick in acoustics would involve a uniform pressure field acting on a rigid body.
Such a procedure allows the coefficient corresponding to this 'self
influence' element to be evaluated without integration, but at the
price of losing a possible check on the accuracy of the numerical
procedure.

Consider the simple time harmonic, two-dimensional acoustic
case as an example. The governing integral equation for a plane
wave scattering problem with a rigid boundary is Eq. (9) using
$d\phi/dn$ equal to zero on S. Taking the surface S to be subdivided
into L linear segments, defined by endpoints $(x(N-1),y(N-1))$
and $(x(N),y(N))$ and taking the potential ϕ to be constant over
each segment with a value $\phi(N)$ equivalent to that at the midpoint
of the segment $(x(N-1/2),y(N-1/2))$, Eq. (9) applied successively to all of the nodal points, $M = 1, \ldots, L$, yields a set of
coupled linear algebraic equations

$$(1/2)\,\phi(M) = \phi_w(M) + \sum_{N=1}^{L} \alpha(M,N)\,\phi(N) \qquad M = 1, \ldots, L \qquad (25)$$

where

$$\alpha(M,N) = (i/4) \int_{S_N} dH_0^{(1)}(K|\bar{r} - \bar{r}_0(M)|)/dn \; dS(\bar{r}) \qquad (26)$$

or equivalently

$$\sum_{N=1}^{L} \beta(M,N)\phi(N) = \phi_w(M) \qquad M = 1, \ldots, L \qquad (27)$$

with $\beta(M,N)$ equal to $(1/2)\delta(M,N) - \alpha(M,N)$. Note that the integration over the linear segment containing the field point M is singular but is to be taken in a Cauchy principal value sense. The influence of the singularity itself has already been accounted for in the reduction of the coefficient of the left-hand side of the integral equation from 1 to 1/2. The remaining integral is zero since the integrand contains a factor dR/dn which is zero since R lies along the linear segment and n is perpendicular to the linear segment. Thus the net 'self influence' coefficient $\beta(M,M)$ is always $1/2\delta(M,N)$. This system of equations may be solved by any standard technique keeping in mind that the coefficient matrix will have a zero determinant at the 'fictitious interior eigenvalues' where this approach fails. An illustration of this is carried out in the appendix. Note that the integration in Eq. (26) can be carried out by simple Gaussian schemes as described in all of the standard texts on boundary element methods, e.g., Banerjee and Butterfield [1981], etc. As the shape functions for both the geometry and the dependent variable become more complicated, the integrations for the influence coefficients become more difficult, but they are essentially still possible to carry out by Gaussian integration techniques, although with some minor modifications. Details of these more complicated schemes will be left for the standard texts mentioned above and papers cited in this study. It must be mentioned, however, that the boundary integral equation itself may be manipulated into many alternative forms which may have different computational advantages, e.g., Schuster and Smith [1985].

EIGENVALUE PROBLEMS

It is clear that eigenvalues may also be obtained from a boundary integral equation approach and this has been the source of some effort. The initial work was done for acoustic media, e.g., Tai and Shaw [1974], etc. Some other authors, e.g., DeMey [1976], have suggested the use of only the singular part of the fundamental solution in these problems; this is unfortunately incorrect since it will provide an additional set of spurious eigenvalues as well as the correct ones. A simple example for an acoustic problem should suffice for illustration.

Consider free oscillations in a rigid boundary two-dimensional acoustic domain. The standard boundary element formulation will provide matrix relations between the unknown nodal values of the potential or displacement and the forcing function which in this case is zero, i.e., Eq. (25) with a zero forcing function,

$$\Sigma \; \beta(M,N)\phi(N) = 0 \qquad M = 1, \ldots, L \tag{28}$$

which will have a nontrivial solution if and only if the determinant of the coefficient matrix $\beta(M,N)$ is zero. Once these eigenvalues are found, it is a straightforward task to return to this underdetermined system to find ratios of the nodal values to one particular value, chosen not to be zero by inspection (not always easily). There are, of course, the eigenfunctions corresponding to this eigenvalue.

DISCUSSION

The main purpose of this presentation is twofold. First, it is to provide a simple introduction to a powerful solution technique to acoustic radiation and scattering problems. Second, and perhaps the more important of the two, is to distinguish between the boundary integral equation method, which has many solution techniques, and the boundary element method which is one of these techniques. Alternative solution techniques have been presented and in some cases illustrated by simple example. However, it must be said that the boundary element method has clearly become the solution technique chosen by most researchers in this field. This method is especially powerful when combined with other numerical methods such as finite elements in applications such as fluid-structure interaction problems. One obvious application for alternative methods, however, is that of multiple scattering problems, especially when simple geometrical scattering shapes are repeated. Here, iteration methods representing first interactions, first reflections, etc., appear to have distinct advantages. A side benefit of this discussion is to draw attention to the 'fictitious interior eigenvalue' problem which exists for all exterior scattering problems based on an integral formulation, including an example for the T matrix method where this difficulty does not appear to be widely recognized. Martin [1982] notes this problem in a reverse sense. It is said not to exist if an appropriate (complex) set of expansion functions is used; this is not the case in common practice.

APPENDIX: A SIMPLE EXAMPLE

Consider the simple example of time harmonic acoustic plane wave scattering by a rigid circular cylinder. In this case, the normal derivative of the dependent variable, the velocity potential ϕ is zero on the surface, $r = a$. This solution to this problem will be described in several different forms. Morse and Feshbach [1953] is a reference for most of the series expansions.

(a) Separation of Variables

The first solution method will be that of classical separation of variables to establish the exact solution. The governing equation is

$$\nabla^2\phi + K^2\phi = 0 \qquad \text{for } r > a \tag{A1}$$

with a boundary condition

$$d\phi/dn = 0 \qquad \text{on } r = a \tag{A2}$$

and an incident wave of the form

$$\phi_w = \exp(iKx - ipt) \tag{A3}$$

where the time harmonic term $\exp(-ipt)$ will be suppressed for convenience.

Separation of variables assumes a solution for the scattered field, $\phi_s = \phi - \phi_w$, in the form

$$\phi_s = R(r)T(\theta) \tag{A4}$$

Substitution of this form into the differential equation on ϕ which is identical to that for ϕ under the conditions of periodicity and symmetry in θ and outgoing waves in r leads to

$$\phi_s(r,\theta) = \sum_{m=0}^{\infty} B_m H_m^{(1)}(Kr)\, \cos(m\theta) \tag{A5}$$

In order to satisfy the boundary condition, it is convenient to expand the incident plane wave as

$$\exp(iKx) = \sum_{m=0}^{\infty} \varepsilon_m i^m J_m(Kr)\, \cos(m\theta) \tag{A6}$$

where ε_m is the Neumann function and is 1 for $m = 0$ and 2 otherwise. The boundary condition Eq. (A2) then requires

$$K \sum_{m=0}^{\infty} B_m \cos(m\theta) H_m^{(1)\,'}(Ka) = -K \sum_{m=0}^{\infty} \varepsilon_m i^m J_m'(Ka)\, \cos(m\theta) \tag{A7}$$

which, because of the orthogonality of the cosine functions over
this boundary, leads to

$$B_m = -\varepsilon_m i^m J_m'(Ka)/H_m^{(1)}{}'(Ka) \tag{A8}$$

The total field is then given by

$$\phi = \sum_{m=0}^{\infty} i^m \varepsilon_m [J_m(Kr) - J_m'(Ka)H_m^{(1)}(Kr)/H_m^{(1)}{}'(Ka)]\cos(m\theta) \tag{A9}$$

This form may be modified to a more standard one using the Wronskian for J_m and $H_m^{(1)}$. This solution, which is very straightforward,
is included to illustrate the differences between this classical
solution method and those based on boundary integral equation methods. In particular, the 'built-in' difficulty of the 'fictitious
interior eigenvalues' may be seen more clearly through this
comparison.

(b)   Boundary Integral Equation Method

This problem is one of the few that can be solved by direct
integration of the boundary integral equation due to the orthogonality of the eigenfunctions over the circular boundary. Consider
the boundary integral equation as

$$\alpha\phi(r_0,\theta_0) = \phi_w(r_0,\theta_0) + (i/4) \oint_S \phi(a,\theta)$$

$$\times [dH_0^{(1)}(K|r - r_0|)/dr]\ dS(\bar{r}) \tag{A10}$$

where the Green's function has an expansion

$$H_0^{(1)}(K|r - r_0|) = \sum_{m=0}^{\infty} \varepsilon_m \cos[m(\theta_0 - \theta)]J_m(Kr_<)H_m^{(1)}(Kr_>) \tag{A11}$$

where $r_>$ is r and $r_<$ is r_0 when $r > r_0$ and vice versa. If the
boundary integral equation is taken at a field point $r = a^-$, i.e.,
inside the cylinder or exterior to the elastic domain, α is zero.
If, furthermore, the total wave field on the surface is taken to
have an expansion

$$\phi(a,\theta) = \sum_{m=0}^{\infty} A_m \cos(m\theta) \tag{A12}$$

the integral equation requires

$$A_m = -(4/Ka\pi)i^{m-1}J_m(Ka^-)/[J_m(Ka)H_m^{(1)}{}'(Ka)] \tag{A13}$$

The solution in the field can then be obtained from the original integral equation using α equal to one. This yields

$$\phi(r_0,\theta_0) = \sum_{m=0}^{\infty} i^m \varepsilon_m [J_m(Kr_0) - J_m(Ka) H_m^{(1)'}(Kr) J_m(Ka^-)/$$

$$[J_m(Ka^-) H_m^{(1)}(Ka)]]\cos(m\theta_0) \qquad (A14)$$

This coefficient will match that from separation of variables if and only if $J_m(Ka)$ is not equal to zero. These values of K are the eigenvalues for the interior Dirichlet problem; at these values, the coefficient A_m is undetermined. Actually in the limit, the correct form would be obtained, but in a numerical scheme this would imply that at frequencies near to these interior natural frequencies the coefficient would have a value determined by round-off errors.

(c)   T Matrix Method

The T matrix method as described in the main text could use

$$\psi_m(\bar{r}) = \sqrt{\varepsilon_m} H_m^{(1)}(Kr)\cos(m\theta) \qquad (A15)$$

as its basis functions, remembering that the choice of expansion set for the surface variable was the real part of this set. Then

$$Q_{pm} = 2 J_p(Ka) H_m^{(1)'}(Ka)_{pm} \qquad (A16)$$

which is diagonal. The corresponding T matrix is also diagonal and is indeterminate whenever the determinant of Q_{pm} vanishes, i.e., at $J_p(Ka) = 0$. Then the relationship between the scattered wave coefficients, f_m, and the incident wave coefficients, a_m, is

$$f_m = -[J_m(Ka) J_m'(Ka)]/[J_m(Ka) H_m^{(1)'}(Ka)] a_m \qquad (A17)$$

The total field solution matches Eq. (A14) and thus 'fails' at the interior Dirichlet eigenvalues just as did the boundary integral solution. As an alternative, the surface potential could be expanded in terms of the derivative of this set as its basis. The major difference is that $J_m(Ka)$ is replaced by $J_m'(Ka)$ and thus the failure in solution occurs at the interior Neumann eigenvalues. Note that the 'fictitious interior eigenvalue' problem has shifted to a different set of eigenvalues but is still present. While this analytic solution is correct at all other values of Ka and indeterminate with the proper limiting value at these eigenvalues, if these coefficients were calculated numerically, there would be nonzero errors in the off-diagonal terms and the solution near

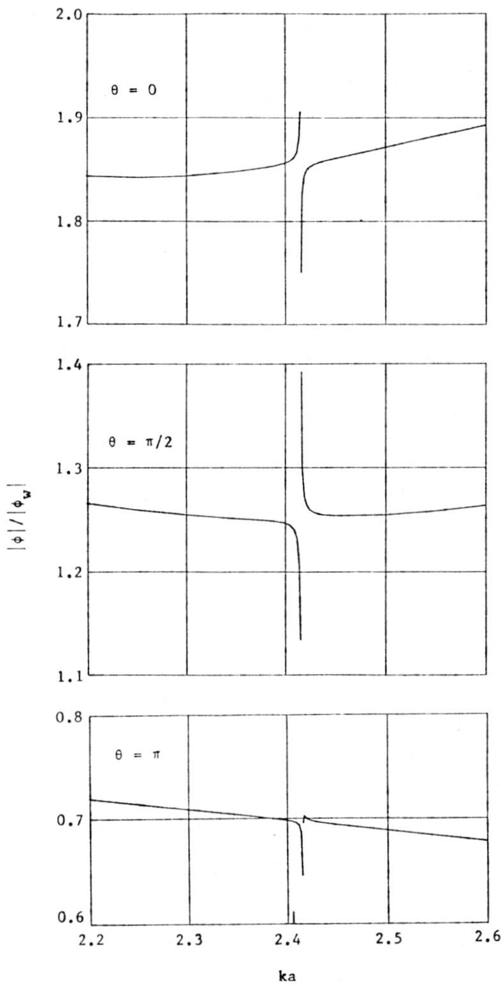

Figure 1

these interior eigenvalues should be strongly affected by this indeterminate form. Such difficulties may be the cause of some of
the discrepancies mentioned by Waterman [1976] for a spherical
geometry.

(d)   Boundary Element Method
     One primary goal of this study is to allow comparison of re-
sults of several approximate and numerical methods to an exact solution. Choices as to the complexity of the shape functions, the
type of numerical integration schemes to be used in determining
the influence coefficients, etc., are required in general but for
the specific problem at hand the simplest scheme is used, i.e.,

straight line boundary elements with constant potentials corresponging to the midpoint values.

Use of a numerical scheme such as that described in the main text with these element and dependent variable shape functions allows a very simple program to be written and run. Results for values of Ka from 2.0 to 3.0 at a field point at $r = a$ and $\theta = 0$, $\pi/2$, π are given in Fig. 1. This range includes the first interior Dirichlet eigenvalue and the solution clearly fails there. Actually, the solution fails at a value of Ka slightly less than the interior Dirichlet eigenvalue for the circle since the actual surface was approximated by an inscribed polygon of 32 sides (elements). Similar results hold for higher interior Dirichlet eigenvalues.

REFERENCES

1. Achenbach, J. D., and M. Kitahara, Reflection and transmission of an obliquely incident wave by an array of spherical cavities, JOURNAL OF THE ACOUSTICAL SOCIETY OF AMERICA (submitted, 1986).

2. Atkinson, K. E., A survey of numerical methods for the solution of Fredholm integral equations of the second kind, SIAM Journal on Numerical Analysis (1976).

3. Baker, B. B., and E. T. Copson, THE MATHEMATICAL THEORY OF HUYGENS PRINCIPLE, Oxford Press (1st ed.), 1939, (2nd ed.), 1944.

4. Banaugh, R. P., and W. Goldsmith, Diffraction of steady acoustic waves by surfaces of arbitrary shape, JOURNAL OF THE ACOUSTICAL SOCIETY OF AMERICA, 35 (1963), 1590-1601.

5. Banerjee, P. K., and R. Butterfield, BOUNDARY ELEMENT METHODS IN ENGINEERING SCIENCE, McGraw-Hill Book Co., London, 1981.

6. Boley, B. A. A method for the construction of Green's functions, QUARTERLY JOURNAL OF APPLIED MATHEMATICS, 14 (1956), 249-257.

7. Brebbia, C. A., and S. Walker, BOUNDARY ELEMENT TECHNIQUES IN LNGINEERING, Butterworth Pub., London, 1980.

8. Chertock, G., Convergence of iterative solutions to integral equations for sound radiation, QUARTERLY JOURNAL OF APPLIED MATHEMATICS, 26 (1968), 268-272.

9.   Cruse, T. A., An improved boundary integral equation method
     for three-dimensional elastic stress analysis, JOURNAL OF
     COMPUTATION AND STRUCTURES, 4 (1974), 745-754.

10. Delves, L. M., and J. Walsh, NUMERICAL SOLUTION OF INTEGRAL
     EQUATIONS, Claredon Press, Oxford, U.K., 1974.

11. DeMey, G., Calculation of eigenvalues of the Helmholtz equa-
     tion by an integral equation, INTERNATIONAL JOURNAL FOR NUM-
     ERICAL METHODS (in English), 10 (1976), 59-66.

12. Everstine, G. C., F. M. Henderson, E. A. Schroeder, and R. R.
     Lipman, A general low frequency acoustic radiation capability
     for NASTRAN, 14TH NASTRAN USERS' COLLOQUIUM, NASA CP-2419,
     NASA, Washington, D.C., 1986, 293-310.

13. Friedman, M. B., and R. P. Shaw, Diffraction of pulses by
     cylindrical obstacles of arbitrary cross section, JOURNAL OF
     APPLIED MECHANICS, 29 (1962), 40-46.

14. Kantorovich, L. V., and V. I. Krylov, APPROXIMATE METHODS OF
     HIGHER ANALYSIS, Interscience Pub., The Netherlands, 1956.

15. Kobayashi, S., and N. Nishimura, Some considerations on the
     improvement of integral equation methods, PROCEEDINGS OF JSCE
     (in Japanese), No. 291 (1979).

16. Martin, P. A., Acoustic scattering and radiation problems and
     the null field method, WAVE MOTION, 4 (1982), 391-408.

17. Morse, P. M., and F. Feshbach, METHODS OF THEORETICAL PHYSICS,
     I AND II, McGraw-Hill, New York, 1953.

18. Schenck, H. A., Improved integral formulation for acoustic
     radiation problems, JOURNAL OF THE ACOUSTICAL SOCIETY OF
     AMERICA, 77 (1985), 850-864.

19. Schuster, G. T., and L. C. Smith, A comparison among four
     direct boundary integral equation methods, JOURNAL OF THE
     ACOUSTICAL SOCIETY OF AMERICA, 77 (1985), 850-864.

20. Schuster, G. T., Hybrid BIE + Born series modeling scheme:
     Generalized Born series, JOURNAL OF THE ACOUSTICAL SOCIETY
     OF AMERICA, 77 (1985), 865-879.

21. Schuster, G. T., Solution of the acoustic transmission prob-
     lem by a perturbed Born series, JOURNAL OF THE ACOUSTICAL
     SOCIETY OF AMERICA, 77 (1985), 880-886.

22.   Shaw, R. P., Integral equation formulation of dynamic acoustic fluid-elastic solid interaction problems, JOURNAL OF THE ACOUSTICAL SOCIETY OF AMERICA, 53 (1973), 514-520.

23.   Shaw, R. P., Time dependent acoustic radiation from a submerged elastic shell defined by nonconcentric circular cylinders, JOURNAL OF THE ACOUSTICAL SOCIETY OF AMERICA, 4 (1978), 311-317.

24.   Shaw, R. P., Boundary integral equation methods applied to wave problems, in DEVELOPMENTS IN BOUNDARY ELEMENT METHODS, I, P. K. Banerjee and R. Butterfield, eds., Applied Science Pub., Essex, U.K., 1979, pp. 121-153.

25.   Tai, G. R. C., and R. P. Shaw, Helmholtz equation eigenvalues and eigenmodes for arbitrary domains, JOURNAL OF THE ACOUSTICAL SOCIETY OF AMERICA, 56 (1974), 796-804.

26.   Varadan, V. K., and V. V. Varadan, eds., ACOUSTIC, ELECTROMAGNETIC AND ELASTIC WAVE SCATTERING—FOCUS ON THE T MATRIX APPROACH, Pergamon Press, New York, 1979.

27.   Varatharajulu (Varadan), V., and Y. H. Pao, Scattering matrix for elastic waves, I: Theory, JOURNAL OF THE ACOUSTICAL SOCIETY OF AMERICA, 60 (1976), 556-566.

28.   Waterman, P. C., Matrix formulation of electromagnetic scattering, PROCEEDINGS OF THE IEEE, 53 (1965), 805-812.

29.   Waterman, P. C., New formulation of acoustic scattering, JOURNAL OF THE ACOUSTICAL SOCIETY OF AMERICA, 45 (1968), 1417-1429.

30.   Waterman, P. C., Matrix theory of elastic wave scattering, JOURNAL OF THE ACOUSTICAL SOCIETY OF AMERICA, 60 (1976), 567-580.

31.   Watson, J. O., Advanced implementation of the boundary element method in two and three dimensional elastostatics, Chap. III, in DEVELOPMENTS IN BOUNDARY ELEMENT METHODS, 1, P. K. Banerjee and R. Butterfield, eds., Applied Science Pub., London, U.K., 1979.

COMPUTATIONAL ACOUSTICS: Wave Propagation
D. Lee, R.L. Sternberg, M.H. Schultz (Editors)
Elsevier Science Publishers B.V. (North-Holland)
© IMACS, 1988

APPROXIMATION BY FINITE DIFFERENCE SCHEMES
OF THE PROPAGATION OF ACOUSTIC WAVES
IN STRATIFIED MEDIA

J. C. Guillot

Département de Mathématiques
Centre Scientifique et Polytechnique
Université Paris Nord
Villetaneuse, France

P. Joly

Institut National de Recherche
en Informatique et Automatique
Domaine de Voluceau
Le Chesnay, France

INTRODUCTION

There has been much recent research concerning the study of
the behavior of finite difference schemes on a uniform grid with
respect to wave propagation phenomena in dimension 2. In the
whole space, in homogeneous media, the two main numerical effects
are the dispersion and the anisotropy, which can be characterized
in different ways (see [2,4,17,18]). Other studies have been concerned with the approximation of reflection and transmission phenomena at plane boundaries of interfaces [7,16]. In [13], we
looked at the approximation of Rayleigh waves in an elastic half
space. This wave is an example of a guided surface wave. In this
paper, we are interested in the approximation of another kind of
guided wave, the Love waves, which generally exist in stratified
media and are of great importance in geophysical prospecting. In
a recent paper, K. R. Kelly [14] presented numerical computations
concerning this problem for the simple case of a layer over an
infinite homogeneous half space. Our purpose here is the theoretical analysis of this model problem.

Our presentation will be essentially descriptive and we shall
not go into the details of the mathematical analysis, for which we
essentially use the techniques of the spectral theory of self-adjoint operators [11,19,20]. The reader who is interested can find
these details in [9,10,12]. Moreover, for the sake of simplicity,
we shall restrict ourselves to the simplest second order variational scheme and shall consider only the spatial semidiscretization.

All the results can be extended without specific difficulty to
other spatial discretizations and to numerical schemes fully discretized in space and time. For these extensions, the results
are of the same qualitative nature as soon as classical stability
criteria are satisfied. An outline of this paper is as follows.
In Section 1, we recall briefly the main numerical phenomena occurring in homogeneous media. In Section 2, we describe our problem and give the exact solution. In Section 3, we present our
numerical scheme and obtain the semidiscrete solution. We compare
this approximate solution to the exact one and distinguish the
parasitic and nonparasitic parts. We give in particular the dispersion equations for the numerical Love waves. In Section 4, we
deal with approximation results and present different dispersion
curves to illustrate these results.

1.   CLASSICAL NUMERICAL PHENOMENA FOR WAVE
     PROPAGATION IN A HOMOGENEOUS MEDIUM

Let us consider the acoustic wave equation in a homogeneous
medium [$x = (x_1, x_2)$ is the space variable, t is the time]

$$\frac{1}{c^2} \frac{\partial^2 u}{\partial t^2} - \Delta u = 0 \tag{1.1}$$

where the velocity c is constant. The plane wave analysis of
(1.1) consists of the study of particular solution in the form:

$$u(x,t) = \exp i(k \cdot x - \omega t) \qquad k = (k_1, k_2) \in \mathbf{R}^2 \qquad \omega \in \mathbf{R} \tag{1.2}$$

A simple calculation shows that the wave vector k and the pulsa-
tion ω must satisfy the dispersion relation

$$\omega^2 = c^2 |k|^2 = c^2 (k_1^2 + k_2^2) \Leftrightarrow \omega = \pm c |k| \tag{1.3}$$

In particular one sees that for each plane wave (1.2) the phase
velocity $c_\phi(k) = \omega k / |k|^2$ is constant and equal to c. It also co-
incides with the group velocity $c_g(k) = \nabla_k \omega$. This means that the
medium is nondispersive. Let us recall that the study of such
particular solutions is justified by the use of the space Fourier
transform which permits us to represent any finite energy solution of (1.1) as the superposition of harmonic plane waves, namely,

$$u(x,t) = \int a^+(k) \exp i(k \cdot x - |k| t) \, dk$$
$$+ \int a^-(k) \exp i(k \cdot x - |k| t) \, dk \tag{1.4}$$

Let us now consider a semidiscretization in space of equation (1.1):

$$\frac{1}{c^2}\frac{d^2 u_{pq}}{dt^2} - \frac{u_{p+1,q} - 2u_{p,q} + u_{p-1,q}}{h^2}$$

$$- \frac{u_{p,q+1} - 2u_{pq} + u_{p,q-1}}{h^2} = 0 \tag{1.5}$$

where $u_{pq}(t)$ denotes the approximation of $u(x,t)$ at point $x = (ph,qh)$, $(p,q) \in \mathbf{Z}^2$, where h is the stepsize of the grid. Denoting by $u_h(x,t) = (u_{p,q}(t),\ x = (ph,qh),\ (p,q) \in \mathbf{Z}^2)$, the approximate solution, the dispersion analysis of scheme (1.5) is the study of numerical harmonic plane waves:

$$u_h(x,t) = \exp i(\omega t - k \cdot x) \tag{1.6}$$

The dispersion relation of scheme (1.5) is:

$$\omega^2 = \frac{4}{h^2} c^2 \left\{ \sin^2\left(\frac{k_1 h}{2}\right) + \sin^2\left(\frac{k_2 h}{2}\right) \right\} = \omega_h(k)^2 \tag{1.7}$$

In this case, the phase velocity $c_\phi(k) = \omega k/|k|^2$ and the group velocity $c_g(k) = \nabla_k \omega$ do depend on the wave vector k. More especially:

> The dependence with respect to $|k|$, i.e., to the wave length $\lambda = 2\pi/|k|$, characterizes the numerical dispersion.
>
> The dependence with respect to $k/|k|$, i.e., the direction of propagation, characterizes the numerical anisotropy.

The dispersion and anisotropy can be illustrated as in [18] by representing the variations of the phase velocity as a function of $1/N = h/\lambda$ (inverse of the number of grid points per wave length) for a fixed direction (Fig. 1.1) or as a function of $k/|k|$ (in polar coordinates) for fixed $|k|$ (Fig. 1.2).

Note that the use of the discrete Fourier transform permits us to write any finite energy solution of (1.5) as the superposition of numerical harmonic plane waves:

$$u_h(x,t) = \int a_h^+(k) \exp i(k \cdot x - \omega_h(k)t)\ dk$$

$$+ \int a_h^-(k) \exp i(k \cdot x + \omega_h(k)t)\ dk \tag{1.8}$$

The comparison between (1.4) and (1.8) shows that dispersion and anisotropy are the only numerical effects introduced by the space

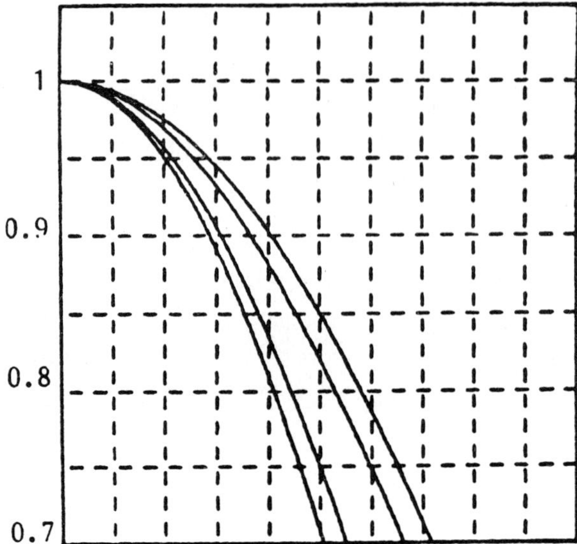

Figure 1.1

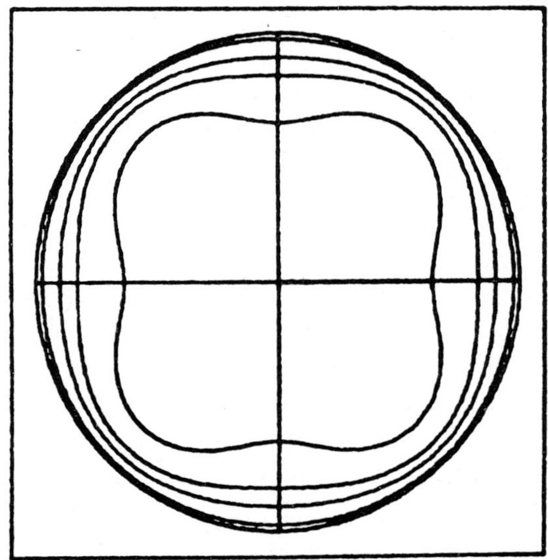

Figure 1.2

discretization. In particular, note that no parasitic phenomenon
appears.

2. THE MODEL PROBLEM - REPRESENTATION OF THE EXACT SOLUTION

Acoustic wave propagation in a layered half space $\mathbf{R}^2_+ =$
$\{(x,y) \in \mathbf{R}^2, \; y > 0\}$ is described by:

$$\rho(y) \frac{\partial^2 u}{\partial t^2} - \mathrm{div}(\mu(y)\,\mathrm{grad}\,u) = 0 \tag{2.1}$$

We impose the Neuman boundary condition on $\Gamma = \{(x,0)\}$:

$$\left(\mu \frac{\partial u}{\partial y}\right)(x,0,t) = 0 \tag{2.2}$$

and study the pure initial value problem corresponding to initial
data:

$$u(x,y,0) = u_0(x,y) \qquad \frac{\partial u}{\partial t}(x,y,0) = u_1(x,y) \tag{2.3}$$

We consider a two-layered medium corresponding to:

$$\rho(y) = \rho_1 \qquad \mu(y) = \mu_1 \qquad \text{for } 0 < y < L$$
$$\rho(y) = \rho_2 \qquad \mu(y) = \mu_2 \qquad \text{for } y > L \tag{2.4}$$

and we assume that the velocities $c_1 = (\mu_1/\rho_1)^{1/2}$, $c_2 = (\mu_2/\rho_2)^{1/2}$
satisfy the following inequality:

$$c_2 > c_1 \tag{2.5}$$

which is the necessary and sufficient condition for the existence
of Love waves [1,15]. The complete mathematical study of (2.1),
(2.2), (2.3), (2.4) is due to Wilcox [19,20]. It is essentially
based on the use of the partial Fourier transform in the x direction and on the spectral theory of selfadjoint second order Sturm-
Liouville operators [6,8]. The result is the following: the solution u can be represented as the sum of a series converging in
$L^2(\mathbf{R}^2_+)$:

$$u(x,y,t) = u_0(x,y,t) + \sum_{j=1}^{+\infty} u_j(x,y,t) \tag{2.6}$$

where $u_0(x,y,t)$ is the body wave and $\sum_j u_j(x,y,t)$ is the guided
part of the solution. Formula (2.6) is a generalized eigenfunction expansion of $u(x,y,t)$.

                        J.C. Guillot and P. Joly

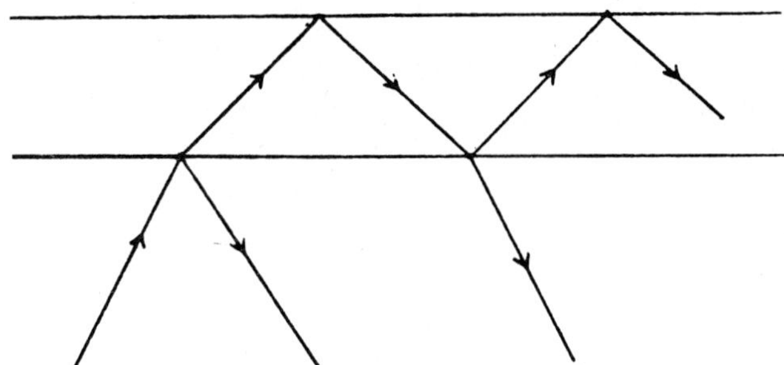

Figure 2.1.   The body wave.

The body wave can be represented as the superposition of non-dispersive harmonic waves depending on two parameters, the frequency ω and the x wave-number k,

$$u_0(x,y,t) = \int a(\omega,k,t)\tilde{\psi}_0(\omega,k;x,y)\ d\omega\ dk \qquad (2.7)$$

where $\tilde{\psi}_0(\omega,k;c,y)$ is a generalized eigenfunction of the operator $A = -\ 1/\rho\ \mathrm{div}(\mu\nabla)$ $(A\tilde{\psi}_0 = \omega^2\tilde{\psi}_0)$. Physically, $\tilde{\psi}_0(\omega,k;\cdot,\cdot)$ can be interpreted as a plane wave being reflected and transmitted at the interface and at the boundary Γ (see Fig. 2.1). The energy of u_0 is distributed in the entire half-space. The amplitude function $a(\omega,k,t)$ can be expressed with the help of the initial data (u_0,u_1) (see [9] for an explicit formula for $\tilde{\psi}_0$ and a).

The function $u_j(x,y,t)$ is the jth Love wave. It appears as the superposition on an infinite interval $[k_j,+\infty)$ of x wave-numbers k, of harmonic plane waves propagating in the x direction, namely,

$$u_j(x,y,t) = \int_{|k|>k_j} a_j(k)\psi_j(k,y)\ \exp\ i(\omega_j(k)t - kx)\ dk \qquad (2.8)$$

The wave number

$$k_j = \frac{(j-1)\pi}{L}\ \frac{c_1}{(c_2^2 - c_1^2)^{1/2}}$$

is the threshold for the apparition of the jth Love wave. The functions $\{\omega_j(k),\ k > k_j\}$, for $j = 1, 2, \ldots,$ are defined in the following way: For $k \in (k_j,k_{j+1})$, $\{\omega_\ell(k),\ 1 \le \ell \le j\}$ are given by

$$\omega_\ell(k)^2 = c_1^2(k^2 + K_1^\ell(k)^2) = c_2^2(k^2 - \xi_2^\ell(k)^2) \qquad (2.9)$$

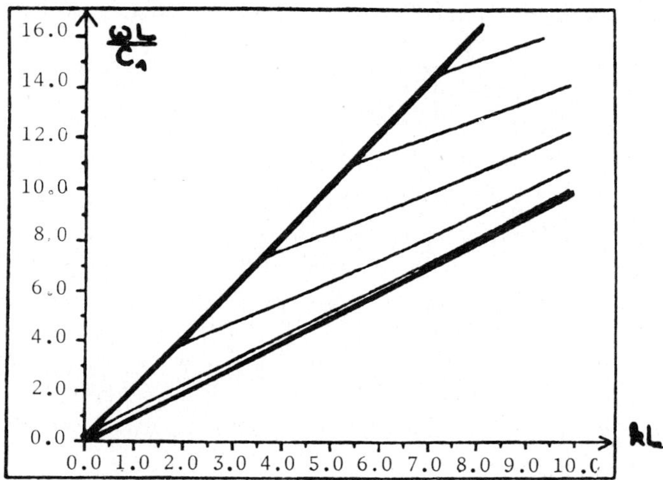

Figure 2.2.   Dispersion waves.

where $\{K_1^{\ell}(k)\xi_2^{\ell}(k))$, $1 \leq \ell \leq j$ are the j solutions of the following nonlinear system (Love wave's equation):

$$c_1^2 K_1^2 + c_2^2 \xi_2^2 = (c_2^2 - c_1^2)k^2$$

$$\mu_2 \xi_2 = \mu_1 K_1 \ tg(K_1 L) \tag{2.10}$$

$$K_1 \in \mathbf{R} \qquad \xi_1 \in \mathbf{R}_*^+$$

Each harmonic wave propagates in the x direction with the phase
velocity $\omega_j(k)/k$. Equation (2.9) is thus the dispersion relation
of the jth Love wave. We illustrate the dispersion properties of
these waves in Fig. 2.2. The amplitude of each wave varies as
the function $\psi_j(k,y)$:

$$\psi_j(k,y) = c_j(k) \ \cos(K_1^j(k)y)e^{-\xi_2^j(k)L} \qquad y < L$$

$$\psi_j(k,y) = c_j(k) \ \cos(K_1^j(k)L)e^{-\xi_2^j(k)y} \qquad y > L \tag{2.11}$$

where $c_j(k)$ is a normalization constant. In particular, this amplitude is exponentially decaying in the layer $y > L$, which means
that the energy of the guided wave u_j is essentially localized in
the layer $0 < y < L$. (The amplitude factor $a_j(k)$ can be expressed
as a function of the initial data (see [9]).) We summarize the
different properties of these waves in the picture below (Fig.
2.3). Note that the oscillations of $\psi_j(k,y)$ in the layer $0 < y < L$
essentially depend on j and not on k.

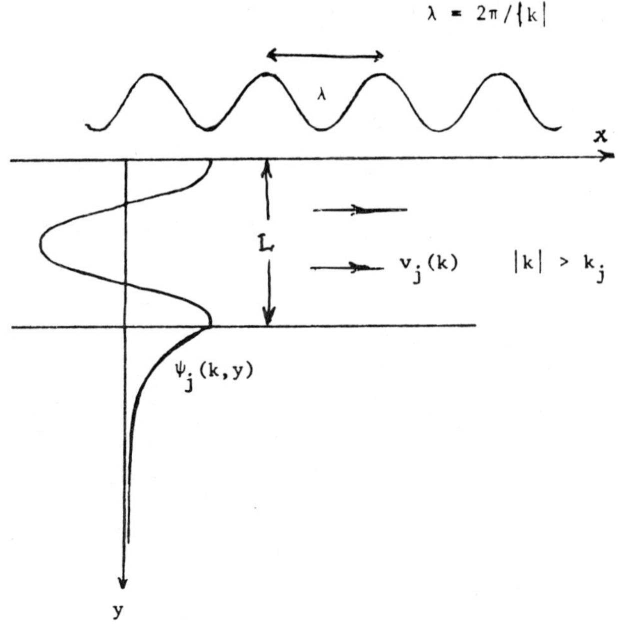

Figure 2.3.   The Love waves.

3.   THE NUMERICAL SCHEME - DESCRIPTION OF THE DISCRETE SOLUTION

The numerical scheme we shall consider can be obtained by standard variational techniques (see [3,5]), which ensure the stability and more particularly the conservation of a discrete energy. For the model problem (2.1), (2.2), (2.4), if we denote by h the stepsize of the uniform grid we use to discretize the half space R_+^2 and if we assume that there exists an integer $N \geq 2$ such that $N = L/h$ (so that the interface coincides with a line of the grid), this scheme can be written as follows:

$$\rho_1 \frac{d^2 u_{pq}}{dt^2} - \mu_1 \frac{u_{p+1,q} - 2u_{pq} + u_{p-1,q}}{h^2}$$

$$- 2\mu_1 \frac{u_{p,q+1} - u_{p,q}}{h^2} = 0 \qquad q = 0$$

$$\rho_1 \frac{d^2 u_{pq}}{dt^2} - \mu_1 \left(\frac{u_{p+1,q} - 2u_{p,q} + u_{p-1,q}}{h^2} \right.$$

$$\left. + \frac{u_{p,q+1} - 2u_{pq} + u_{p,q-1}}{h^2} \right) = 0 \qquad 0 < q < N$$

$$(3.1)$$

$$\frac{\rho_1 + \rho_2}{2} \frac{d^2 u_{pq}}{dt^2} - \frac{1}{h}\left[\mu_2 \frac{u_{p,q+1} - u_{p,q}}{h} - \mu_1 \frac{u_{p,q} - u_{p,q-1}}{h}\right]$$

$$- \frac{\mu_1 + \mu_2}{2}\left(\frac{u_{p+1,q} - 2u_{pq} + u_{p-1,q}}{h^2}\right) = 0 \qquad q = N$$

$$\rho_2 \frac{d^2 u_{pq}}{dt^2} - \mu_2\left(\frac{u_{p+1,q} - 2u_{p,q} + u_{p-1,q}}{h^2}\right.$$

$$\left. + \frac{u_{p,q+1} - 2u_{pq} + u_{p,q-1}}{h^2}\right) = 0 \qquad q > N$$

(3.1)
cont.

We have used the same notation as in Section 1 except that the index q varies in the set $\mathbf{N}$. The line of nodes q = 0 coincides with the boundary (Γ) and the line q = L with the interface. In particular, we denote by $u_h(x,y,t) = u_{pq}(t)$ for x = ph, y = qh the discrete solution of (3.1). For completeness, we have to prescribe for $u_h(t)$ initial conditions which are approximations of (2.3). Problem (3.1) can be rewritten in an abstract form

$$\frac{d^2 u_h}{dt^2} + A_h u_h = 0 \tag{3.2}$$

where A_h is a self adjoint operator (approximation of A) on a weighted L^2 space on the grid. The method we use for solving (3.2) follows closely the one adopted by Wilcox in the continuous case. Using a discrete Fourier transform in the x-direction leads us to study the spectral theory of discrete Sturm-Liouville operators for which we adapt the techniques of [6] and [8]. The details of the mathematics can be found in [9] and [12].

The result of our analysis is the obtention of a decomposition of $u_h(t)$ which will be the discrete equivalent of the decomposition (2.6) of u(t). To describe these results we have to distinguish two cases:

$$(1) \quad c_1 < c_2 \le c_1\sqrt{2}$$
$$(2) \quad c_1\sqrt{2} < c_2 < +\infty \tag{3.3}$$

The value $c_1\sqrt{2}$ has no physical meaning. It only depends on the numerical scheme we considered. In both cases (1) and (2), the solution $u_h(t)$ can be written as

$$u_h(t) = \bar{u}_h(t) + \varepsilon_h(t) \tag{3.4}$$

where $\bar{u}_h(t)$ denotes the <u>nonparasitic part</u> of the solution and $\varepsilon_h(t)$ the <u>parasitic part</u>. Moreover the structure of $\bar{u}_h(t)$ and $\varepsilon_h(t)$ can be described as follows:

$$\bar{u}_h(t) = u_0^h(t) + \sum_{j=1}^{n(h)} u_j^h(t) \tag{3.5}$$

$$\varepsilon_h(t) = \begin{cases} \varepsilon_0^h(t) & \text{in case (1)} \\[2ex] \varepsilon_0^h(t) + \varepsilon_h^*(t) & \text{in case (2)} \end{cases} \tag{3.6}$$

Analysis of the Nonparasitic Part

The component $u_0^h(t)$ is the numerical body wave. It is the discrete equivalent of $u_0(t)$. As for the continuous wave, it can be decomposed as the superposition of (dispersive) numerical harmonic waves

$$u_0^h(x,y,t) = \int a_h(\omega,k,t)\,\tilde{\psi}_0^h(\omega,k;x,y)\ d\omega\ dk \tag{3.7}$$

where $\tilde{\psi}_0^h(\omega,k;x,y)$ is a generalized eigenfunction of A_h which can be interpreted, as for the continuous case, as a numerical plane wave reflecting and transmitting at the interface and the boundary, according to <u>numerical Snell's laws</u>. The function $\tilde{\psi}_0^h(\omega,k;x,y)$ is known explicitly and can be shown to be a second-order approximation of $\tilde{\psi}_0(\omega,k;x,y)$ when h tends to zero (see [9] or [12] for further details).

The energy of this wave is distributed quasi-uniformly in the entire half space.

The integer n(h) is the number of numerical Love waves. Contrary to the continuous case, this number is finite. Its value depends on the case one considers:

$$n(h) = N = \frac{L}{h} \qquad\qquad \text{in case (1)}$$

$$n(h) = \left[\frac{2L}{h}\ \text{Arcsin}\left(\left(\frac{c_2^2}{c_1^2} - 1 \right)^{1/2} \right) \right] \qquad \text{in case (2)} \tag{3.8}$$

However, note that in both cases (1) and (2) n(h) tends to infinity as 1/h when h tends to 0 but that it is strictly smaller than N in case (2).

The component $u_j^h(t)$ is the jth numerical Love wave (for $1 \le j \le h$). It appears as the superposition of numerical evanescent plane waves propagating in the x direction. Such a wave exists if and only if the x-wave number k exceeds a numerical threshold $k_j(h)$,

$$k_j(h) = \frac{2}{h} \text{Arcsin}\left[\frac{c_1}{(c_2^2 - c_1^2)^{1/2}} \sin\left(\frac{(j-1)\pi h}{2L}\right)\right] \tag{3.9}$$

Thus, one has the formula (for $1 \le j \le n(h)$):

$$u_j^h(x,y,t) = \int_{|k|>k_j(h)} a_j^h(k)\psi_j^h(k,y) \exp i(\omega_j(k,h)t - kx) \, dk \tag{3.10}$$

The equation $\omega = \omega_j(k,h)$, for $|k| > k_j(h)$ is the numerical dispersion relation of the jth numerical Love wave. The functions $\omega_j(k,h)$ are defined in the following way: When $|k|$ belongs to $(k_j(h),k_{j+1}(h))$, $\{\omega_\ell(k,h), 1 \le \ell \le j\}$ are given by

$$\omega_\ell(k,h)^2 = \frac{4}{h^2} c_1^2\left[\sin^2\left(\frac{kh}{2}\right) + \sin^2\left(\frac{K_1^\ell(k,h)h}{2}\right)\right]$$

$$= \frac{4}{h^2} c_2^2\left[\sin^2\left(\frac{kh}{2}\right) - \sinh^2\left(\frac{\xi_2^\ell(k,h)h}{2}\right)\right] \tag{3.11}$$

where $\{(K_1^\ell(k,h),\xi_2^\ell(k,h)), 1 \le \ell \le j\}$ denote the j solutions of the following nonlinear system (numerical Love wave's equation):

$$\frac{4}{h^2} c_1^2 \sin^2\left(\frac{K_1 h}{2}\right) + \frac{4}{h^2} c_2^2 \sin^2\left(\frac{\xi_2 h}{2}\right) = \frac{4}{h^2}(c_2^2 - c_1^2)\sin^2\left(\frac{kh}{2}\right)$$

$$\frac{\mu_2}{h} \sinh(\xi_2 h) = \frac{\mu_1}{h} \sin(K_1 h)\tan(K_1 L) \tag{3.12}$$

$$K_1 \in \left[0,\frac{\pi}{h}\right] \qquad \xi_2 > 0$$

Note that (3.12) is clearly a second-order approximation for small h of (2.10). Each harmonic numerical Love wave propagates in the x direction with the phase velocity $v_j(k,h) = \omega_j(k,h)/k$. We shall illustrate its dispersion in Section 4. It has the structure of the continuous Love wave. In particular, it is exponentially decaying in the y direction in the layer $y > L$, as can be seen on the expression of the amplitude function $y \to \psi_j^h(k,y)$:

$$\psi_j^h(k,y) = c_j^h(k) \cos(K_1^j(k,h)y)e^{-\xi_2^j(k,h)L} \qquad y < L$$

$$\tag{3.13}$$

$$\psi_j^h(k,y) = c_j^h(k) \cos(K_1^j(k,h)L)e^{-\xi_2^j(k,h)y} \qquad y > L$$

We shall give in Section 4 convergence results for the thresholds $k_j(h)$ and the dispersion relations $\omega = \omega_j(k,h)$. One can also show that $\psi_j^h(k,y)$ converges to $\psi_j(k,y)$ which implies that the numerical component $u_j^h(t)$ tends to the continuous one $u_j(t)$. Note that as $n(h)$ goes to infinity when $h \to 0$, one can show that:

$$\sum_{j=1}^{n(h)} u_j^h(t) \to \sum_{j=1}^{+\infty} u_j(t) \qquad \text{in } L^2(\mathbb{R}_+^2) \tag{3.14}$$

(We associate to each discrete solution $u_h(t)$ the corresponding piecewise constant function.) In some sense, the guided part of the numerical solution can be seen as a partial sum of the series which constitutes the guided part of the exact solution.

Analysis of the Parasitic Part

$\varepsilon_h(t)$ always contains a parasitic body wave $\varepsilon_0^h(t)$ which can be written in the same form as the nonparasitic one:

$$\varepsilon_0^h(x,y,t) = \int_{D_h} \alpha_h(\omega,k,t)\tilde{\theta}_0^h(\omega,k;x,y) \, d\omega \, dk \tag{3.15}$$

for a domain of values of (k,ω) for which ω is greater than $2/h\, c_1$ (see [9]) $\tilde{\theta}_0^h(\omega,k;x,y)$ is a generalized eigenfunction of the operator A_h which has no equivalent in the exact solution. It can be interpreted as a numerical plane wave giving rise to parasitic transmitted and reflected numerical waves at the interface. One shows that $\tilde{\theta}_0^h$ is highly oscillating in the y direction. $\varepsilon_0^h(x,y,t)$ is a body wave since its energy is distributed in the entire half space. However, one shows that:

$$\varepsilon_0^h(x,y,t) \to 0 \qquad \text{in } L^2(\mathbb{R}_+^2) \text{ when } h \to 0 \tag{3.16}$$

When $c_2 > c_1\sqrt{2}$, i.e., in case (2), $\varepsilon_h(t)$ also contains a numerical guided wave $\varepsilon_h^*(t)$. This wave is the superposition of plane harmonic waves which travel in the x direction and are evanescent in the y direction. The threshold for the apparition of this wave is:

$$k^*(h) = \frac{c}{h} \text{Arcsin}\left(\frac{c_1}{(c_2^2 - c_1^2)^{1/2}}\right) \tag{3.17}$$

(Note that $k^*(h) \to +\infty$ when $h \to 0$.) Thus, one can write

$$\varepsilon_h^*(x,y,t) = \int_{|k|>k^*(h)} \alpha_h^*(k)\psi_h^*(k,y) \exp i(\omega^*(k,h)t - kx) \, dk \tag{3.18}$$

The dispersion relation of this parasitic guided wave is

$$\omega = \omega^*(k,h) \qquad |k| \geq k^*(h) \tag{3.19}$$

where $\omega^*(k,h)$ is defined by:

$$\omega^*(k,h)^2 = \frac{4}{h^2}\, c_1^2 \left\{ \sin^2\left(\frac{kh}{2}\right) + \cosh^2\left(\frac{\xi_1^*(k,h)h}{2}\right) \right\}$$

$$= \frac{4}{h^2}\, c_2^2 \left\{ \sin^2\left(\frac{kh}{2}\right) - \sinh^2\left(\frac{\xi_2^*(k,h)h}{2}\right) \right\} \tag{3.20}$$

where $(\xi_1^*(k,h), \xi_2^*(k,h))$ is the unique solution of:

$$c_1^2 \sinh^2\left(\frac{\xi_1 h}{2}\right) + c_2^2 \sinh^2\left(\frac{\xi_2 h}{2}\right) = (c_2^2 - c_1^2)\left\{ \sin^2\left(\frac{kh}{2}\right) \right.$$

$$\left. - \sin^2\left(\frac{k^*(h)h}{2}\right) \right\} \tag{3.21}$$

$$\mu_2 \sinh(\xi_2 h) = \mu_1 \sinh(\xi_1 h)\, \tanh(\xi_1 L)$$

$$(\xi_1, \xi_2) \in \mathbb{R}_*^+ \times \mathbb{R}_*^+$$

The amplitude function $\psi_h^*(k,y)$ is exponentially decreasing for $y > L$ and highly oscillating in the layer $0 < y < L$:

$$\psi_h^*(k,y) = (-1)^q a_h^*(k)\cosh(\xi_1^*(k,h)y)\, e^{-\xi_2^*(k,h)L} \qquad \text{for } y = qh \leq L$$

$$\psi_h^*(k,y) = (-1)^N a_h^*(k)\cosh(\xi_1^*(k,h)L)\, e^{-\xi_2^*(k,h)y} \qquad \text{for } y = qh \geq L \tag{3.22}$$

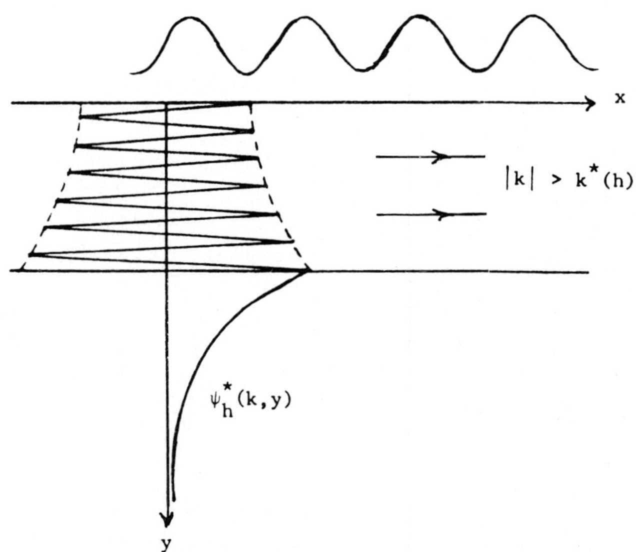

Figure 3.1. The parasitic guided wave.

When h tends to 0, the parasitic guided wave can be shown to tend
to 0 because of (3.17).

4.   CONVERGENCE AND APPROXIMATION RESULTS

It is possible to make a general study of the convergence of
the appropriate solution u_h to the exact one and more precisely
to show that:

Each nonparasitic component of the discrete solution converges
to the corresponding component of the exact solution.
The parasitic part of the discrete solution tends to 0.

In our case, one can prove that for each t > 0:

$$u_0^h(t) \to u_0^h(t) \qquad\qquad \text{in } L^2(\mathbb{R}_+^2)$$

$$\forall\, j \in \mathbb{N} \quad u_j^h(t) \to u_j(t) \qquad\qquad \text{in } L^2(\mathbb{R}_+^2)$$

$$\sum_{j=1}^{n(h)} u_j^h(t) \quad \sum_{j=1}^{+\infty} u_j(t) \qquad\qquad \text{in } L^2(\mathbb{R}_+^2)$$

$$\varepsilon_h(t) \to 0 \qquad\qquad \text{in } L^2(\mathbb{R}_+^2)$$

It is also interesting to have more precise results, namely, error
estimates, on the different characteristics of the components of
the solution. That is what we are going to give now for the guided
waves. More precisely, we shall be interested in the approximation
of:

The thresholds k_j (Section 4.1)
The dispersion relation $\omega = \omega_j(k)$ (Section 4.2)

4.1   Approximation of the Thresholds

For a given j in $\mathbb{N}^*$, the jth Love wave exists as soon as
$n(h) \geq j$, i.e.,

$$\varepsilon = \frac{h}{L} < \varepsilon_j \tag{4.1}$$

In this case we are able to compare the exact threshold k_j and
the approximate one $k_j(h)$.

One can show that:

$$k_j(h) = k_j \qquad \text{if } c_2 = c_1\sqrt{2}$$

$$k_j(h) > k_j \qquad \text{if } c_2 < c_1\sqrt{2} \qquad [\text{case } (1)]$$

$$k_j(h) < k_j \qquad \text{if } c_2 > c_1\sqrt{2} \qquad [\text{case } (2)]$$

Moreover, an asymptotic analysis leads to

$$k_j(h) = k_j \left[1 + \left(\frac{2c_1^2 - c_2^2}{c_2^2 - c_1^2} \right) \frac{(j-1)^2 \Pi^2 h^2}{24L^2} + 0 \frac{(j-1)^4 h^4}{L^4} \right) \right] \quad (4.2)$$

which shows that $k_j(h)$ is a second order approximation of k_j with respect to $\varepsilon = h/L$ and that the error one makes increases with the index j and the quantity $|(2c_1^2 - c_2^2)/(c_2^2 = c_1^2)|$. We illustrate these properties in Figs. 4.1 and 4.2.

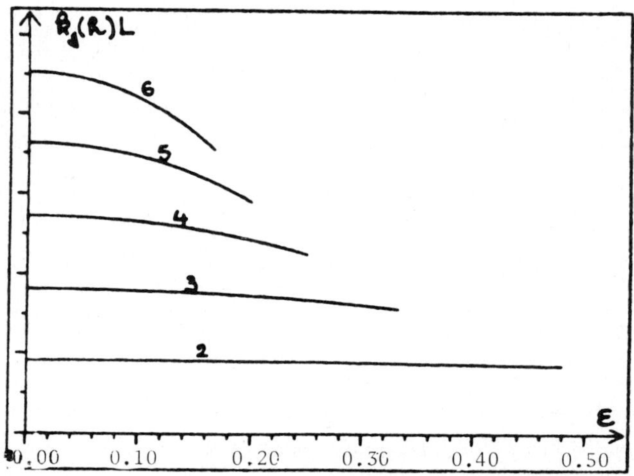

Figure 4.1

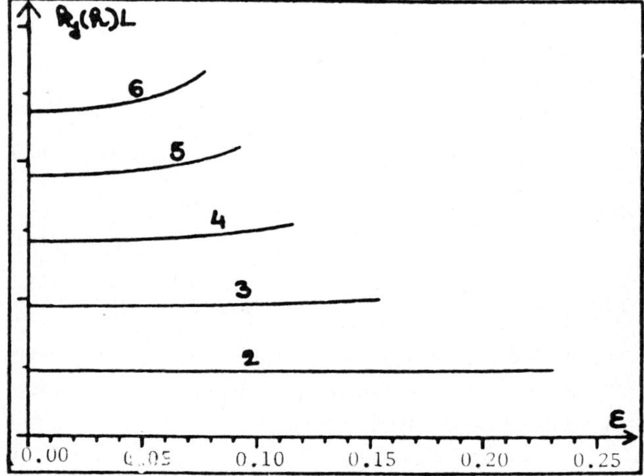

Figure 4.2

4.2 Approximation of the Dispersion Relations

The exact and discrete dispersion relations are respectively given by:

$$\omega^2 = \omega_j(k)^2 = c_1^2(k^2 + K_1^j(k)^2) = c_2^2(k^2 - \xi_2^j(k)^2)$$

$$\omega^2 = \omega_j(k,h)^2 = \frac{4}{h^2} c_1^2\left[\sin^2\left(\frac{kh}{2}\right) + \sin^2\left(\frac{K_1^j(k,h)h}{2}\right)\right] \tag{4.3}$$

Thus the comparison of $\omega_j(k)$ and $\omega_j(k,h)$ is clearly based on the comparison of $(K_1^j(k),\xi_2^j(k))$ [solution of (3.12)] and $(K_1^j(k,h),\xi_2^j(k,h))$ [solution of (2.10)].

Such an analysis can be made with the help of the implicit function theorem. One shows that:

$$\omega_j(k,h)^2 = \omega_j(k)^2 - c_1^2 k^2\left\{1 + \left(\frac{K_1^j(k)}{k}\right)^4\right.$$

$$\left. - 2\alpha_1^j(k)\left(\frac{K_1^j(k)}{k}\right)^2\right\}k^2h^2 + 0(k^4h^4) \tag{4.4}$$

where $\alpha_1^j(k)$ is known explicitly and bounded with respect to k (see [9]). Equation (4.4) shows in particular that $\omega_j(k,h)$ is a second-order approximation of $\omega_j(k)$ with respect to kh $= 2\pi/N$, where N denotes the number of grid points per x wavelength.

4.3 Dispersion Curves

In Figures 4.3 and 4.4, we give the dispersion curves for the numerical guided waves, when kh varies in the interval $[0,\Pi]$, in the two following examples:

$$\text{Example 1:} \quad \frac{\mu_2}{\mu_1} = 1 \qquad \frac{c_2}{c_1} = 2 \qquad N = 5$$

$$\text{Example 2:} \quad \frac{\mu_2}{\mu_1} = 1 \qquad \frac{c_2}{c_1} = 1.2 \qquad N = 5$$

In example 1, we are in case (2). There exists a parasitic guided wave (represented by the dotted curve in Fig. 4.3) and the number n(h) of nonparasitic Love waves is equal to 5. In example 2, we are in case (1). The number n(h) of nonparasitic guided waves is equal to 2 ($<$ N = 5).

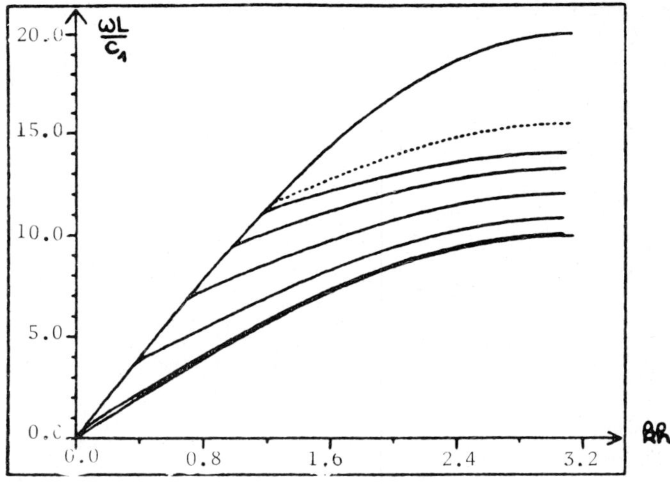

Figure 4.1.   Example 1.

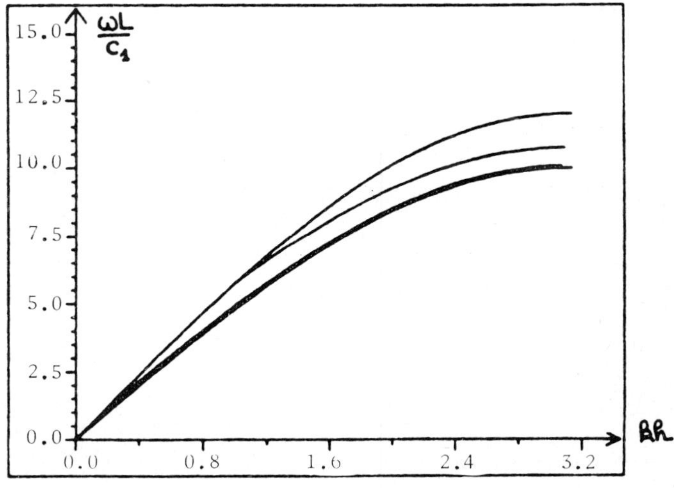

Figure 4.2.   Example 2

Finally, to analyze the accuracy of our scheme with respect
to the propagation of Love waves, we present in Figures 4.5 through
4.7 different dispersion curves corresponding, on each figure, to
a same value of the index j (i.e., for a given Love wave) and different values of h (i.e., different values of N), the adimensional
wave number kL varying in the interval [0,10]. We can illustrate
in this way, the convergence of the numerical method.

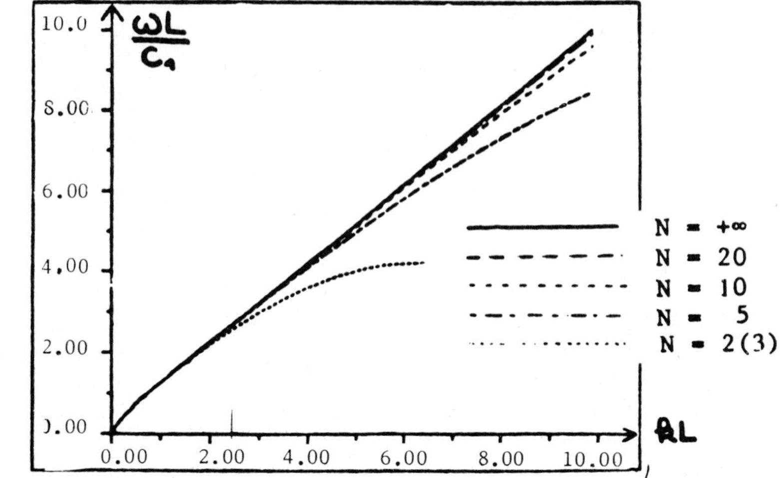

Figure 4.5.   Corresponds to j = 1, N = 2, 5, 10, 20, +∞.

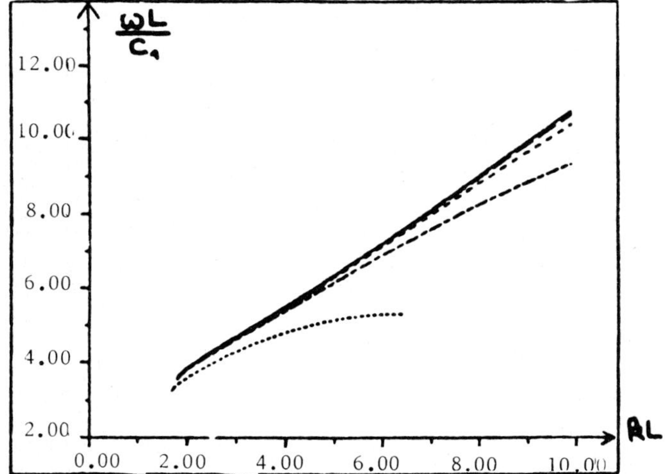

Figure 4.6.   Corresponds to j = 2, N = 2, 5, 10, 20, +∞.

Note in particular that it clearly appears that the error increases with j and kL and that for N = 20 this error does not exceed 1% for j = 1, 2, 3, when kL varies between 0 and 10.

Of course, the accuracy of the scheme can also be measured by looking at other quantities. For example it is interesting to compare the discrete amplitude functions $\psi_j^h(k,y)$ [see formula (3.13)] and the continuous ones $\psi_j(k,y)$ [formula (2.11)]. This comparison is made in [9].

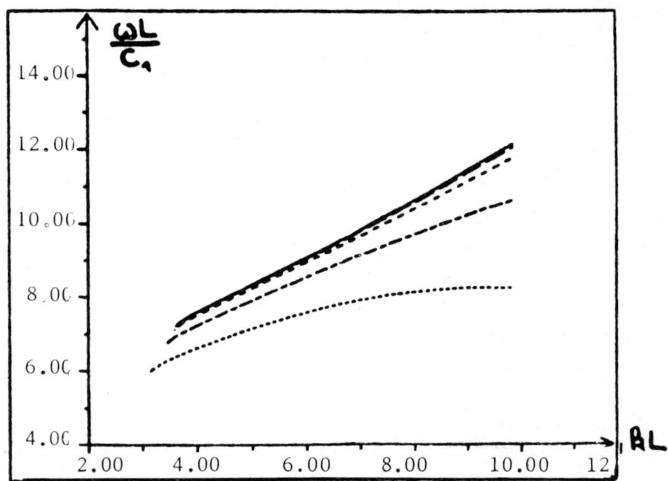

Figure 4.7.   Corresponds to j = 3, N = 3, 5, 10, 20, +∞.

5. CONCLUSION

From a theoretical point of view, our work points out the
existence of new numerical phenomena, as soon as one considers
propagation in heterogeneous media. In particular, we show, for
our model problem, the existence of purely numerical parasitic
waves and have shown that the countable set of guided waves existing in the exact solutions is approximated by a finite number
of numerical ones. Moreover, this model example demonstrates
that the introduction of the techniques of the spectral theory
in the domain of the numerical analysis is very useful for the
analysis of the properties of the approximate solution and for
obtaining sharp results of convergence.

From a practical point of view, the dispersion curves we obtain are a useful guide for the choice of the discretization parameters as a function of the physical parameters.

REFERENCES

1.   Achenbach, J. D., WAVE PROPAGATION IN ELASTIC SOLIDS, North
     Holland, 1973.

2.   Alford, R. M., R. Kelly, and D. M. Boore, Accuracy of finite
     difference modeling of the acoustic wave equation, GEOPHYSICS,
     vol. 6 (1974).

3.   Bamberger, A., J. C. Guillot, and P. Joly, Etude de schémas
     numériques des équations de l'élastodynamique linéaire,

RAPPORT INSTITUT NATIONAL DE RECHERCHE EN INFORMATIQUE ET
AUTOMATIQUE 41 (1980).

4. Bamberger, A., J. C. Guillot, and P. Joly, Numerical diffrac-
    tion by a uniform grid, SIAM JOURNAL ON NUMERICAL ANALYSIS
    (to appear).

5. Ciarlet, P. G., THE FINITE ELEMENT METHOD FOR ELLIPTIC PROB-
    LEMS, North Holland, 1978.

6. Coddington, E. A., and N. Levinson, THEORY OF ORDINARY DIF-
    FERENTIAL EQUATIONS, McGraw Hill, New York, 1955.

7. Daudier, S., and L. Nicoletis, Study of reflection and trans-
    mission properties of the variational scheme with finite dif-
    ferences, RAPPORT INSTITUT FRANCAIS DU PETROLE, 1984.

8. Dunford, N., and J. J. Schwartz, LINEAR OPERATORS, PART II:
    SPECTRAL THEORY, Interscience, New York, 1983.

9. Guillot, J. C., and P. Joly, Approximation par différences
    finies de la propagation d'ondes acoustiques en milieu
    stratifié (I), RAPPORT INSTITUT NATIONAL DE RECHERCHE EN
    INFORMATIQUE ET AUTOMATIQUE (à paraître).

10. Guillot, J. C., and P. Joly, Approximation numérique de la
    propagation des ondes acoustiques en milieu stratifié, Note
    aux Comptes Rendus de l'ACADEMIE DES SCIENCES DE PARIS, t.
    303, Serie I, No. 5 (1986).

11. Guillot, J. C., and C. H. Wilcox, Spectral analysis of the
    Epstein operator, PROCEEDINGS OF THE ROYAL SOCIETY OF EDIN-
    BURGH, Sect. A80 (1978), 85-98.

12. Joly, P., Analyse numérique et mathématique de modéles de
    propagation d'ondes acoustiques, élastiques et électromagné-
    tiques, Thèse d'Etat, Paris Dauphine, 1986.

13. Joly, P., Analyse numérique des ondes de Rayleigh, Thèse de
    3eme Cycle, Université de Paris Dauphine, 1983.

14. Kelly, R., Numerical study of Love waves, GEOPHYSICS, vol.
    48, No. 7 (1983).

15. Miklowitz, J., THE THEORY OF ELASTIC WAVES AND WAVE-GUIDES,
    North Holland, 1980.

16. Nicoletis, L., Simulation numérique de la propagation d'
    ondes sismiques, Thèse de Docteur Ingenieur, Université
    Pierre et Marie Curie, 1981.

17. Trefethen, L. N., Group velocity in finite difference schemes,
     SIAM REVIEW, Vol. 24, No. 2 (April 1983).

18. Vichnevetsky, R., and J. B. Bowles, Fourier analysis of
     numerical approximations of hyperbolic equations, SIAM STUD-
     IES IN APPLIED MATHEMATICS (1982).

19. Wilcox, C. H., Spectral analysis of the Pekeris operators,
     ARCHIVE FOR RATIONAL MECHANICS AND ANALYSIS, Vol. 69 (1976),
     259-300.

20. Wilcox, C. H., Sound propagation in stratified fluids,
     APPLIED MATHEMATICAL SCIENCES, vol. 50, Springer-Verlag,
     Berlin, 1984.

COMPUTATIONAL ACOUSTICS: Wave Propagation
D. Lee, R.L. Sternberg, M.H. Schultz (Editors)
Elsevier Science Publishers B.V. (North-Holland)
© IMACS, 1988

THE INFLUENCE OF THE ELASTICITY OF THE
OCEAN BOTTOM ON WAVE PROPAGATION

John T. Kuo

Aldridge Laboratory of Applied Geophysics
Columbia University
New York, New York

ABSTRACT

The influence of elasticity on wave propagation in a marine
environment is an extremely complex phenomenon and has slowly
gained its appreciation in underwater acoustics during the most
recent years.

A marine environment may be simplistically characterized by
acoustic (water), viscoelastic and poroelastic (particularly unconsolidated sediments), and elastic (consolidated sediments, possibly volcanics, and the basement rocks) media. The influence of
the elasticity of the ocean bottom and subbottom on wave propagation, deviating from being an acoustic medium, is virtually all
explicitly expressed in the stress term of the classic Euler equations of motion in the framework of the infinitesimal strain theory
of continuum mechanics.

The accumulated evidence from laboratory physical experiments,
analytic solutions, numerical simulations, and field observations
reveals that the elasticity of the ocean bottom and subbottom must
be taken into account, particularly at relatively low frequencies
and long-range acoustic transmissions in the oceans. Apparently,
the elasticity of an acoustic/elastic (poroelastic) coupled medium
has a profound and complex influence on both wave propagation and
attenuation.

INTRODUCTION

When the late Professor Maurice Ewing and I were studying the
surface-wave dispersion in the Pacific Ocean nearly twenty years
ago, we then recognized the importance of the influence of the
seemingly low rigidity of the sediments on the surface dispersion
[9]. If one wishes to examine the problem of wave propagation in
a marine environment more realistically, it is unavoidable that
one not only needs to take into account the elasticity of the

underlying sediments both unconsolidated and consolidated, but
also that of the basement for long-range transmission at relatively low frequencies.

The influence of elasticity on wave propagation in a marine
environment has slowly gained its entry in the most recent years.
The phenomenon of such an influence of elasticity on wave propagation is extremely complex and is a function of relative elastic-
parameter contrast in terms of relative contrast of P and S velocities in the layers, thickness of the elastic, viscoelastic or
poroelastic layers, conversion of P to S and S to P and multiples,
vertical and lateral inhomogeneities, generation of Stoneley-type
interface waves, frequencies, velocity gradients, porosity, and
permeability, etc. This paper naturally is not intended to be a
review of the subject but it highlights some of the recent findings related to the subject that may illuminate our further appreciation of the influence of elasticity on wave propagation, particularly in a marine environment.

A marine environment may be simplistically characterized by
a water layer which is underlain by unconsolidated and consolidated sediments, and the basement, which may all be classified
under acoustic (water), viscoelastic and poroelastic (particularly unconsolidated sediments), and elastic (consolidated sediments,
possibly volcanics, and the basement rocks) media. Wave propagation in all these media in the framework of infinitesimal strain
theory is governed by the classic Euler equations of motion, neglecting the advective nonlinear term, in continuum mechanics as
follows:

$$\nabla \cdot \underset{\sim}{\tau}(\underset{\sim}{x},t) + \rho \underset{\sim}{f}(\underset{\sim}{x},t) = \rho \frac{\partial^2 \underset{\sim}{u}(\underset{\sim}{xt})}{\partial t^2} \tag{1}$$

or

$$\frac{\partial p_{ij}}{\partial x_j} + \rho f_i = \rho \frac{\partial^2 u_i}{\partial t^2}$$

The influence of the elasticity on wave propagation is virtually all explicitly expressed in the stress $\tau(x,t)$ or P_{ij} (see
Appendix I for the details and notations).

For an acoustic medium,

$$p_{ij} = k\Theta\delta_{ij}$$

For an elastic medium,

$$p_{ij} = (k - \frac{2}{3}\mu)\Theta\delta_{ij} + 2\mu e_{ij} = \lambda\Theta\delta_{ij} + 2\mu e_{ij}$$

For an ideal viscous fluid,

$$p_{ij} = k\Theta\delta_{ij} + 2\nu \frac{dE_{ij}}{dt}$$

For an ideal imperfectly behaved elastic solid,

$$p_{ij} = k\Theta\delta_{ij} + 2\mu E_{ij} + 2\nu \frac{dE_{ij}}{dt}$$

For a poroelastic solid,

$$p_{ij} = (k + \frac{1}{3}\mu)\Theta - N\Omega + 2\mu e_{ij}$$

$$(p_{ij})_f = -N\Theta + M\Omega$$

Therefore, as expected, the influence of the elasticity of an elastic medium on wave propagation, by assuming the elastic medium to be an acoustic medium, is explicitly controlled by the complexity of the rigidity and the imperfect behavior of the medium.

1. LABORATORY EXPERIMENTS COMPARED WITH FINITE-
    ELEMENT SIMULATION FOR ACOUSTIC/ELASTIC
    COUPLED MEDIA

Kuo et al. (1985) reported at the last IMACS Symposium on Computational Acoustics on the significance of the influence of the elasticity of an elastic bottom on wave propagation in an acoustic/elastic medium through physical experiments and finite-element modeling carried out at the Aldridge Laboratory of Applied Geophysics (ALAG), Columbia University. The physical modeling experiment of the wave propagation in such an acoustic/elastic coupled medium was a three-dimensional composite elastic model consisting of a sloping-faced marble block cemented in mortar mix submerged in a large water tank. The finite-element modeling of an acoustic/elastic coupled medium couples the acoustic wave equation to the elastic wave equation by means of the double-node technique. Comparison of the laboratory physical modeling experiment and the finite-element numerical simulation of the physical modeling experiment of wave propagation in an acoustic/elastic coupled medium revealed that the elasticity of an acoustic/elastic coupled medium, as expected, has a profound influence on the shear wave conversion and propagation; in turn, the characteristics of wave forms and arrival events involve both compressional and shear waves.

In order to further substantiate the complexity of the influence of the elasticity of an acoustic/elastic coupled medium on wave propagation, another comparison of the finite-element and the Roever and Vining experiment results was made by Teng and Kuo [22]. Strick, Roever, and Vining [21] have experimentally and analytically investigated the problem of propagation of elastic waves from an impulsive source along a liquid/solid interface. Their dual attack of the problem led to a better understanding of the complex nature of the influence of elasticity on the wave motion in two different acoustic/elastic coupled media. In their laboratory physical experiments, Roever and Vining used a water layer overlying an elastic solid for the cases of (1) water/pitch, where the shear-wave velocity of the pitch is lower than the wave velocity of water, while the compressional-wave velocity of the pitch is greater than the wave velocity of water. Consequently, when the source-receiver distance is sufficiently large, the responses include critically refracted compressional waves, direct waves, reflected waves, and Stoneley-type interface waves; and (2) water/plaster-of-paris, where both the compressional and shear wave velocities of plaster-of-paris are greater than the velocity of water. Consequently, the responses include critically refracted compressional waves, direct waves, reflected waves, critically refracted shear waves, and Stoneley-type interface waves. Results of comparisons among the finite-element method of Teng and Kuo, the physical experiment of Roever and Vining, and the analytic solutions of Strick have substantiated the complexity of the influence of elasticity on wave propagation in an acoustic/elastic coupled medium. For details, the reader is referred to Teng and Kuo [22].

2. NUMERICAL AND OBSERVATIONAL RESULTS OF ACOUSTIC/
    ELASTIC-LAYERED COUPLED MEDIA

2.1 Numerical Study

The propagation of low-frequency waves in the oceans is complicated by the acoustic interaction with the ocean bottom and subbottom structures. In recent years, it has been recognized that the conversions of compressional-(P) to shear-(S), shear-(S) to compressional-(P), and Stoneley-type interface waves generated in the sediment-subbottom structure are one of the principal energy-loss mechanisms at low-frequency acoustic propagation in the oceans. For the large angles of incidence appropriate in

long-range propagation problems, the high-frequency response of
the ocean bottom obeys geometric ray theory well and the ocean
bottom may be modeled as a fluid. However, wave conversions occur
at the sediment-basement interface; the bottom must be modeled as
an elastic medium. Shear-wave conversions will occur even for
rays approaching turn-over depth within the sedimentary column,
except the column is thick compared to the wavelength. In this
case, high-frequency asymptotic solutions to the wave equation are
adequate to find the response. Ray theories break down for acoustic-wave propagation at low frequencies. At still lower frequencies, the concept of compressional-(P) and shear-(S) decoupling
also becomes inappropriate, because the P and S motions are intrinsically coupled [10].

Fryer [10] and Vidmar [23], for example, investigated the
effect of sediment rigidity on the plane-wave reflection coefficient for a typical deep-sea subbottom structure, composed of a
single inhomogeneous turbidite layer overlying a homogeneous
basalt substratum with an emphasis on large angles of incidence
(or low grazing angles) and low frequencies. A typical comparison of the effect of sediment rigidity on the reflection loss for
a solid sediment and a fluid sediment is shown in Figure 1. In
this figure, Vidmar [23] shows the reflection loss as a function
of grazing angle at 20 Hz for the hypothetical turbidite layer of
518-m thickness in comparison with that of 650 m used in the Fryer
model [10]. Computational methods adopted by Vidmar and by Fryer
were different. While Fryer used the method of the delta-matrix
extension of Thomson-Haskell, Vidmar used expansion theory for
reflection and transmission coefficients. In Figure 1, the three
curves represent (1) a fluid sediment and fluid substrate (dotted
line), (2) a fluid sediment and solid substrate (dashed line),
and (3) a solid sediment and solid substrate (solid line) as given
by Vidmar. Interference effects within the sediments give rise to
absorption peaks between 20° and 40°. In the range of less than
40° grazing angle, the effects of sediment rigidity on reflection
loss are negligible. Again, at large grazing angles, the predominant energy loss is in the form of compressional waves in the
lower half-space; the effects of sediment rigidity on reflection
loss are insignificant. The most pronounced effects of the bottom
rigidity on wave propagation at 20 Hz lies in the critical angles
of P and S waves in the present case in the range of 50° to 70°.

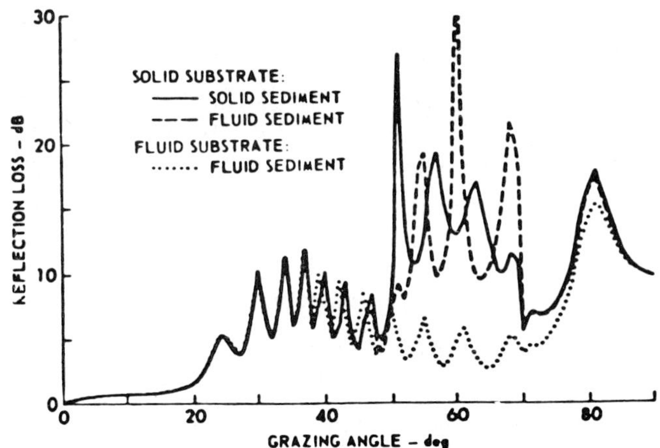

Fig. 1. Reflection loss versus grazing angle for a 518 m thick hypothetical turbidite layer at 20 Hz (Vidmar [23]).

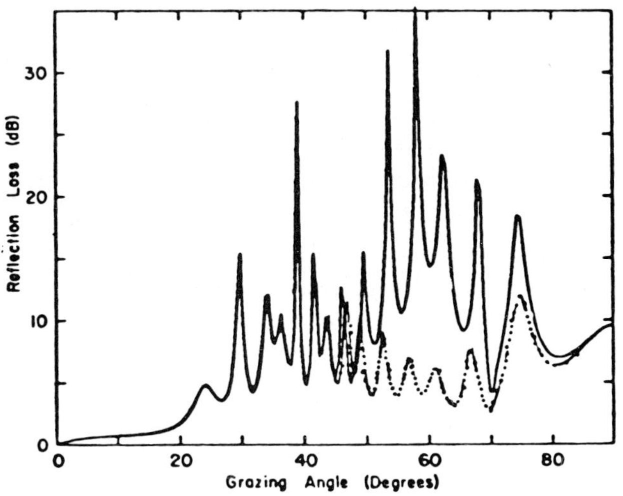

Fig. 1a. Bottom reflection loss at 20 Hz for the turbidite model (Fryer [10]).

Comparison of Figure 1 with that of Fryer (Figure 1a) shows essentially the same characteristics, and qualitative agreement for a fluid sediment and fluid substrata and a solid sediment and solid substratum (the dotted and solid lines, respectively).

2.2 Observations from Real Field Data of ESPs

Alsop and Diebold [1] made use of marine reflection data obtained by the Multichannel Seismic Group of the Lamont-Doherty Geological Observatory of Columbia University to observe converted

shear waves. The particular type of data they used are expanding
spread profiles (ESP), in which two ships move opposite from each
other with one ship shooting and the other recording in such a
manner that a common-mid-point (CMP) configuration is maintained
between the two ships. The data acquired are a common-mid-point
gather with the offset range of almost 0 to 60 or 70 km. The
traces are "binned" in 50-km offset bins, resulting in redundan-
cies of about 15 traces per bin.

Because converted shear waves are most conveniently observed
in tau-p seismograms as events with large slowness occurring
deeper within the seismogram than, say, compressional events, the
ESP data in the X-t (distance-time) domain are Radon-transformed
into the tau-p domain through the following relation:

$$u(z,\tau,p) = \int_{-\infty}^{\infty} p(x,z;\ \tau - px)\ dx$$

where p is the horizontal slowness or ray parameter, and τ is the
vertical delay time or intercept time.

However, in order for the converted shear waves to be observed,
these waves must be reflected upwards and then converted to a compressional wave. In fact, as might be expected from reciprocity,
the energy converted from shear to compressional as a function of
slowness for the upcoming shear wave is the same for compressional-
to-shear conversion for the downgoing compressional wave. If
shear-wave conversion takes place in a given range of slowness,
one can expect to see it in the tau-p seismogram. The predominant
events of shear-wave conversion take place at the sediment-basement interface, where the shear wave-velocity of the basement is
generally higher than the compressional-wave velocity of the overlying sediments. Thus, the shear-wave branch for reflections
within the basement joins the compressional branch for reflection
from the sediment-basement interface. However, within the sediment sequence, the shear-wave velocity of the underlying layers
is generally less than the compressional-wave velocity of the overlying layer, and the shear-wave branch in the tau-p seismogram is
not connected to a compressional branch to be identical in appearance for a low-velocity layer.

Figure 2 shows the real data of ESP 5 of the North Atlantic
Transect Experiment, in which the converted shear waves are present. Figure 3 is the tau-p seismogram obtained from the real data
as shown in Figure 2. In Figure 3, a compressional-wave branch

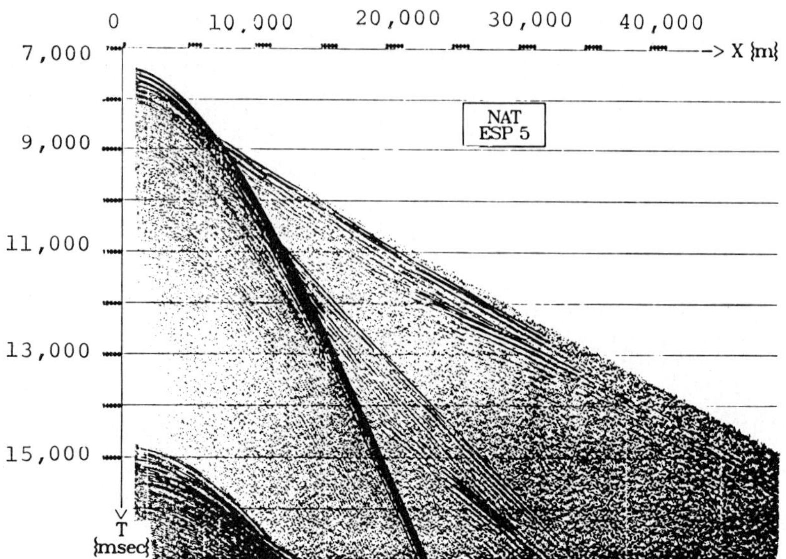

Fig. 2. Real data of ESP 5 of the North Atlantic Transect Experiment [1].

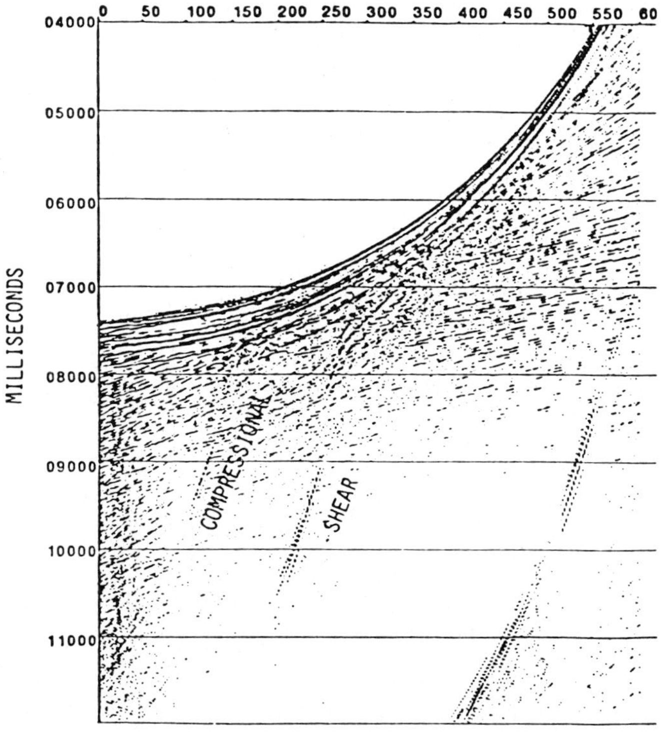

Fig. 3. The tau-p seismogram of ESP5/NAT shows the converted shear waves [1].

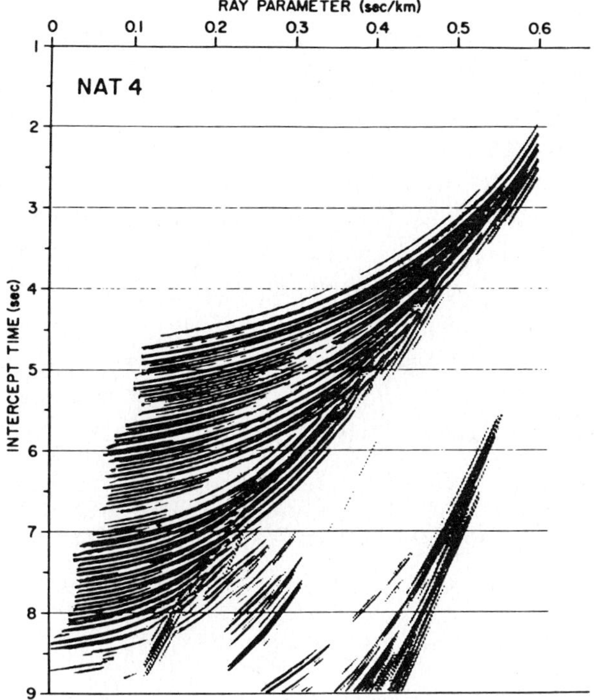

Fig. 4. The tau-p seismogram of ESP4/NAT(NAT4) data [Ref. 1].

due to refraction into the basement and reflection from the mantle
occurs between 100 and 180 msec/km. A similar branch for the converted shear waves from the sediment-basement interface occurs between 180 and 335 msec/km and is clearly shown [1].

As expected, one of the complications involved in observing
converted shear waves is the problem of dissipation of energy. It
is not necessary that they are always observable, even for these
converted waves from the sediment-basement interface. Laboratory
experiments [20] have indicated that in the unconsolidated or partially consolidated sediment, particularly in the upper section
of the sedimentary sequence, shear waves are very dissipative,
corresponding to low values of Q and assume very low velocity in
comparison to that of the velocity of water. Therefore, converted
shear waves in the upper sedimentary sequence of any of the tau-p
seismograms obtained from ESPs have not been observed [1]. Figure
4 is the tau-p seismogram for the real data of ESP4 of the North
Atlantic Transect Experiment, in which the converted shear waves
are absent not only in the upper layers but also at the sediment-

basement interface. In order to explain the absence of the converted shear waves in ESP4/NAT, Alsop and Diebold [1] generated
synthetic tau-p seismograms designated as NAT4 MODEL A with typical values of Q for shear waves in the consolidated sediment
based on ray theories. In order to enhance the appearance of converted shear waves, they assumed a Poisson's ratio of 0.25 for the
sediments and used Q values for both compressional and shear
waves slightly higher than generally accepted. The compressional velocities used were those obtained previously from an inversion of the ESP X-t and tau-p seismograms. The converted shear
waves, indicated by arrows in Figure 5, are present in the top
layer of the synthetic seismogram in a range of slowness between
0.5 and 0.6. The absence of the converted shear waves in the tau-
p seismogram can always be attributed to the high rate of dissipation. Thus, Alsop and Diebold generated another synthetic tau-p
seismogram designated as NAT4 MODEL C, the model of which is similar to that of NAT4 MODEL A, except (1) the Q for shear

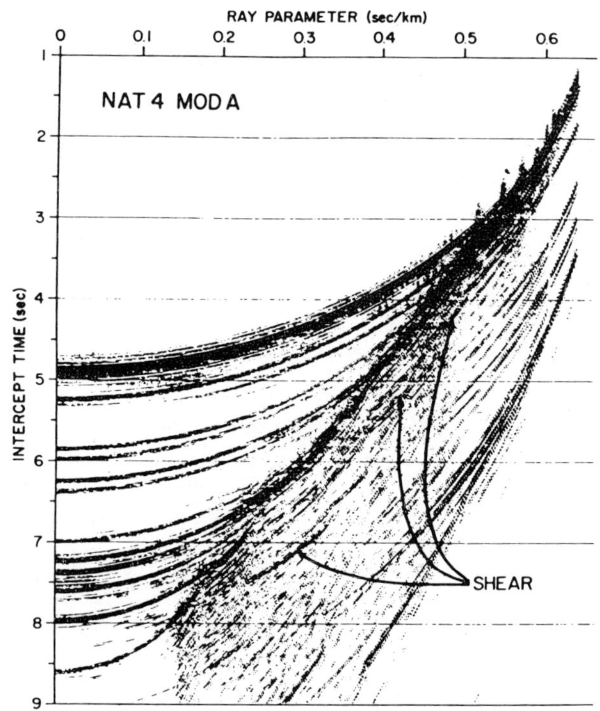

Fig. 5. The tau-p synthetic seismogram of NAT4 MODEL A (Ref. 1).

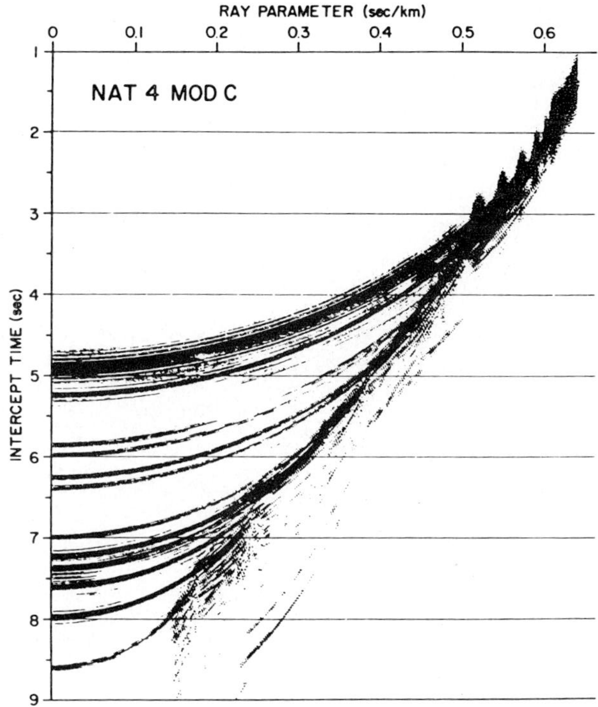

Fig. 6. The tau-p synthetic seismogram for NAT4 MODEL C (Alsop
and Diebold [1]).

waves in the upper sedimentary sequence has been assigned low values in accord with the results of laboratory experiments, (2) the
velocities of shear waves in the top of two layers are those of
the results of laboratory experiments, and (3) the compressional
wave Q's are lowered. In the resulting tau-p seismogram for NAT4
MODEL C (Figure 6), all the converted shear waves that appeared in
Figure 5 for NAT4 MODEL A are no longer present. Despite the fact
that Figure 6 confirms the tau-p seismogram for the real data of
ESP4/NAT, it does not rule out the possibility that the velocity
structure for the ESP4/NAT is more complicated than has been determined from inversion.

3. POROELASTIC MEDIA

3.1 The Plane-Wave Response of a Water Layer
       Over a Poroelastic Half-Space

As the water depth of an ocean environment becomes comparable to the wavelength, the influence of the bottom sediment, and
in turn, of elasticity on wave propagation becomes increasingly

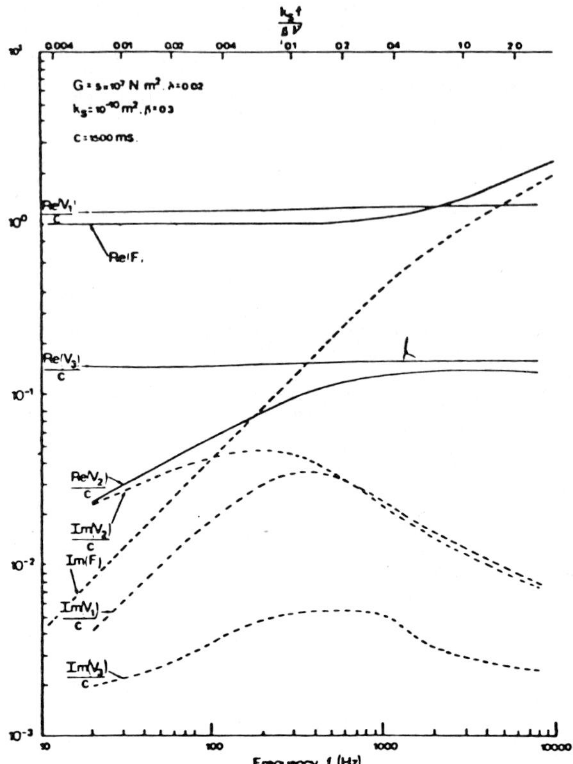

Fig. 7. The real and imaginary part of V_1, V_2, V_3, and F versus frequency: A fine sand for $k_s = 10^{-10}$ m^2 [24].

important. Existing wave-propagation models generally treat the bottom sediments as either a dissipative fluid, an elastic, or a dissipative elastic medium [10,2,17].

In a marine environment, the sediments generally are multi-phased media, viz., solid bodies permeated with fluid-filled pores. This is further complicated by the imperfect elasticity of the skeletal frame and the heterogeneity of sediment properties. To account for the elastic behavior of fluid-saturated materials, Biot [4,5] and Gassmann [11] introduced a semiphenomenological formulation of the equations of elasticity for porous aggregates. Later, Biot [6] presented a general theory of elastic wave propagation in porous media. Subsequent applications of the Biot theory (see, for example, Domenico [8], Stoll [18], Berryman [3], Johnson and Plona [14], etc.) have generally confirmed the predictions of the Biot model.

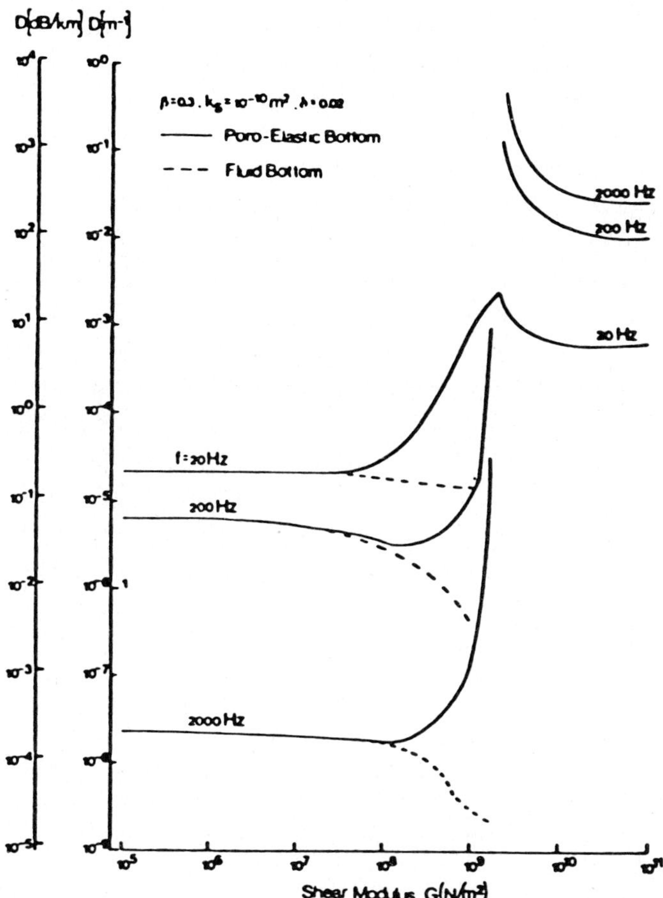

Fig. 8. Acoustic attenuation versus shear modulus (see Yamamoto [24].

The effect of consolidated and unconsolidated sediments and the pore fluid on acoustic-wave propagation may be modeled through introducing the concept of a poroelastic medium, i.e., the Biot model. Yamamoto [24] is among the first to incorporate Biot's theory into the normal-mode analysis of acoustic-wave propagation, in terms of harmonic plane waves for the case of a homogeneous ocean overlying a homogeneous half-space, which is considered a poroelastic medium.

In a poroelastic medium, there are three kinds of elastic waves propagated as shown in Appendix I. Yamamoto demonstrated the effect of the poroviscous frequency number $k_s f/\phi \nu_f$ on the complex velocities of the three kinds of elastic waves for a fine sand (Figure 7). As the figure shows, the phase velocities of the compressional waves V_1 and particularly of the shear waves V_3 are virtually independent of frequency. The phase velocity of the compressional waves V_2, however, increases like $f^{1/2}$ for small values of the poro-viscous-frequency number and slowly approaches a constant value as the frequency increases. The attenuation, or the imaginary part of V_1, V_2, and V_3, reaches a maximum at $k_s f/\phi \nu_f \simeq 0.1$.

In order to investigate the effects of physical properties of the bottom sediments, mainly shear modulus, porosity, and permeability, on the propagation of acoustic waves, Yamamoto adopted the values of the physical properties of the two typical bottom materials to approximately represent a fine sand and a porous rock (such as sandstones and corals). Figure 8 shows the acoustic attenuation as a function of shear modulus for $\phi = 0.3$, $k_s = 10^{-10}$ m^2, and $\delta = 0.02$, where δ is the Coulomb specific loss in the frame. Because fast compressional waves dominate in a soft sediment layer, while slow compressional waves dominate in a hard porous layer, Yamamoto found that the attenuation coefficient decreases with frequency in the soft sediment with the shear modulus smaller than the critical value of 2.5×10^9 N/m^2, while the opposite is true for the porous hard rocks with the shear modulus larger than the critical value. It is of interest to compare the poroelastic-bottom model with the fluid-bottom model, in which the sediment is modeled to be a fluid with the complex sound velocity equal to the fast compressional-wave velocity V_1 as in the case of a poroelastic medium. For the fluid-bottom model, the attenuation coefficient and the admittance are closely correlated with the imaginary part of the fast compressional-wave velocity V_1. For the shear modulus less than 5×10^7 N/m^2, the attenuation coefficients are nearly constant, corresponding to those for the poroelastic-bottom model. However, the fluid-bottom model deviates from the poroelastic-bottom model drastically for the shear modulus larger than 5×10^7 N/m^2.

In the cases for porous hard rocks and for soft sediments, the maximum attenuation of acoustic waves for a given frequency occurs at the poroviscous frequency number $k_s f/\phi \nu$ about equal to 0.1. However, as this number deviates from the critical value of 0.1, in the case of soft sediments, the error by a fluid-bottom model becomes increasingly large.

As in the laboratory experiments of Roever and Vining, the contrast between compressional- and shear-wave velocities of

the underlying half space and the overlying water velocity is
another critical factor in the influence of elasticity on wave
propagation.

3.2 The Transient Response of a Poroelastic
      Half-Space

An adequate description of the acoustic and elastic transient
response of fluid-saturated sediments requires the use of more
sophisticated numerical procedures such as finite-element or finite-difference methods. Because the finite-element method has
great flexibility in modeling irregular boundaries, inhomogeneity
of sediments, as well as the high contrast of acoustic and elastic properties of media, Hassanzadeh et al [13] developed the
finite-element Biot model, which incorporates the Biot equations
of poroelasticity into the Aldridge finite-element algorithm.
They initially considered the propagation of low-frequency acoustic waves in a macroscopically isotropic, homogeneous, porous
medium.

Although at low frequencies compressional waves of the second
kind have never been observed, studies have shown that they are
generated at an interface between dissimilar fluid-saturated porous aggregates as well as any interface between a fluid and a
poroelastic medium (see, for example, [7], [12], and [19]). Thus,
the existence of the compressional wave of the second kind and its
generation at interfaces between different porous sediments, as
predicted by the Biot theory, can influence the propagation of
acoustical elastic waves in porous media.

Hassanzadeh et al [13] presented their initial finite-element results of the transient-wave response in a poroelastic half-
space with model parameters as follows. The velocities of the
compressional and shear waves for solid grains are 2700 m/sec and
1500 m/sec, respectively; the velocity of the fluid is 1500 m/sec,
and the densities of solid grains and fluid are 2.4 g/cm^3 and 1.0
gm/cm^3, respectively; the forcing function used is the first derivative of a Gaussian wavelet with a center frequency of 50 Hz.

Figure 9 shows the finite-element synthetic seismogram with
a porosity of the medium of 40% for the said poroelastic half
space. The source is located 100 m below the surface at x = 0,
and the receiver array is located 50 m below the surface uniformly. The presence of direct and reflected compressional waves
of the second kind marked by P_2 and P_{22}, respectively, is distinct.
For comparison of the direct and reflected compressional waves of

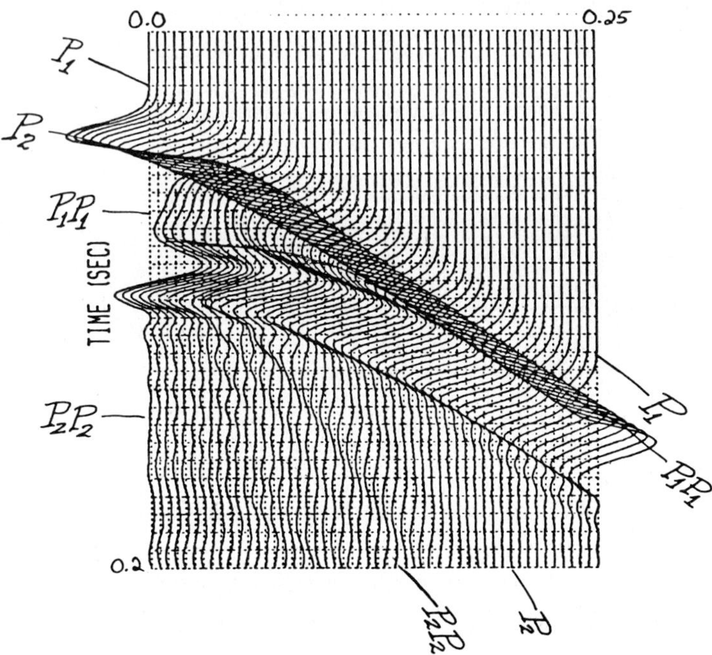

Fig. 9. The transient acoustic wave response in a poroelastic
half space [13].

the first kind, marked by P_1 and P_1P_1, with these compressional
waves for a nonporous elastic medium, Figure 9a shows the direct
and reflected compressional waves, marked by P and PP, in a nonporous elastic medium. The arrivals of the compressional waves
of the first kind are delayed, owing to the influence of poroelasticity. As all the amplitudes of the calculated wave fields
have been normalized according to those of the nonporous elastic
case, it can be seen that the amplitudes of the compressional
waves of the second kind are much smaller that those of the first
kind. For the present case of 40% porosity, it is about an order
of magnitude smaller. These two compressional waves of the first
kind and of the second kind are propagated from the same source
point, but they diverge independently according to their respective propagational velocities. In the absence of dissipation,
more energy is converted to the compressional wave of the second
kind with increase in porosity. The compressional wave of the

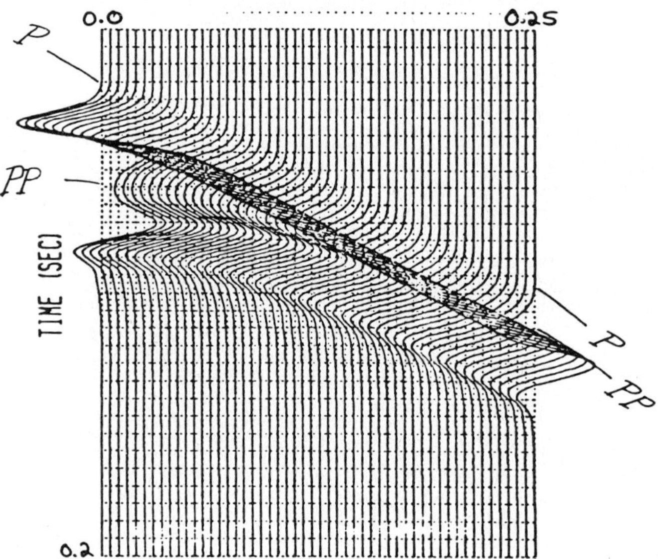

Fig. 9a. The transient acoustic wave response in an elastic half-space [13].

first kind, corresponding to solid and fluid moving in phase, is
similar to the ordinary compressional wave in the nonporous solid
matrix. In contrast, the compressional wave of the second kind,
corresponding to solid and fluid moving out of phase, is generally
diffusive.

4. DISCUSSION AND CONCLUSIONS

The influence of elasticity in terms of rigidity on wave
propagation is virtually all explicitly expressed in the stress
term of the classic Euler's equations of motion. The rigidity is
naturally of the same order of the effect on wave propagation as
the bulk modulus of the medium.

Converted shear waves, converted compressional waves, and
Stoneley-type interface waves are believed to be more abundantly
present in marine seismic-data recordings, particularly recogniza-
ble in a large-aperture recording array than one realizes. These
waves are principally generated due to the complexity of geological media, including stratification, contrast of elastic parameters and densities, inelasticity, discontinuity, and a variety of
lateral and vertical inhomogeneities.

In studying the importance of shear waves, Kuo and Dai [16] have developed a P and S wave simultaneous migration method for the source and receiver noncoincidence case, named the Kirchhoff elastic wave migration. Kuo and Dai then concluded that shear and converted shear waves must be considered as signals in order to have enhanced migrated images in the seismic depth section, opposing that these waves have been treated as background noise in the conventional Kirchhoff acoustic wave migration, resulting in poorer migrated images in the seismic depth section.

A marine environment consists in general of a water layer, which is underlain by unconsolidated sediments, the compressional and at least shear wave velocities for which are lower than the velocity of water, consolidated sediments, the basement and the subbasement. The relative contrasts of the compressional and shear waves of the unconsolidated sediment, the consolidated sediment sequence, and the basement and subbasement rocks are of fundamental importance in determining how the elasticity of the ocean bottom and subbottom influences wave propagation. Physical experiments of Roever and Vining and the field observations of ESPs particularly have shed light on these aspects of the complexity of wave phenomena.

Naturally, the problem of imperfectly elastic behavior of the sediments is another serious problem. If the information content of shear waves will ever be used to determine the elastic parameters, which are indicative of stratigraphy, lithology, porosity, and strength, the problem of energy dissipation of converted shear waves, together with compressional waves and Stoneley-type interface waves, must be addressed. The adoption of Biot theory may well be a good beginning. However, it is unavoidable that one is faced with one of the most fundamental problems in marine geophysics, that is, how to be sure that the velocity functions, both compressional and shear waves, of the unconsolidated, consolidated sequence and the basement and subbasement are correctly determined.

APPENDIX I

The classic Euler equations of motion governing wave propagation in acoustic, elastic, viscoelastic, and poroelastic media, neglecting the advective nonlinear term, are as follows:

$$\nabla \cdot \underset{\sim}{\tau}(\underset{\sim}{x},t) + \rho \underset{\sim}{f}(\underset{\sim}{x},t) = \rho \frac{\partial^2 \underset{\sim}{u}(\underset{\sim}{x},t)}{\partial t^2} \tag{1}$$

or

$$\frac{\partial p_{ij}}{\partial x_j} + \rho f_i = \rho \frac{\partial^2 u_i}{\partial t^2} \tag{1a}$$

where $\tau(x,t)$ or p_{ij} is the stress, ρ is the density, $\rho \underset{\sim}{f}(x,t)$ or ρf_i is the body force of the medium or applied force, and $\underset{\sim}{u}(x,t)$ or u_i is the displacement.

In the case of a perfectly elastic and isotropic medium,

$$p_{ij} = (k - \frac{2}{3} \mu)\theta\delta_{ij} + 2\mu e_{ij} = \lambda\theta\delta_{ij} + 2\mu e_{ij} \tag{2}$$

where k is the bulk modulus, λ and μ are the Lamé constants, and

$$e_{ij} = \frac{1}{2}\left(\frac{\partial u_i}{\partial x_j} + \frac{\partial u_j}{\partial x_i}\right)$$

In the case of an ideal fluid for which $\mu = 0$, (2) reduces to

$$p_{ij} = k\theta\delta_{ij} \tag{3}$$

or in the case of a pseudo-acoustic medium,

$$p_{ij} = \lambda\theta\delta_{ij} \tag{4}$$

If one introduces the deviatoric stress P_{ij} defined by

$$p_{ij} = p_p\delta_{ij} + P_{ij} \tag{5}$$

where p_p is the mean of the three principal stresses defined as $p_p = (1/3) \Sigma p_{ii}$, and the deviatoric strain E_{ij} defined by

$$e_{ij} = \frac{1}{3} \theta\delta_{ij} + E_{ij} \tag{6}$$

then it can be easily shown that

$$P_{ij} = 2\mu E_{ij} \tag{7}$$

On the basis of (7), one may introduce one of the simplest rigorous expressions of a deviation from perfectly elastic behavior, that is, an ideal viscous fluid such that

$$p_{ij} = p_p\delta_{ij} + 2\nu \frac{dE_{ij}}{dt} \tag{8}$$

resulting in the dissipation of energy, where ν is the viscocity.

For $\nu = 0$, (8) reduces to that of an ideal fluid of (3), i.e.,

$$p_{ij} = p_p\delta_{ij} = k\theta\delta_{ij}$$

Moreover, one may combine the perfectly elastic behavior of the solid with that of an ideal viscous fluid to give a mathematically rigorous type of imperfectly elastic behavior of an elastic solid as follows:

$$p_{ij} = p_p \delta_{ij} + 2\mu E_{ij} + 2\nu \frac{dE_{ij}}{dt} \tag{9}$$

Therefore, the value of μ measures the elasticity, or the rigidity of the medium, discriminating between ideal fluids for which as a lower limit $\mu = 0$, and solids, for which as an upper limit $\mu = \infty$, a perfectly rigid solid. The value of ν measures the viscosity, introducing the dissipation of energy. It is apparent that the value of μ plays an important role in the influence of elasticity on wave propagation when the medium is a nonliquid. The value of ν and its related parameters plays an important role so far as the imperfectly elastic behavior of an elastic solid is concerned.

From the above, one may build any sophisticated expressions for imperfect elastic behavior of an elastic medium such as a poroelastic medium, which will be discussed as follows.

For a poroelastic medium, Biot [6] defined the stress p_{ij} as follows:

$$\begin{aligned}
p_{ij} &= (k + \tfrac{1}{3}\mu)\Theta + N\Omega + 2\mu e_{ij} \\
&= (H - \mu)\Theta + N\Omega + 2\mu e_{ij} \tag{10}
\end{aligned}$$

$$(p_{ij})_f = -N\Theta + M\Omega$$

Equations (10) are used to form two coupled equations, in the absence of the body force or applied force, such that the modified Euler equations of motion are:

$$\frac{\partial p_{ij}}{\partial x_j} = \frac{\partial^2}{\partial t^2}(\rho u_i + \rho_f w_i)$$

$$\frac{\partial (P_{ij})_f}{\partial x_i} = \frac{\partial^2}{\partial t^2}\left(\rho_f u_i + \frac{\rho_f}{\phi} w_i\right) \tag{11}$$

Thus the effects of the elasticity on wave propagation in fluid-saturated porous elastic media are embedded in the stress expressions of equations (10) and the equations of motion (11).

Upon substituting equations (10) into equations (11), Biot's equations of motion for a homogeneous Coulomb-damped poroelastic medium are:

$$\mu \nabla^2 \underset{\sim}{u} + (H - \mu) \nabla\Theta - N\nabla\Omega = \frac{\partial^2}{\partial t^2} (\rho \underset{\sim}{u} + \rho_f \underset{\sim}{w})$$

(12)

$$\nabla(N\Theta - M\Omega) = \frac{\partial^2}{\partial t^2} (\rho_f \underset{\sim}{u} + m\underset{\sim}{w}) + \rho_f \frac{\nu_f}{k_s} \frac{\partial \underset{\sim}{w}}{\partial t}$$

where

$$H = (K_r - K_s)^2 / (D_r - K_s) + K_s + \frac{4}{3}\mu$$

$$N = K_r(K_r - K_s)/(D_r - K_s)$$

$$M = K_r^2 / (D_r - K_s)$$

$$D_r = K_r[1 + \phi(K_r/K_f - 1)]$$

where K_r is the bulk modulus of the solid grain, K_f is the apparent
bulk modulus of the pore fluid, ρ is the bulk density of sediment,
ρ_f is the density of water, ν_f is the kinematic viscosity of fluid,
k_s is the hydraulic coefficient of permeability, and the quantity
m is the virtual mass of the skeletal frame in the accelerated
flow field (Stoll [18]; Yamamoto [24]).

The volumetric strain Ω of the pore fluid is defined as

$$\Omega = -\nabla \cdot \underset{\sim}{w} \qquad \underset{\sim}{w} = \phi(\underset{\sim}{U} - \underset{\sim}{u})$$

(13)

where $\underset{\sim}{U}$ is the displacement vector for the pore fluid, $\underset{\sim}{w}$ is the
seepage displacement vector, and ϕ is the porosity of sediment,
and the volumetric strain e of the solid grains is defined as

$$\Theta = \nabla \cdot \underset{\sim}{u}$$

(14)

It can be shown that the three kinds of elastic waves propagated in a homogeneous poroelastic medium are (1) two kinds of
complex compressional waves with velocities V_1 and V_2

$$V_1, V_2 = \left(\frac{2A}{B \mp (B^2 - 4AC)^{1/2}} \right)^{1/2}$$

(15)

where

$$A = M(H) - N^2$$

$$B = \rho M + \left[\frac{s\rho_f}{\phi} - \frac{i\rho_f \nu_f}{k_s \omega} \right] H - 2\rho_f N$$

$$C = \left(\frac{s\rho_f}{\phi} - \frac{i\rho_f \nu_f}{k_s \omega} \right) \rho - \rho_f^2$$

where the expression $(B^2 - 4AC)$ in (15) must be nonnegative; and
(2) complex shear waves with velocity V_3:

$$V_3 = \left(\frac{\mu}{\rho}\right)^{1/2}\left[1 - \frac{\rho_f^2}{\left(\frac{s\rho_f}{\phi} - \frac{i\rho_f \nu_f}{k_s \omega}\right)}\right]^{-1/2} \tag{16}$$

Equations (12), without the damping term, may be decoupled to yield three wave equations as follows [7,13]:

$$\nabla^2\begin{pmatrix}\Theta_1\\\Theta_2\end{pmatrix} = \begin{pmatrix}V_1^2\\V_2^2\end{pmatrix}^{-1}\frac{\partial^2}{\partial t^2}\begin{pmatrix}\Theta_1\\\Theta_2\end{pmatrix} \tag{17}$$

and

$$\nabla^2\underset{\sim}{\Psi} = (V_3^2)^{-1}\frac{\partial^2}{\partial t^2}\underset{\sim}{\Psi} \qquad \text{where} \quad \underset{\sim}{\Phi} = -\phi\underset{\sim}{\Psi} \tag{18}$$

$$\underset{\sim}{\Psi} = \nabla\times\underset{\sim}{u} \qquad \underset{\sim}{\Phi} = \nabla\times\underset{\sim}{w}$$

Moreover, the relative dilatation between the solid and the fluid can be expressed as

$$\Omega = \Omega_1 + \Omega_2 = \delta_1\Theta_1 + \delta_2\Theta_2 \tag{19}$$

where

$$\begin{pmatrix}\delta_1\\\delta_2\end{pmatrix} = \left[\begin{pmatrix}V_1^2\\V_2^2\end{pmatrix}^{-1}\left(H - \frac{N^2}{M}\right) + NM^{-1}\rho_f - \rho\right]\left(\frac{NM^{-1}s\rho_f}{\phi} - \rho_f\right)^{-1}$$

The two kinds of complex compressional waves are referred to as the compressional waves of the first kind and of the second kind, or the fast compressional and the slow compressional waves, respectively.

ACKNOWLEDGMENTS

The author is most indebted to Dr. Ding Lee for his encouragement and understanding. He is grateful to Drs. L. Alsop and Y. C. Teng and Mr. S. Hassanzadeh for fruitful discussions. The support of the sponsors of Project MIDAS is appreciated.

The manuscript of this paper was completed at the Institut für Geophysik, Technische Universität Clausthal, FRG, while the author was granted the Humboldt Award for Distinguished Senior U. S. Scientist by the Alexander-von-Humboldt Stiftung. The author wishes to acknowledge the genuine kindness of the German hosts, particularly Professor and Mrs. Otto Rosenbach of Clausthal and Dr. H. Hanle of Humboldt. He is grateful to Columbia University for granting him sabbatical leave and to the Alexander-von-Humboldt Stiftung for the Award.

REFERENCES

1. Alsop, L. E., and J. B. Diebold, Determination of shear velocities from converted phases in marine sediment data, TRANSAC-
   TIONS, AMERICAN GEOPHYSICAL UNION (abstract) 67 (1986), 1230.

2. Beebe, J. H., S. T. McDaniels, and L. A. Ruano, Shallow-water
   transmission loss prediction using the Biot sediment model,
   JOURNAL OF THE ACOUSTICAL SOCIETY OF AMERICA 71 (1982), 1417-
   1426.

3. Berryman, J. G., Confirmation of Biot's theory, APPLIED
   PHYSICS LETTERS 37 (1980), 382-384.

4. Biot, M. A., Theory of propagation of elastic waves in a
   fluid-saturated porous solid; Part I: Low frequency range;
   Part II: Higher frequency range; JOURNAL OF THE ACOUSTICAL
   SOCIETY OF AMERICA 28 (1956), 168-191.

5. Biot, M. A., General theory of three-dimensional consolidation, JOURNAL OF APPLIED PHYSICS 12 (1941), 155-165.

6. Biot, M. A., Generalized theory of acoustic propagation in
   porous dissipative media, JOURNAL OF THE ACOUSTICAL SOCIETY
   OF AMERICA 34 (1962), 1254-1264.

7. Deresiewicz, H., The effect of boundaries on wave propagation
   in a liquid-filled porous solid: I. Reflection of plane
   waves at a free plane boundary (non-dissipative), BULLETIN OF
   THE SEISMOLOGICAL SOCIETY OF AMERICA 50 (1960), 599-607.

8. Domenico, S. N., Effect of water saturation on seismic reflectivity of sand reservoirs encased in shale, GEOPHYSICS 39
   (1974), 759-769.

9. Ewing, M., J. Brune, and J. T. Kuo, Surface wave studies of
   the Pacific crust and mantle, crust of the Pacific basin,
   GEOPHYSICS Monograph No. 6 (1962), AMERICAN GEOPHYSICAL UNION,
   30-40.

10. Fryer, G. J., Reflectivity of the ocean bottom at low fre-
    quency, JOURNAL OF THE ACOUSTICAL SOCIETY OF AMERICA 63
    (1978), 35-42.

11. Gassmann, F., Über die Elastizität poröser Medien, VIERTEL-
    JAHRSSSHRIFT DER NATURFORSCHENDEN GESELLSCHAFT in Zürich
    96 (1951), 1-23.

12. Geertsma, J., and D. C. Smit, Some aspects of elastic wave propagation in fluid-saturated porous solids, GEOPHYSICS 26 (1961), 169-181.

13. Hassanzadeh, S., Y. C. Teng, and J. T. Kuo, Finite element analysis of marine sediment acoustics, JOURNAL OF THE ACOUSTICAL SOCIETY OF AMERICA, December 1986 (abstract).

14. Johnson, D. L., and T. S. Plona, Acoustic slow waves and the consolidation transition, JOURNAL OF THE ACOUSTICAL SOCIETY OF AMERICA 72 (1982), 556-565.

15. Kuo, J. T., Y. C. Teng, P. Pecholcs, and J. Blair, A note on the influence of the elasticity on wave propagation in an acoustic/elastic coupled medium, COMPUTATION AND MATHEMATICS WITH APPLICATIONS 11 (1985), 887-896.

16. Kuo, J. T., and T. F. Dai, Kirchhoff elastic wave migration for the case of source and receiver noncoincidence, GEOPHYSICS 49 (1984), 1223-1238.

17. Kupermann, W. A., and F. B. Jenson (ed.), Bottom-interacting ocean acoustics, NATO CONFERENCE SERIES, Vol. 5, Plenum, New York, 1980.

18. Stoll, R. D., Acoustic waves in ocean sediments, GEOPHYSICS 42 (1977), 715-725.

19. Stoll, R. D., and T. K. Kan, Reflection of acoustic waves at a water-sediment interface, JOURNAL OF THE ACOUSTICAL SOCIETY OF AMERICA 70 (1981), 149-156.

20. Stoll, R. D., Acoustic waves in marine sediments, PROCEEDINGS OF CONFERENCE ON OCEAN SEISMO-ACOUSTICS, NATO ASW Centre, La Spezia, Italy, June, 1985.

21. Strick, E., W. L. Roever, and T. F. Vining, Propagation of elastic wave motion from an impulsive source along a fluid/ solid interface; I. Experimental pressure response; II. Theoretical pressure response; III. The pseudo-Rayleigh wave, PROCEEDINGS OF THE ROYAL SOCIETY, A, 251 (1959), 445-523.

22. Teng, Y. C., and J. T. Kuo, A finite-element algorithm for solving the dynamic problem of a fluid/solid interface, BULLETIN OF THE FIRST IMACS SYMPOSIUM ON COMPUTATIONAL ACOUSTICS (in press), 1987.

23. Vidmir, P. J., Ray path analysis of sediment shear wave effects, JOURNAL OF THE ACOUSTICAL SOCIETY OF AMERICA 68 (1980), 639-648.

24. Yamamoto, T., Acoustic propagation in the ocean with a poro-elastic bottom, JOURNAL OF THE ACOUSTICAL SOCIETY OF AMERICA 73 (1983), 1587-1596.

COMPUTATIONAL ACOUSTICS: Wave Propagation
D. Lee, R.L. Sternberg, M.H. Schultz (Editors)
Elsevier Science Publishers B.V. (North-Holland)
© IMACS, 1988

FINITE DIFFERENCE METHODS FOR
BOTTOM INTERACTION PROBLEMS

R. A. Stephen

Woods Hole Oceanographic Institution
Woods Hole, Massachusetts

ABSTRACT

The finite difference method is a powerful technique for studying seismo-acoustic problems on the ocean bottom. A finite difference method is presented which is stable for sharp, rough liquid-
solid interfaces. Arbitrary variations of compressional and shear
wave velocities and density in two dimensions (for either a point
source with cylindrical symmetry or a line source with Cartesian
symmetry) are allowed. The method solves the elastic wave equation
(a second-order system of hyperbolic partial differential equations
in horizontal and vertical displacement) in the time-space domain
using an explicit, centered-difference scheme.

In flat lying media the "snapshot" displays show clearly the
partitioning of compressional and shear energy at the sea floor into
reflected waves (P), interference head waves (P and S), diving waves
(P + S), and the pseudo-Rayleigh wave (Scholte wave). All evanescent wave effects are demonstrated. Scattering from deterministic
models of the sea floor including hills, valleys, and strong lateral
velocity variations can also be modelled and agree well with observed data from experiments in the Gulf of California and in the
'ROSE' area. In shallow water models the technique shows clearly
the evolution of the normal mode pattern and Stoneley waves from a
point, impulsive source. The behavior of these waves in range dependent wave guides including into and out of wedges such as the
continental margin can also be studied.

INTRODUCTION

The method of finite differences is applied to solving the
elastic wave equation for problems of sound propagation at range-
dependent fluid-solid interfaces. In analyzing sound propagation
at the sea floor it is usually assumed that either the wavelengths
are short relative to the roughness or that the sea floor is flat.
Unfortunately, these assumptions are rarely valid in the field.

The finite difference method makes it possible to obtain solutions in cases where these assumptions break down. Although the method has the potential for generating accurate synthetic seismograms in complex media, it is typically restricted by computational limitations to two-dimensional problems with ranges of a few tens of wavelengths.

The application of finite difference methods to the elastic wave equation is not new (Alterman and Loewenthal, 1972; Boore, 1972; Kelly et al., 1976). However, Stephen (1983) pointed out that traditional finite difference methods can be unstable at sharp liquid-solid interfaces and can yield inaccurate results at solid-solid interfaces. The examples shown in this paper were generated using the schemes of Stephen (1983) and Nicoletis (1981). Both methods are based on centered finite-difference approximations to the elastic wave equation in space and time (explicit schemes) with point, impulsive sources and both methods allow for general variation of compressional and shear velocity and density within the grid. The method of Stephen (1983) is unstable at sharp, liquid-solid boundaries, but its accuracy for gradients in elastic parameters was clearly demonstrated. The method of Nicoletis (1981) is stable at liquid-solid interfaces and its accuracy for wave guide problems was discussed by Stephen (1985).

Many papers on computational ocean acoustics consider the details of the numerical algorithms or attempt to analyze mathematically the stability and accuracy of given schemes. The finite difference method is no exception (e.g., Boore, 1972; Alford et al., 1974) but in this paper we try to demonstrate the application of the method to some real world problems. This is important and necessary for at least three reasons. First, most analyses of finite difference methods for wave equations assume either constant coefficients (homogeneous media) or zero shear modulus. Both these assumptions are inapplicable to sea-floor scattering problems. Second, because the finite difference method is limited computationally to small models, it is criticized for not being able to handle realistic problems. Most of the examples discussed in this paper were run to explain phenomena observed in field data. Third, the finite difference method can be used to study deterministically scattering in range-dependent environments. The method has been criticized for producing results which are too complicated to be interesting or to be useful in providing insight into the wave propagation process. The snapshot format, which the finite difference

method provides, is at least, if not more, as informative as ray
diagrams or dispersion curves and is generally more physically
realistic.

In this paper we present a number of examples of finite difference synthetic seismograms. Two of the examples involve scattering of sound from sea-floor hills in deep water. In one case,
multipathing causes the appearance of a 'double head wave' and in
the other case the hill acts as a secondary source (or diffractor)
for head wave energy causing a "refraction branch diffraction."
Propagation of sound in shallow water and in wedges is also demonstrated.

ENERGY PARTITIONING AT A FLAT SEA FLOOR

Before considering range-dependent problems it is useful to
study results for flat lying media. A point, impulsive source (10
Hz) is fired 2 km above the top of the grid at zero range (Fig. 1).
Since only homogeneous water separates the source from the sea

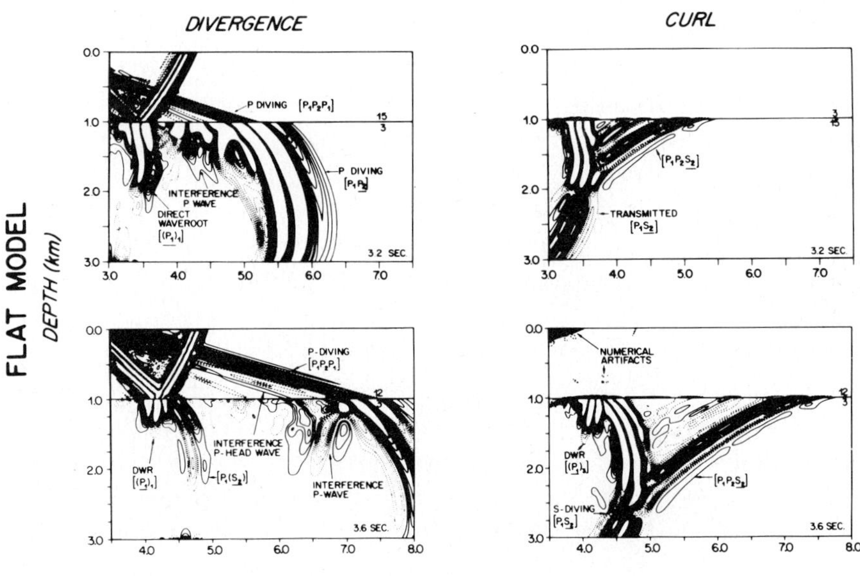

Fig. 1. Snapshots of the wavefield generated by the finite difference method for a flat sea floor 3.2 and 3.6 seconds after the
source was initiated. The source is located 2 km above the computational grid at 0 km range and has a dominant frequency of 10 Hz.
The plotting scale differs above and below the sea floor as indicated by the numbers on the right-hand side. Temporal, spatial and
energy relationships among the various wave types scattered by sea
floor interaction are readily observed. (Figure from Dougherty
and Stephen, submitted).

floor in this model we can reduce the amount of computation by introducing the effect of the point source as a time-dependent boundary condition on the top edge of the grid. The left-hand side is
an axis of symmetry and the lower and right sides are absorbing
boundaries (Clayton and Engquist, 1977; Emerman et al., 1982).
This example is taken from Dougherty and Stephen (submitted) and
uses the method of Necoletis (1981) to compute the interior wave
field. In order to simplify the snapshots we compute the divergence and curl of the displacement field which separates the field
into compressional and shear components.

The velocity-depth model for this example consists of uniform
water above the sea floor. Within one grid interval of the sea
floor the compressional wave velocity increases to 3.2 km/sec and
the Poisson's ratio decreases to 0.25. The velocity profile con-
tinues to increase with depth along two linear gradients until it
reaches 6 km/sec at about 1.5 km below the sea floor.

On interaction with the sea floor the incident compressional
wave splits into a number of compressional and shear waves. The
location and relative amplitude of these waves at 3.2 and 3.6 sec
are shown in Fig. 1. Some of the waves closely approximate the
waves predicted for interaction at boundaries between fluid and
solid half-spaces (Brekhovskikh, 1960) and are indicated using this
notation (for example, $P_1P_2P_1$ for the compressional head wave).
However, because of the gradient in the solid below the interface
new wave types arise such as the 'interference head wave' (Cerveny
and Ravindra, 1971). Evanescent waves such as the 'direct wave
root' (Stephen and Bolmer, 1985) are also apparent in the snapshots.
The ability of the finite difference method to handle homogeneous
blocks of material and regions of continuous velocity gradient with
equal ease is one of its strong points. Most bottom interaction
problems are best modeled by velocity gradients below the sea floor.

The snapshot format (Fig. 1) shows clearly the temporal, spatial and energy relationships between the various wave types which
occur for a spherical (or cylindrical) wavefront incident on the
sea floor. The pedagogical importance of this format should not
be underestimated. However, the time series response of pressure
or displacement at point receivers is more useful for relating results to real observations. Receivers can be placed anywhere in
the grid, permitting simple comparison of deep or shallow hydrophones and even buried receivers. Figure 2 compares the time
series results for horizontal lines of receivers at two depths in

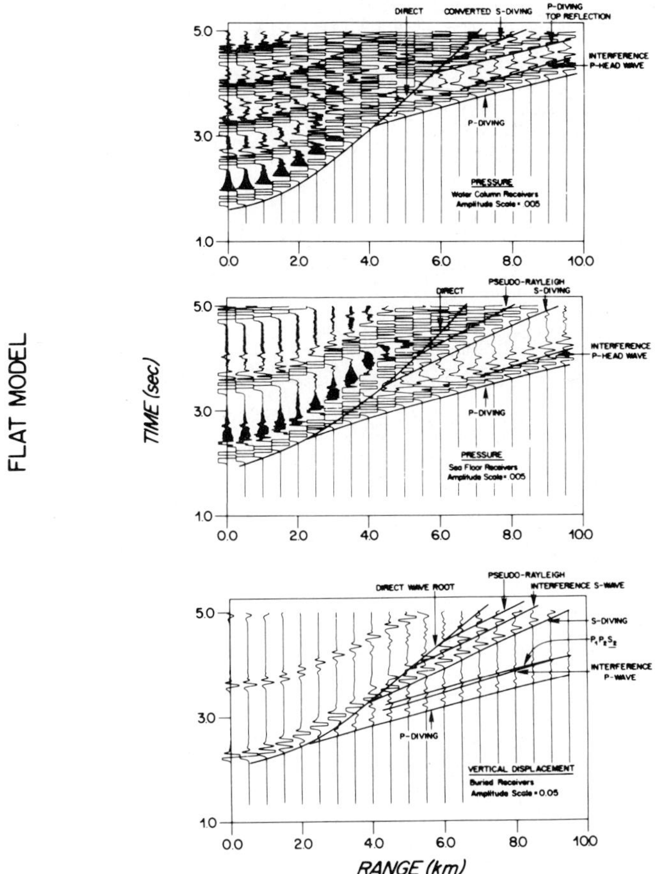

Fig. 2. Time series representation for receivers 520 m above the
sea floor, 20 m above the sea floor, and 480 m below the sea floor
for the model of Fig. 1. The wave types which one would expect to
see in a field experiment are indicated. In practice, the interference P- and S-waves, the pseudo-Rayleigh wave, and $P_1P_2S_2$ are
rarely if ever identified in field data. These waves are very sensitive to properties at the interface and could be attenuated
rapidly due to scattering from small-scale roughness. (Figure
from Dougherty and Stephen, submitted).

the water column and one depth in basement for the same model as
Fig. 1. Features which can be identified from the synthetics but
which are not generally recognized in field data are the interference P-head wave and the pseudo-Rayleigh wave.

Model Parameters

The flat sea floor example described above was computed on a
1000 × 300 grid for 3000 time steps. The space increment was 0.010
km in both range and depth and the time increment was 0.002 sec.
The source waveform in pressure was the third derivative of a

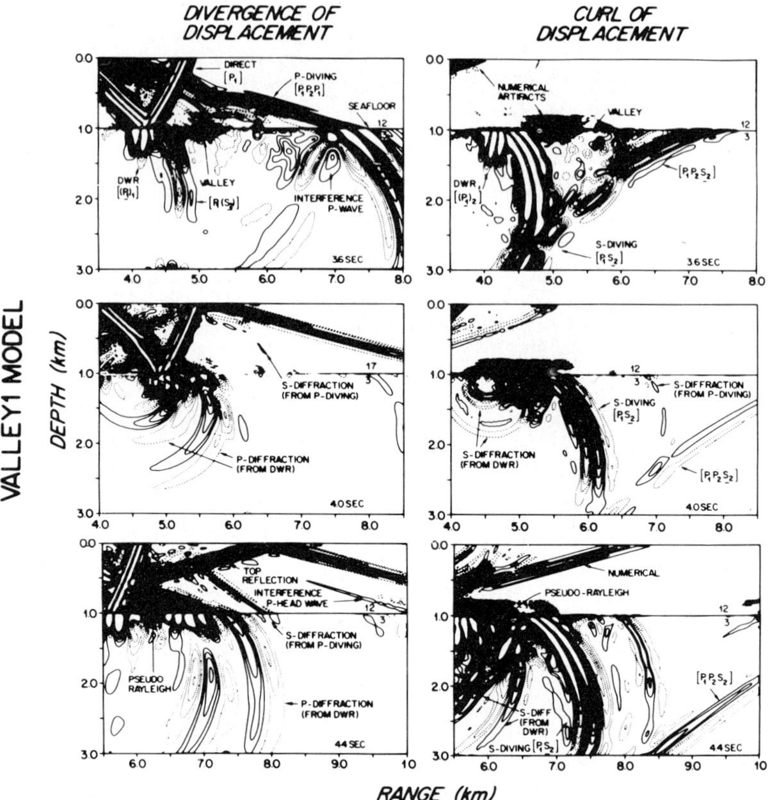

Fig. 3. Snapshots for the same velocity-depth model of Fig. 1 but
with a valley at 5 km range. Distortions can be observed in the
usual, flat model waves as well as numerous diffractions of both
compressional (divergence of displacement) and shear (curl of displacement) waves. (Figure from Dougherty and Stephen, 1986.)

Gaussian curve with a dominant frequency of 10 Hz (see Stephen et
al., 1985). This wavelet was propagated through a region 10 km
long by 3 km deep for a duration of 6 sec. Computationally, the
model required 10.5 megabytes of memory and 45 hours of CPU time
on a VAX 11-780 with the VMS operating system and the optimized
Fortran compiler. All calculations were done with single precision. Similar models on a Cyber 205 (a Class VI supercomputer)
require about 15 minutes of CPU time.

Scattering from a Valley at Super-Critical Ranges

The flat sea floor model discussed above sets the stage for
more complicated, range-dependent examples. Dougherty and Stephen
(submitted) consider a number of shapes and sizes of scatterers.
As an example here we show the effect on the wavefield of a valley
at 5.0 km range from the source. The valley, which is 750 m wide

and 250 m deep, has dimensions comparable to the seismic wavelengths in water (150 m) and basalt (~400 m). This problem is not
amenable to solution by high frequency methods.

   Inspection of the snapshot for the valley model (Fig. 3) shows
how the normal waves for a flat sea floor have been altered by the
valley and shows new wave types not normally present. The P-diving
wavefront bends due to the inflection in the sea floor and the
continuity of the $P_1P_2S_2$ wave is disrupted. There are four types
of diffraction from the valley: two each (compressional and shear)
for the P-diving wave and the direct wave. It is quite remarkable
how complicated the scattered wavefield from such a simple (but
realistic) model can be.

   The time series corresponding to ocean bottom hydrophones for
the valley model is shown in Fig. 4. The shear diffraction arising
from the P-diving wave and the compressional diffraction from the
direct wave are distinct, large amplitude arrivals. The compressional diffraction from the direct wave was observed quite clearly
in the field data from the ROSE experiment in the eastern Equator-
ial Pacific (Dougherty and Stephen, submitted).

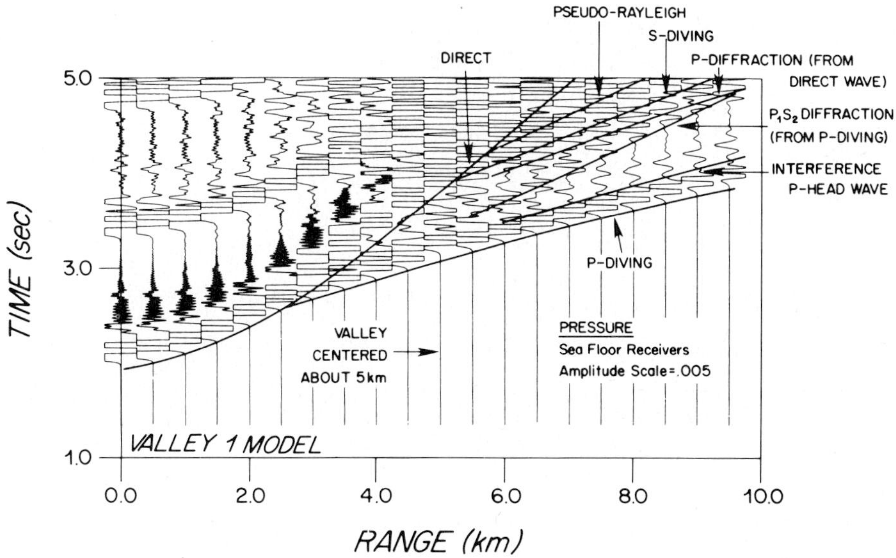

Fig. 4. Time series results for the valley model (Fig. 3) with
receivers 20 m above the sea floor. The diffractions from the
valley have comparable energy to other ground wave arrivals.
(Figure from Dougherty and Stephen, 1986.)

Scattering from a Hill at Sub-Critical Ranges

     The finite difference method was used by Stephen (1984) to
explain a diffraction phenomenon from a sea floor hill on borehole
seismic data from the Gulf of California. In this study the method of Stephen (1983) was used and the sea floor is represented by
a continuous change in velocity. Figure 5 shows the snapshots for
the flat sea floor example and Fig. 6 shows the snapshots in the
presence of a hill directly below the source. Comparison of the
two figures shows clearly the distortions in the wave field due to
the hill. The most obvious effect is the generation of a second
head wave (or diffraction) from the side of the hill. Since the
hill was located at sub-critical ranges the traditional head wave

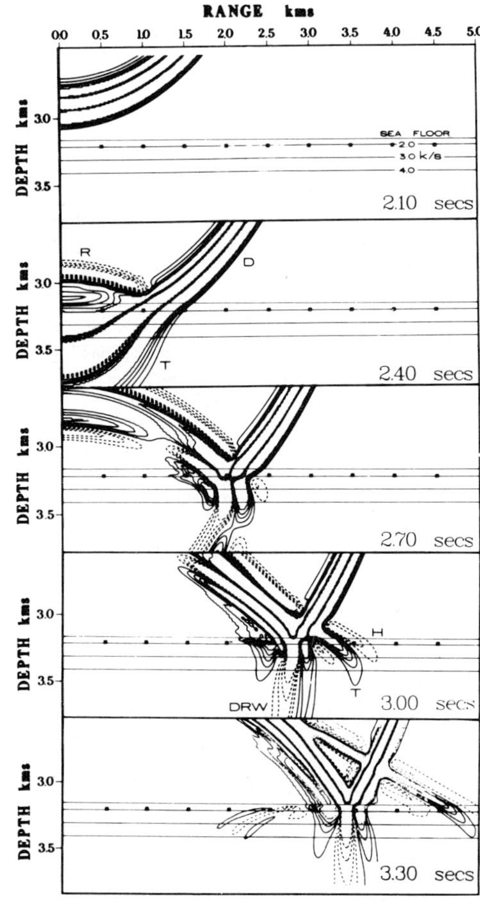

Fig. 5. Snapshots for a flat sea floor model without a velocity
discontinuity at the sea floor. These results should be compared
with the hill model in Fig. 6. (Figure from Stephen, 1984.)

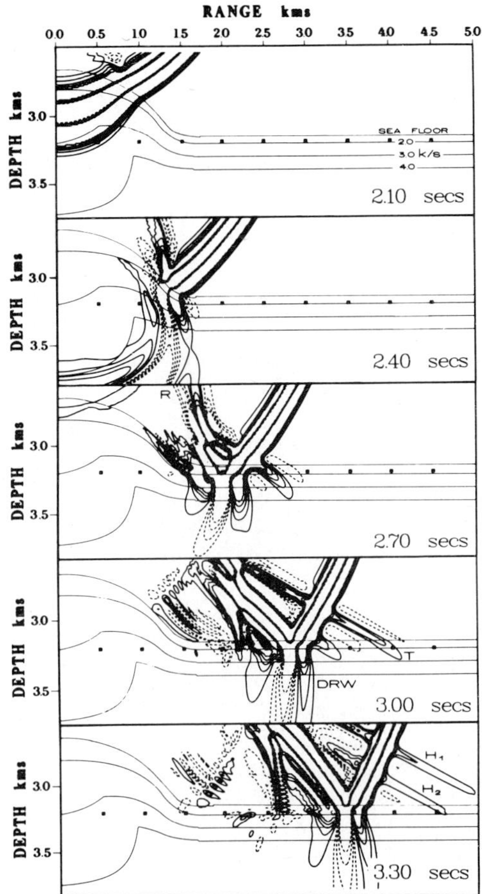

Fig. 6. Snapshots corresponding to a model with a hill directly below the source. The most obvious effects of the hill are the double head wave (H1 and H2) and the diffracted wave to the left of the sea floor reflection. Compare with Fig. 5. (Figure from Stephen, 1984.)

is still present although it is somewhat reduced in amplitude. The reflected wave field is distorted in shape and is reduced in amplitude due to the decreased velocity gradient under the hill. The 'tail' just to the left of the reflected wave is the reflection from the side of the hill. As expected the direct wave root is not affected by the hill in this example.

Shallow Water Waveguides

The finite difference synthetic seismogram method can also be applied to wave propagation in shallow water (Fig. 7). In this case a point, compressional source (with peak energy at 10 Hz) is fired 40 m below the sea surface in a channel 97.5 m deep.

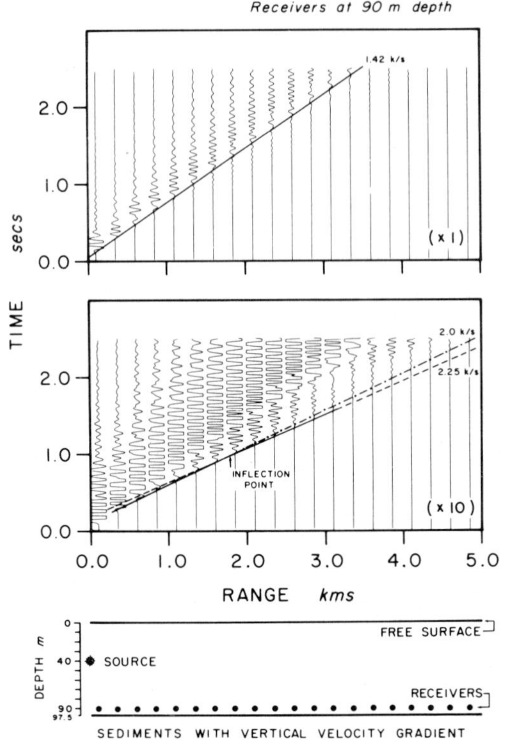

Fig. 7. Time series for a 10 Hz impulsive source in a waveguide
97.5 m thick over a sedimentary bottom with nonzero shear modulus
and a velocity gradient with depth below the sea floor. Ground
and water waves can be identified. The effect of the velocity
gradient below the sea floor is to increase velocities with range
and to enhance the amplitudes of the first arrivals up to 3 km
range.

Hydrophone receivers are located 10 m above the sea floor. The
bottom has properties representing sediment and consists of a gradient from 2.0 km/sec for P, 0.68 km/sec for S and a density of
2.0 gm/cc to 2.25 km/sec for P, 0.76 km/sec for S and a density of
2.1 gm/cc at 250 m below the sea floor. The compressional diving
wave (or ground wave) is the first arrival, followed by a packet
of P wave multiples (or leaky PL modes). The amplitude of the
ground wave decreases dramatically beyond 3.0 km range. This is
a shadow zone in which first order rays do not return to the surface and is due to the transition from velocity gradient to homogeneous half-space at 250 m depth below the sea floor. Subsequent
multiples of the compressional diving wave have higher amplitudes
than the primary as explained by Stephen et al. (1985).

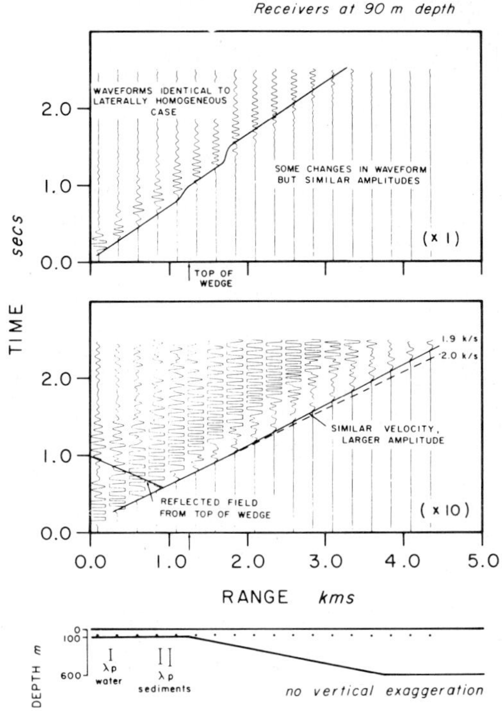

Fig. 8. This model is similar to Fig. 7 but water depth increases
linearly from 97.5 m at 1.25 km to 597.5 m at 3.75 km range. On
comparison with Fig. 7 the key new features are a backward reflected wave from the wedge, a large amplitude first compressional wave
with little change in apparent velocity and the persistence of the
water wave packet out to 2.0 km over the wedge.

The water wave arrives with a velocity of 1.4 km/sec. (The
water velocity for homogeneous media is 1.5 km/sec.) This wave
packet is composed of modes due to constructive interference of up
and down going waves in the water column. The dispersive nature
of the packet is apparent. It broadens with range, with lower frequencies arriving later in the packet.

Shallow Water Waveguide Energy Incident on a
Fluid Wedge from Above

Figure 8 uses the same model as Fig. 7 except the water column becomes progressively thicker starting at 1.25 km range. The
velocity function below the sea floor is the same as in Fig. 7.
The model represents the continental margin off New Jersey. There
is a weak reflection from the top of the wedge seen at about 1.0
second on the 0.1 km trace. There are two other effects, however,
which are not intuitively obvious. Both the compressional headwave

and later parts of the direct water wave packet propagate over 3 km into the wedge with little or no change in velocity or waveform.

One would expect the arrival times of the first compressional arrival to be delayed by up to approximately 0.5 sec. compared to the flat, waveguide case because of the increased length of the water path. However, at 4 km range, where the water depth is 400 m deeper than in the channel the delay of the compressional wave is only 0.14 sec. Even more dramatic is the amplitude of the compressional wave which is 20 db greater than for the laterally homogeneous case at ranges greater than 3.0 km. The frequency of the first arrival is also somewhat lower. The combination of velocity gradient with depth and shape of the wedge focuses more ground wave energy to larger ranges and less delayed arrival times than expected.

In Fig. 8 we have marked the region of the water wave packet which is identical for the laterally homogeneous and wedge examples. The later arriving modes are unaltered out to 2 km beyond the onset of the wedge. Apparently the modal pattern does not adapt immediately to changes in water depth. This phenomenon is analogous to the tunneling of seismic energy into shadow zones. The analysis of these "persistent wave effects" is still in progress.

CONCLUSIONS

The finite difference synthetic seismogram method can be used to solve a broad spectrum of range dependent problems in bottom-interacting ocean acoustics. We have the capability to quantitatively study problems which a decade ago were regarded as too complicated for consideration. Energy partitioning around hills and valleys on the sea floor can be quite complex, but the snapshot format, which the finite difference method provides, is extremely useful in identifying propagation paths and concentrations of energy density.

Two new effects predicted by the method are the enhancement of first compressional waves over a wedge and the persistence of the water wave modes as they tunnel beyond the waveguide into the top of a wedge.

ACKNOWLEDGMENTS

This work was supported by the Office of Naval Research under Contract N00014-85-C-001. Woods Hole Oceanographic Institution contribution No. 6294.

1. R. M. Alford, K. R. Kelly, and D. M. Boore, Accuracy of finite-difference modelling of the acoustic wave equation, GEOPHYSICS, 39 (1974), 834-842.

2. Z. Alterman and D. Loewenthal, Computer generated seismograms, in METHODS IN COMPUTATIONAL PHYSICS, 12 (A. B. Bolt, ed.), Academic Press, New York, 1972, pp. 35-164.

3. D. M. Boore, Finite difference methods for seismic waves, in METHODS IN COMPUTATIONAL PHYSICS, 11 (A. B. Bolt, ed.), Academic Press, New York, 1972, pp. 1-37.

4. L. M. Brekhovskikh, WAVES IN LAYERED MEDIA, Academic Press, New York, 1960, pp. 318-324.

5. V. Cerveny and R. Ravindra, THEORY OF SEISMIC HEAD WAVES, University of Toronto Press, Toronto, 1971, pp. 235-250.

6. R. Clayton and B. Engquist, Absorbing boundary conditions for acoustic and elastic wave equations, BULLETIN SEISMOLOGY SOCIETY OF AMERICA, 67 (1977), 1529-1540.

7. M. E. Dougherty and R. A. Stephen, Seismic scattering from sea-floor features in the ROSE area, submitted to JOURNAL OF THE ACOUSTICAL SOCIETY OF AMERICA.

8. S. H. Emerman, W. Schmidt, and R. A. Stephen, An implicit finite-difference formulation of the elastic wave equation, GEOPHYSICS, 47 (1982), 1521-1526.

9. K. R. Kelly, R. W. Ward, S. Treitel, and R. M. Alford, Synthetic seismograms: A finite-difference approach, GEOPHYSICS, 41 (1976), 2-27.

10. L. Nicoletis, Simulation numérique de la propagation d'ondes sismiques dans les milieux stratifiés à deux et trois dimensions; contribution à la construction et à l'interprétation des sismogrammes synthétiques, Ph.D. thesis, L'université Pierre et Marie Curie, Paris, 1981, VI.

11. R. A. Stephen, A comparison of finite difference and reflectivity seismograms for laterally homogeneous marine models, GEOPHYSICAL JOURNAL OF THE ROYAL ASTRONOMICAL SOCIETY, 72 (1983), 39-57.

12. R. A. Stephen, Finite difference seismograms for laterally varying marine models, GEOPHYSICAL JOURNAL OF THE ROYAL ASTRONOMICAL SOCIETY, 79 (1984), 184-198.

13.   R. A. Stephen, Finite difference synthetic acoustic logs with
      sharp, rough interfaces, Full Waveform Acoustic Logging Con-
      sortium Annual Report, Earth Resources Laboratory, MIT, 1985,
      101-123.

14.   R. A. Stephen and S. T. Bolmer, The direct wave root in mar-
      ine seismology, BULLETIN SEISMOLOGY SOCIETY OF AMERICA, 75
      (1985), 57-67.

15.   R. A. Stephen, F. Pardo-Casas, and C. H. Cheng, Finite differ-
      ence synthetic acoustic logs, GEOPHYSICS, 50 (1985), 1588-
      1609.

COMPUTATIONAL ACOUSTICS: Wave Propagation
D. Lee, R.L. Sternberg, M.H. Schultz (Editors)
Elsevier Science Publishers B.V. (North-Holland)
© IMACS, 1988

A FINITE ELEMENT ALGORITHM FOR SOLVING THE
TRANSIENT PROBLEM IN A FLUID-SOLID COUPLED MEDIUM

Yu-Chiung Teng and John T. Kuo

Alridge Laboratory of Applied Geophysics
Columbia University
New York, New York

ABSTRACT

On the basis of the double-node-layer technique in finite
element method, we have investigated two fluid/solid coupled models to compare the results with the Roever-Vining experimental
results as well as Strick's analytical solutions. In model I, the
case of the water/pitch interface, the shear wave velocity of the
pitch is lower than the compressional wave velocity of water,
while the compressional wave velocity of pitch is greater than
the wave velocity of water. In model II, the case of the water/
plaster-of-paris interface, both the shear and compressional wave
velocity of the plaster-of-paris are greater than the compressional velocity of water. Consequently, when the source-receiver distance is sufficiently large, the responses include critically refracted compressional waves, direct waves, reflected waves, and
Stoneley interface waves for the first case. As for the second
case, we expect one more critically refracted shear wave arrival.
Owing to the slowly rising bell-shaped forcing function used in
the finite element modeling, the numerical results do not exactly
depict the Roever-Vining experimental results. However, all these
arrivals can be clearly identified in the synthetic seismograms.
From these two numerical studies, we found that slightly shifted
in the vertical distance of the receiver location from the fluid/
solid interface will cause drastic changes in waveform of the
pressure fields; that means the response of the pressure field is
highly sensitive to the interface, particularly the receiver which
is located close to the interface. On the contrary, change in the
horizontal source-receiver distance causes very little change in
the waveforms of the pressure field.

INTRODUCTION

The present paper is motivated by the work of Roever and
Vining, and Strick (1959), in which they have experimentally and

analytically investigated the problem of wave propagation from an
impulsive source along a fluid/solid interface. Their dual attack
of the problem certainly led to a better understanding of the solution and the nature of complexity of the influence of the elasticity on the wave motion in an acoustic/elastic coupled medium.

It might be of interest to simulate their problem numerically
by means of the finite element method. Such a finite element simulation may be accomplished by modeling the small acoustic disturbance in the fluid medium with pressure type fluid elements,
and the small elastic deformation in the solid medium by elastic
type solid elements. A double-node technique then can be introduced to cope with the fluid-solid interface. Unfortunately, the
Roever-Vining experimental data provided only a single location of
the receiver, making it difficult to study the temporal and spatial
dependence of the wave field throughout this acoustic/elastic
coupled media. In our finite element results, in addition to comparing our numerical results at the location of their experimental data, we provide the results from various profiles about the
interface, leaning more toward the physical understanding of the
problem of the wave propagation in an acoustic/elastic coupled
medium.

It is hoped that favorably comparing the results of their
experiment with the finite element simulation would not only give
us the confidence of the finite element solutions to the problem,
but also enable us to extend the finite element method to solving
physically realistic problems of more complex fluid/solid inter-
faced media arising in ocean acoustics.

FINITE ELEMENT FORMULATION OF WAVE PROPAGATION
FOR THE FLUID-SOLID COUPLED MEDIUM

In the finite element formulation, the dynamic problem of
small disturbances in an acoustic medium can be solved by using
either pressure-type fluid elements to model the fluid field
directly, or displacement-type fluid elements. In this paper, we
adopt the direct approach of modeling the acoustic wave field by
using the pressure-type fluid elements.

For a compressible fluid, the equation governing the pressure
distribution of a small disturbance, neglecting the viscous term,
is

$$\kappa \nabla^2 p - \rho \frac{\partial^2 p}{\partial t^2} + S = 0 \qquad (1)$$

where p is the pressure, ρ is the density, κ is the fluid bulk
modulus, and S is the external source.

The functional formulation, which is equivalent to equation
(1) and obeys the same boundary conditions, is

$$X = \int \frac{\kappa}{2} \left\{ \left(\frac{\partial p}{\partial x}\right)^2 + \left(\frac{\partial p}{\partial y}\right)^2 + \left(\frac{\partial p}{\partial z}\right)^2 - 2\left[(S - \rho)\frac{\partial^2 p}{\partial t^2}\right]p \right\} dV \qquad (2)$$

Equation (2) must be minimized with respect to a set of nodal
pressure field. If the unknown function $\{p\}$ is defined element
by element in the usual manner with [N] being the shape function
as follows,

$$\{p\} = [N]\{p\}* \qquad (3)$$

where $\{p\}*$ is the nodal pressure column matrix, an approximate
minimization can be performed.

Assembly of the complete set of minimizing equations follows
the usual rules. Thus, for the whole region, we have

$$\frac{\partial X}{\partial \{p\}} = 0 \qquad (4)$$

The final system of equations is:

$$[A]\{\ddot{p}\} + [E]\{p\} + [S] = 0 \qquad (5)$$

with

$$A_{ij} = \Sigma \; a^e_{ij} \qquad (6)$$

$$E_{ij} = \Sigma \; e^e_{ij} \qquad (7)$$

$$a^e_{ij} = \int N_i \rho N_j \; dV \qquad (8)$$

$$e^e_{ij} = \int \kappa \left[\frac{\partial N_i}{\partial x}\frac{\partial N_j}{\partial x} + \frac{\partial N_i}{\partial y}\frac{\partial N_j}{\partial y} + \frac{\partial N_i}{\partial z}\frac{\partial N_j}{\partial z}\right] dV \qquad (9)$$

For an elastic solid medium, similarly neglecting the damping term, we have:

$$[M]\{\ddot{u}\} + [K]\{u\} + \{F\} = 0 \qquad (10)$$

where [M] and [K] are the global mass matrix and global stiffness
matrix, respectively. The forcing function $\{F\}$ includes two
terms, one is the external part $\{F1\}$ and the other $\{F2\}$ is due to
the fluid interface pressure. By principle of virtual work, $\{F2\}$
can be expressed as:

$$\{F2\} = \int_S [N]^T p \; ds \qquad (11)$$

Similarly, $\{S\}$ in equation (5) also includes two terms: $\{S1\}$ is the independently specified external part, and $\{S2\}$ is due to the solid interface interaction. Thus, $\{S2\}$ has the following form:

$$\{S_2\} = -k[G]\,\frac{\partial \ddot{u}}{\partial n} \tag{12}$$

$$[G] = \int_S [N_n]^T \rho_{solid} [N_n]\, ds \tag{13}$$

Finally, we obtain the matrix equation for the fluid/solid coupled system as:

$$[A]\{\ddot{p}\} + [E]\{p\} - k[G]\,\frac{\partial}{\partial n}\{\ddot{u}\} + \{S_1\} = 0 \tag{14}$$

$$[M]\{\ddot{u}\} + [K]\{u\} + \int_S [N_n]^T p\, ds + \{F_1\} = 0 \tag{15}$$

TWO-DIMENSIONAL FINITE ELEMENT CHARACTERISTICS

For the present Aldridge two-dimensional fluid-solid finite element code, we use the averaging 2-CST (constant strain triangles) quadrilateral elements in order to avoid the space-grid skewness (Figure 1).

Figure 2 shows typical triangular elements. For the displacement element, each nodal point has two specified degrees of freedom; for the pressure element, each nodal point has one degree of freedom.

Referring to Zienkiewicz (1977), the function of the displacement field u = $\{u,v\}$ for the solid, and the pressure field $\{p\}$ for the fluid, are approximated by three linear polynomials:

$$u = \alpha_1 + \alpha_2 x + \alpha_3 y$$
$$v = \beta_1 + \beta_2 x + \beta_3 y \tag{16}$$
$$p = \gamma_1 + \gamma_2 x + \gamma_3 y$$

The nine constants α_i, β_i, γ_i (i = 1, 2, 3) can be evaluated by solving three sets of simultaneous linear equations. These

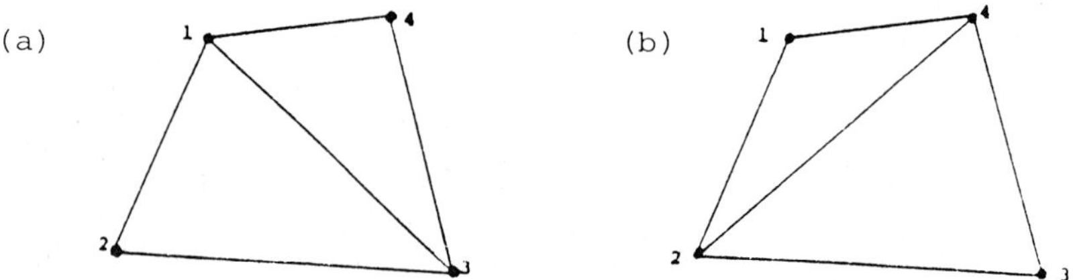

Fig. 1. A quadrangle comprised of two triangles: (a) left-divided, (b) right-divided.

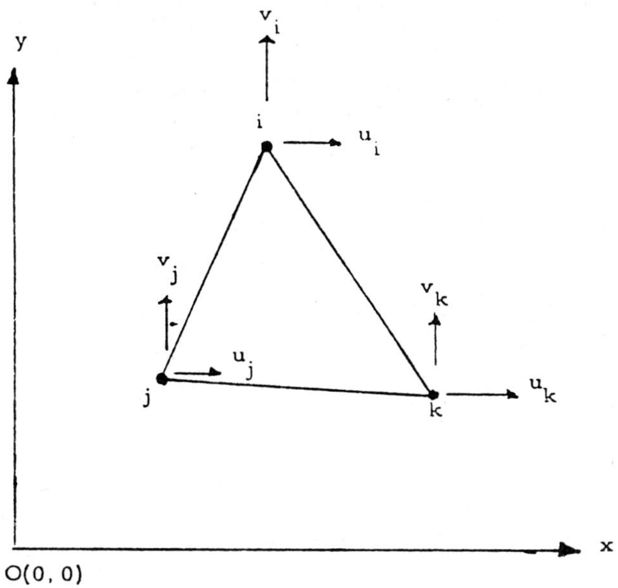

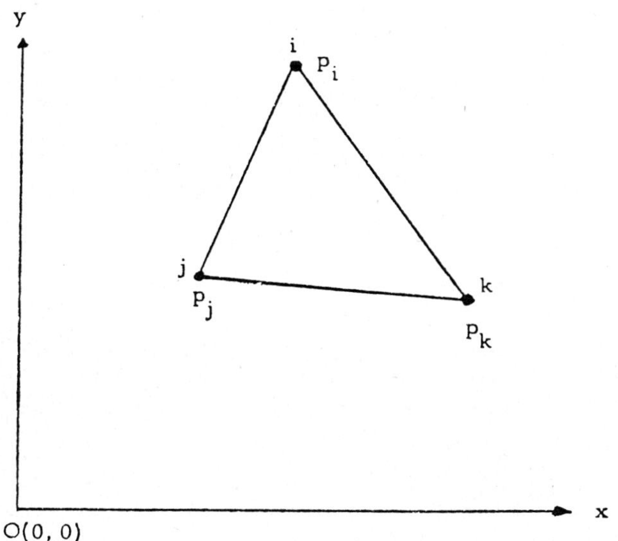

Fig. 2. (a) Triangular displacement element; (b) triangular pressure element.

simultaneous equations can be obtained by equating the values of
the displacement (or pressure) at the triangular nodes, i, j, k
(Figure 2). The coefficients α_i and β_i (γ_i) can be solved in
terms of the nodal displacements (or pressure) and nodal coordinates. The functions [N] so obtained have the form:

$$[N] = [IN_i, IN_j, IN_k] \tag{17}$$

I = 2 × 2 unit matrix

$$N_m = (a_m + b_m x + c_m y)/2 \qquad m = i, j, k \tag{18}$$

$$a_i = x_i y_k - x_k y_j \qquad b_i = y_j - y_k \qquad c_i = x_k - y_j \tag{19}$$

$$\Delta = \frac{1}{2} \begin{vmatrix} 1 & x_i & y_i \\ 1 & x_j & y_j \\ 1 & x_k & y_k \end{vmatrix} \tag{20}$$

with the other coefficients in (19) obtained by a cyclic permuta-
tion of subscripts in the order i, j, k.

DOUBLE-NODE INTERFACE TECHNIQUE

   For a two-dimensional finite element, there is one nodal degree of freedom for fluid elements in contrast to two nodal degrees of freedom for solid elements in which the displacements
are the basic unknowns (Figure 2). Consequently, in the finite
element discretization, boundary conditions are no longer automatically continuous at the fluid/solid interface, whereas those
of the fluid/fluid and solid/solid interfaces are. We thus need
to suppose a double-node fluid-solid interface (see Figure 3) in
which each nodal pair occupies the same position but each node has
its own individual characteristics, that is, the scalar nodal
pressure field on interface A and the vector displacement field
on interface B.

   For time integration, we use lumped mass and central finite-
difference scheme in order to avoid the large in-core storage
requirement

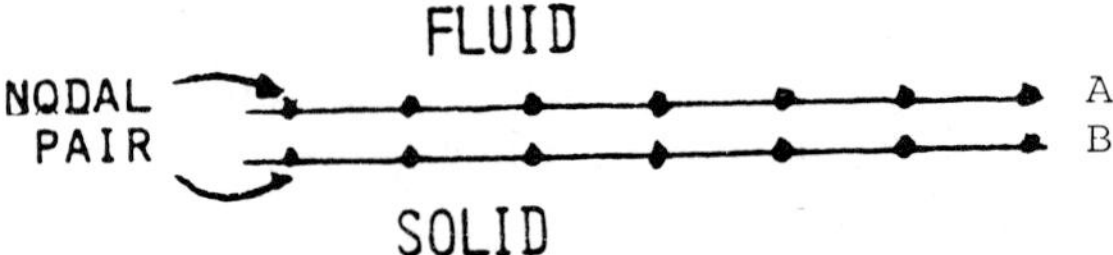

Fig. 3. Double-node interface.

COMPARISON OF THE EXPERIMENT WITH THE
FINITE ELEMENT RESULTS

1. Brief Description of Roever-Vining Experiments

In the classic experiments of Roever and Vining, the acoustic/
elastic coupled medium consists of a water layer overlying an elastic half-space. Two cases were investigated:

A. The case of water/pitch interface, where pitch is very
viscous. The shear wave velocity of pitch is v_s = 1,000 m/sec,
which is lower than the compressional wave velocity of water, v_w =
1,500 m/sec, while the compressional wave velocity of pitch is
v = 2.443 cm/sec greater than the wave velocity of water. The densities of water and pitch are ρ_w = 1.03 gm/cm^3 and ρ_p = 1.309 gm/
cm^3, respectively. Consequently, when the source-receiver distance is sufficiently large, critically refracted compressional
waves, direct waves, reflected waves, and Stoneley interface waves
are present.

B. The case of water/plaster-of-paris interface, where both
the shear and compressional wave velocities, as well as density
(v_s = 1,843 m/sec, v_p = 3.372 m/sec, ρ_{pp} = 1.9657 gm/cm^3) are
greater than the velocity and density of the water. In addition
to critically refracted compressional waves, direct waves, reflected waves, Stoneley interface waves, and critically refracted
shear waves are also present. In their experiments, a spiker
source is used and is located at a vertical distance of 0.5 cm
above the fluid/solid interface. The only observational point is
located at a horizontal distance of 10 cm away from the source and
a same vertical distance of 0.5 cm above the interface. The experimental data of the pressure response as a function of time as
observed for Case I is shown in Figure 6(a) and for Case II in
Figure 11(a).

2. Finite Element Results

A. Case of water/pitch interface (v_s < v_w < v_p). In this
case, the finite element size is 0.25 cm with a time step interval Δt = 0.2 μsec. The forcing function is simulated by a bell-
shaped time function as shown in Figure 5(b).

Figure 6(a) shows the experimental data of the pressure response at the observational point P (10 cm from the source and
0.5 cm from the interface). The first arrival is identified as
the critical refracted P-wave marked by t_p; the second arrival is
the direct wave marked by t_D; the third one is the Stoneley wave

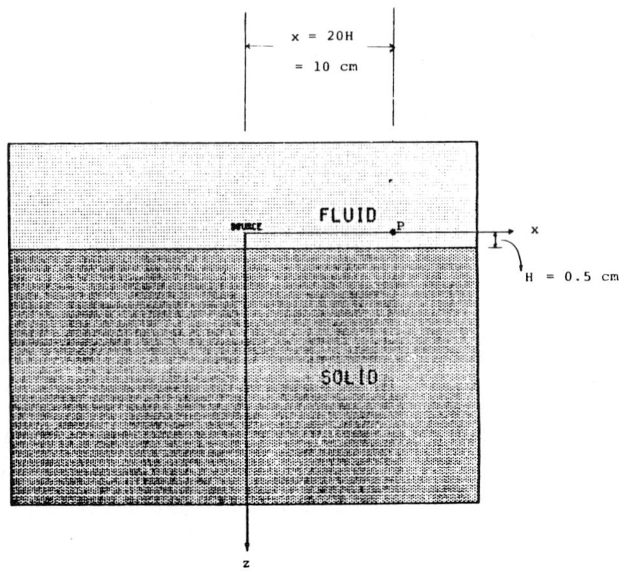

Fig. 4. Geometry of the demonstration problems. v_w = 1,500 m/
sec; ρ_w = 1.03 gm/cm^3. Model I: V_p = 2,443 m/sec; V_s = 1,000 m/
sec; ρ_p = 1.309 gm/cm^3. Model II: V_p = 3,372 m/sec; V_s = 1,843
m/sec; ρ_{pp} = 1.958 gm/cm^3.

marked by t_{St}. Figure 6(b) is the pressure response calculated
based on Strick's analytical solution. The match between the experimental data and the theoretically calculated is rather good,
except the high-frequency of the direct wave is absent in the
theoretically calculated, and the characteristics of the initial
pulse like direct wave which appeared in the experimental data
are rather different from these calculated.

Figure 6(c) is the pressure response of the finite element
modeling, using a 20 sec duration bell-shaped forcing function
as shown in Figure 5. It is to be compared with the experimental data as shown in 6(a) and the calculated results of Strick in
6(b). From the following travel-time formula,

$$t_p = \frac{x}{v_p} + \frac{\partial H - z}{v_w} \sqrt{1 - \frac{v_w}{v_p}^2}$$

the theoretically expected value of t_p is 46.2 μsec, that of t_D
is 66.7 μsec, and that with the Stoneley wave velocity for the
water/pitch case of 818 m/sec (Stonely, 1924; Strick and Ginzbark,

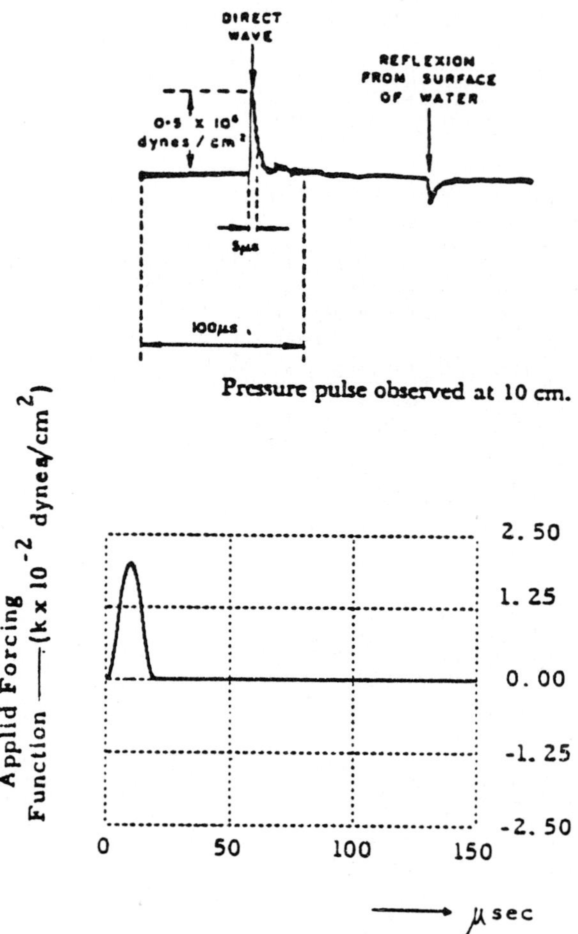

Fig. 5.   Comparison of the experimental and the finite element
source functions.

1956), t_{St} is 122.2 μsec.   The arrival time of t_p, t_D, and t_{St}
for the finite element modeling results are virtually duplicated.
The every portion of the arrival of the critical refracted $P(t_p)$
is magnified in Figure 7 to show the exact arrival of the pulse
at 46.2 μsec.   Because of the width of the bell-shaped source
function about 20 μsec is still too wide, in comparison with the
sharp rise of the experimental spiker source, which is about 7 sec
observed at 10 cm as shown in Figure 5.   Although the wave forms
of all the arrivals of the finite element solution do not exactly
depict the Roever-Vining experimental arrivals, as shown in Figure
6(a), and the analytical result in Figure 6(b), all the t_p, t_D,
and t_{st} arrivals can be clearly identified.   The high frequency

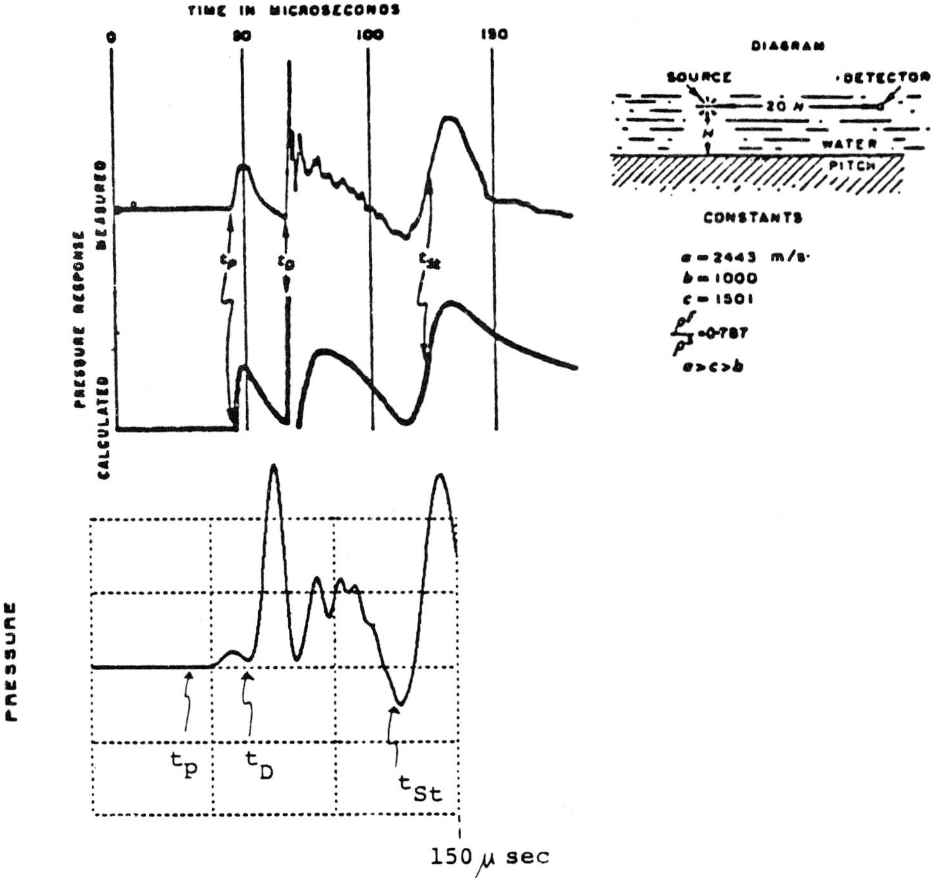

Fig. 6.   Case I.   Comparison of (a) experimental, (b) analytical, and (c) finite element solutions.

portion of the direct waves, which is absent in the theoretically calculated, is preserved in the finite element results.   Apparently, the waveform is very sensitive to the respective location from the fluid/solid interface.   Figure 8 shows the results of the effects of the vertically slight shift of the location of the detector at the receiver, on the wave forms of these arrivals. The closer the receiver is to the interface, the larger the magnitude of the Stoneley wave is.   Using the same bell-shaped forcing function, locations P1, P, and P2, are 0.75, 0.5, and 0.25 cm from the interface, respectively (see Figure 9).   Figure 10 shows the seismogram as received along a profile parallel to the interface, i.e., the profile is 0.5 cm above the fluid/solid interface. The horizontal distances between the source and the observational points are 7.5 to 10 cm (AP in Figure 9).

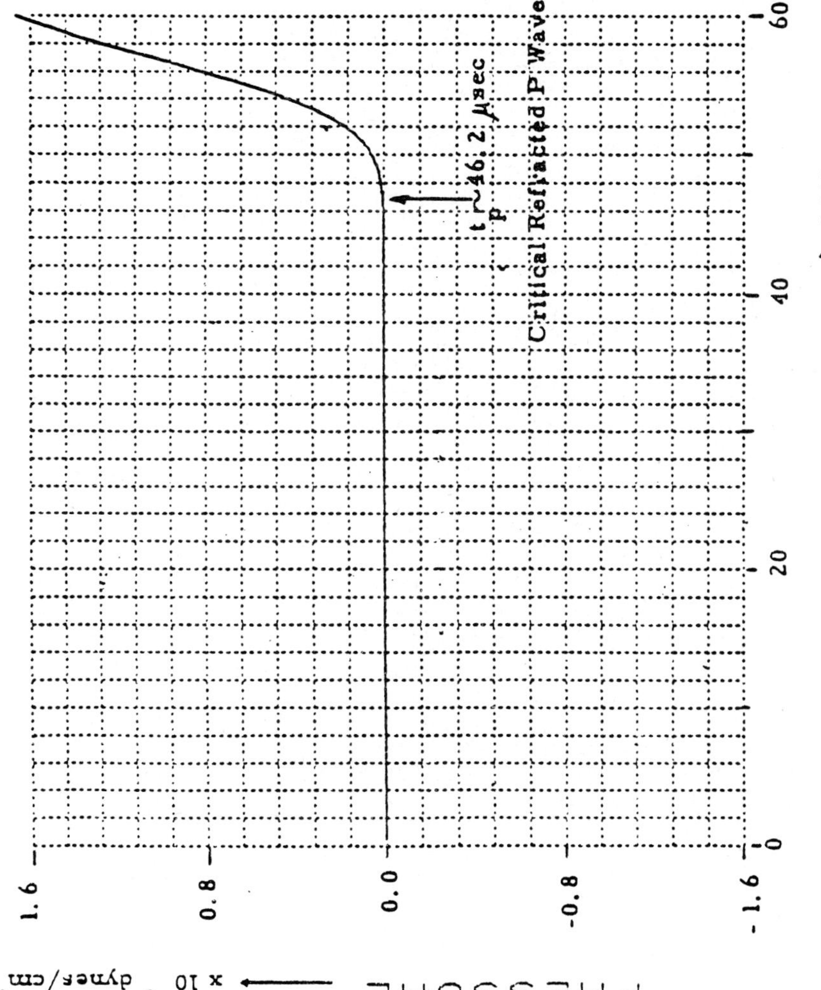

Fig. 7. Pressure field at observational point P with 10 times larger scale than that in Figure 2(c).

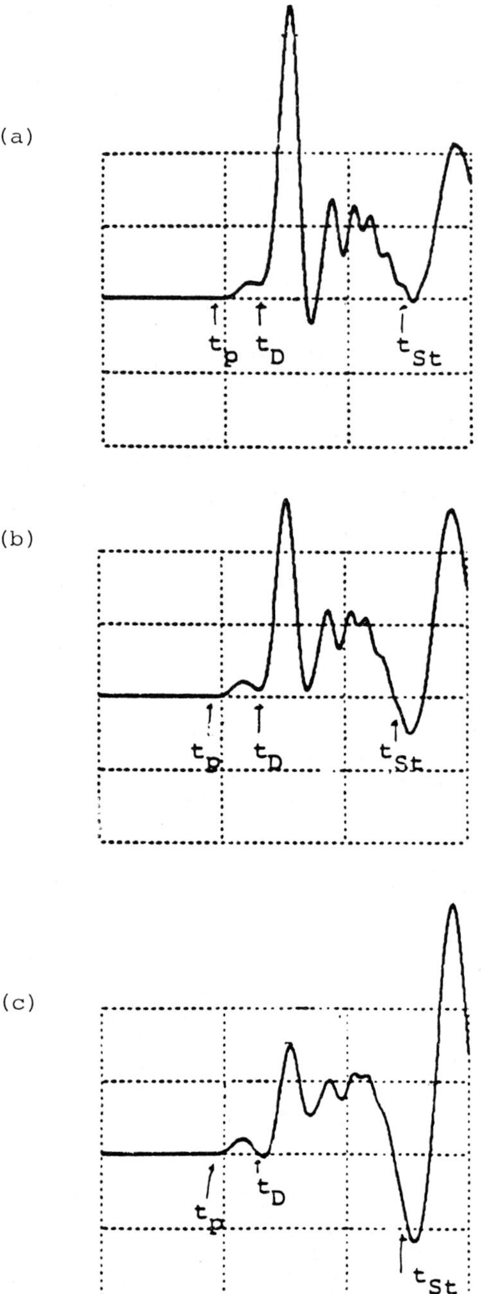

Fig. 8. Pressure field at three different locations: (a) 0.75 cm from the interface, (b) 0.5 cm from the interface; (c) 0.25 cm from the interface.

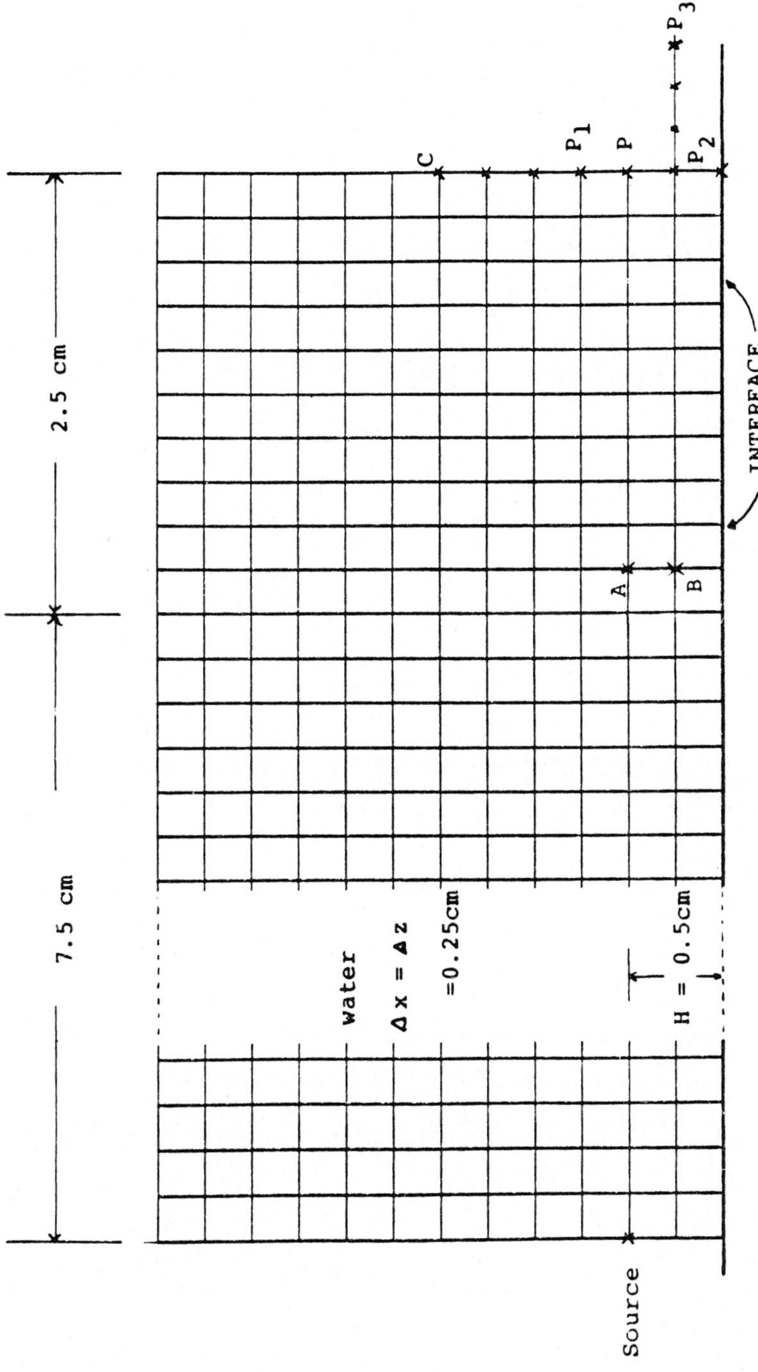

Fig. 9. Blow up of the partial fluid medium of Figure 4.

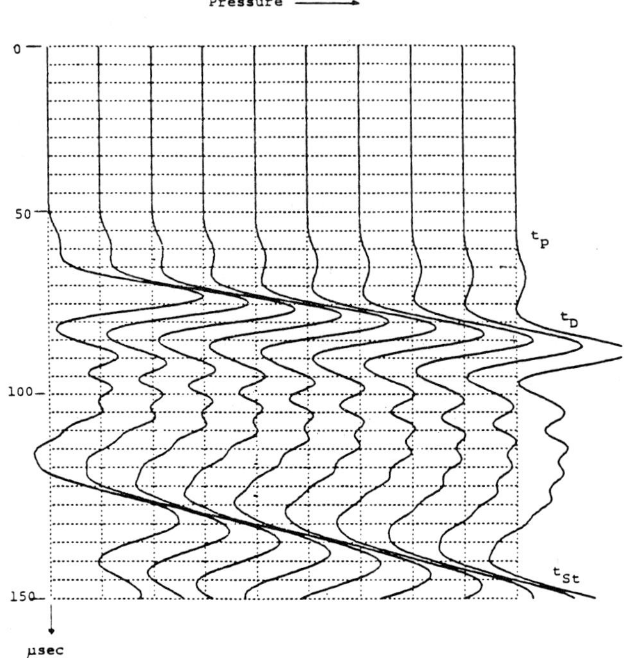

Fig. 10. Seismogram of the pressure field along AP as shown in
Figure 9.

In the seismogram shown in Figure 10, the waveforms of all the
critically refracted compressional waves, direct waves, and Stoneley waves are virtually identical, although the duration of the
arrivals are lengthened.

   B. Case of water/plaster-of-paris interface ($v_w < v_s < v_p$):
In this case, the element size is 0.25 cm and time step interval
Δt = 0.15 μsec. Again, we use the bell-shaped forcing function
with 60 time steps. In addition to the three types of waves for
the water/pitch case, now the responses should include the critically refracted shear waves. At the observational point P, the
theoretically calculated arrival times of t_p, t_s, t_D, and t_{St}
have the values of 35.71 μsec, 58.3 μsec, 66.6 μsec, and 74.3
μsec, respectively. The velocity of the Stoneley wave for this
case is 1.365 m/sec. In the finite element results, the responses
at point P are quite different from the Roever-Vining experiment
results. However, we discovered that the responses at point P2,
one node below point P (see Figure 9), is very close to the experimental results. Figures 11(a) and 11(b) are the comparison
of the Roever-Vining experiment and the finite element results.

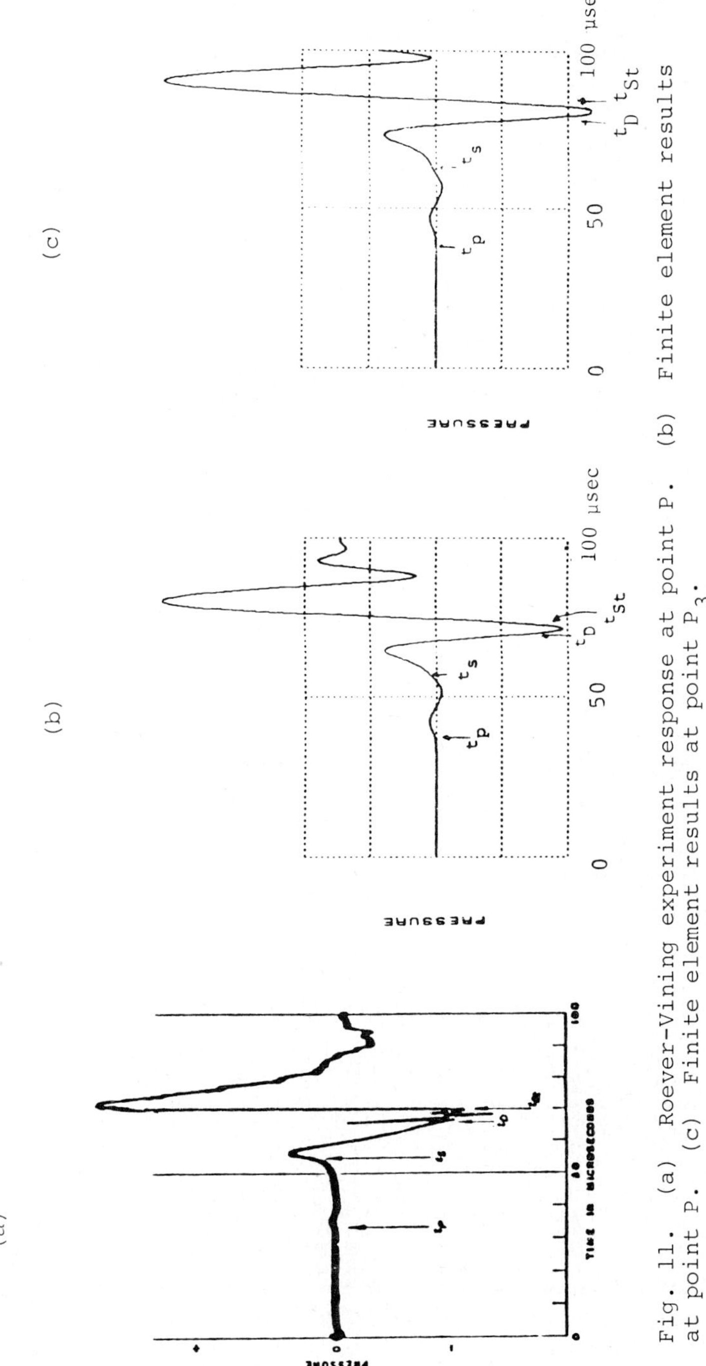

Fig. 11. (a) Roever-Vining experiment response at point P. (b) Finite element results at point P. (c) Finite element results at point P_3.

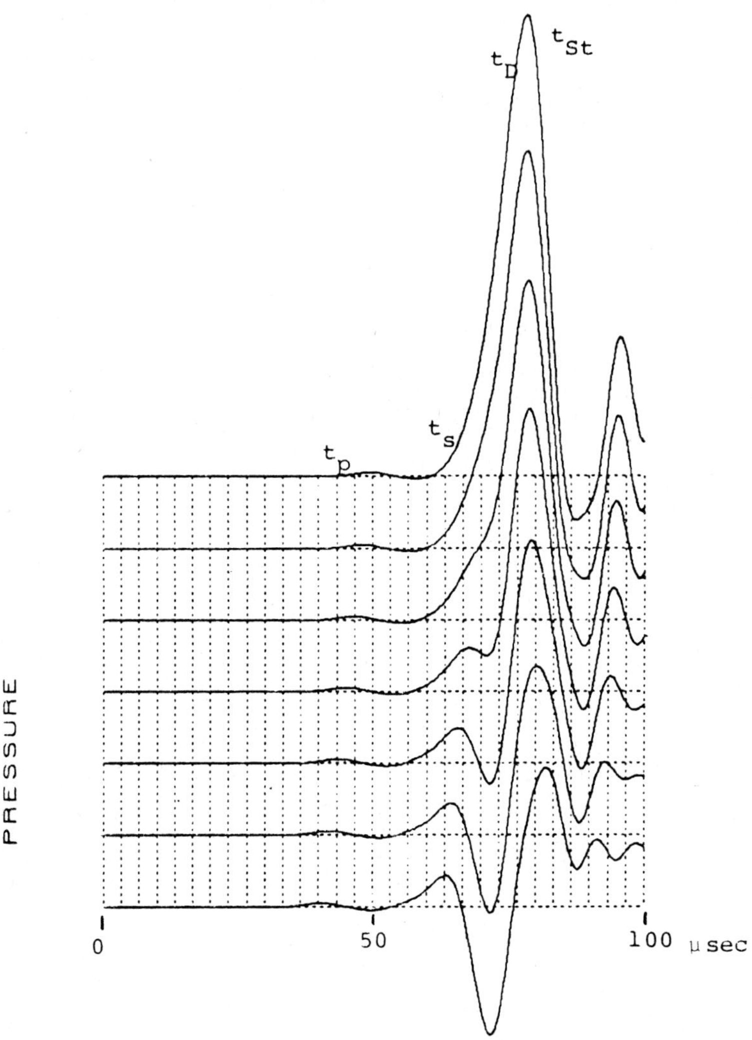

Fig. 12.   Seismogram of pressure field along the vertical plane CD.

In the finite element results, the drastic change in the waveform
of the pressure fields at different vertical locations are shown
in Figure 12, in which the seismograms are along the vertical
plane CD.   The closest similarity with the Roever-Vining experiment results in the waveforms is the response at point P3 in Figure 9 (0.75 cm from point P2).   Figure 11(c) is the response at
point P3 (0.75 cm from point P2).   Figures 13 and 14 are the seismograms along the horizontal planes AP and BP.   The critically refracted compressional and shear waves can be clearly identified.
The direct waves and Stoneley waves interfere with each other,

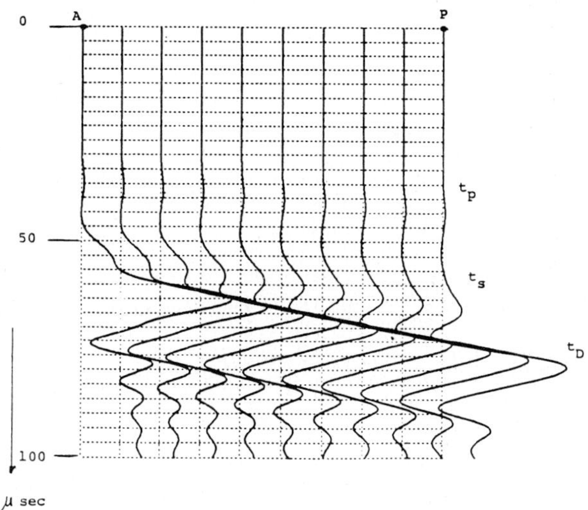

Fig. 13. Seismogram of the pressure field along the horizontal plane AP.

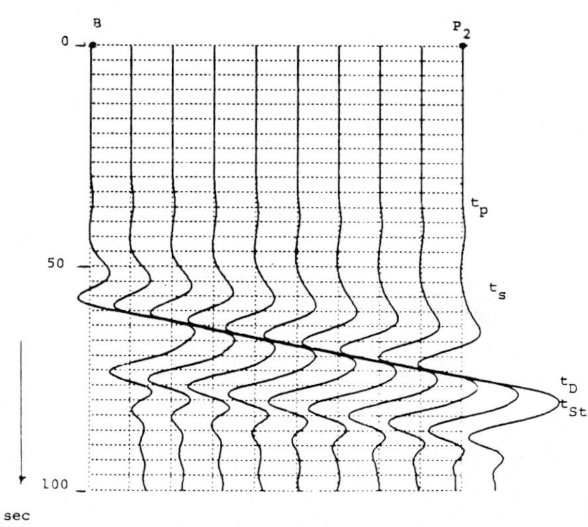

Fig. 14. Seismogram of the pressure field along the horizontal plane BP_2.

since the two velocities of the Stoneley wave and of the water
are virtually the same. Again, these horizontal seismograms show
very little change in waveform from trace to trace.

DISCUSSION AND CONCLUSIONS

We have studied two cases for the acoustic/elastic coupled
medium by the finite element method. The first case is difficult
because of the high contrast in the compressional and shear velocities for the pitch; the Poisson's ratio is almost 0.4 and it was
found that the stabilities in numerical calculations were very
difficult to achieve. In the second case, the Stoneley wave
velocity (v_{St} = 1,345 m/sec) is too close to the compressional
velocity of the water. Consequently, it is very difficult to
match the numerical results with the Roever-Vining experiment results, particularly when the exact source function is unknown.
In comparison of the numerical results and the Roever-Vining experiment results, the finite element results do not exactly depict
the Roever-Vining results at the particular location of the receiver point P. The reasons for this disagreement may be:

1. In the Roever-Vining experiments, the exact forcing function corresponding to the spiker source at the source location is
not known exactly. What we know is the pressure response in a
large tank of water at a distance of 10 cm away from the source
location. Besides, the width of the slowly rising bell-shaped
source function used in the finite element modeling is too wide
in comparison with the spiker responses as shown in Figure 5(a).

2. The effect of attenuation in the Roever-Vining experiment
results is not taken into account in the finite element modeling.

3. The location of the detector might be slightly lower
than the spiker source in their experiments. Unfortunately, the
experiment of Roever and Vining involved only one single receiver,
while in our finite element models there are only two grids between point P and the interface and the grid size is too coarse
for comparison (see Figure 9).

4. The concept of using double-layer nodes to simulate the
fluid/elastic interface does have a certain degree of approximation. Further research in this area is still needed.

From the finite element studies of the above two cases, we
found that the waveform of the pressure field in the fluid is
very sensitive to the fluid/solid interface; the closer the

receiver location is to the interface, the more the Stoneley wave dominates.

ACKNOWLEDGMENTS

This research is supported by Project MIDAS. The authors wish to express their appreciation to Dr. Chung Gong for many fruitful discussions and Dr. Walter Brown for his kind advice in developing software.

REFERENCES

1. Stoneley, R., Elastic waves at the surface of separation of two solids, PROCEEDINGS OF THE ROYAL SOCIETY OF AMERICA, 106 (1924), 416-428.

2. Strick, E., and A. S. Ginzbarg, Stoneley wave velocities for a fluid-solid interface, BULLETIN OF THE SEISMOLOGICAL SOCIETY OF AMERICA, 46 (1956), 281-292.

3. Strick, E., W. L. Roever, and T. F. Vining, Propagation of elastic wave motion from an impulsive source along a fluid/solid interface, I. Experimental pressure response; II. Theoretical pressure responses; III. The pseudo-Rayleigh wave, PROCEEDINGS OF THE ROYAL SOCIETY OF AMERICA, 251 (1959), 445-523.

4. Zienkiewicz, O. C., THE FINITE ELEMENT METHOD, McGraw Hill, third edition, 1977.

COMPUTATIONAL ACOUSTICS: Wave Propagation
D. Lee, R.L. Sternberg, M.H. Schultz (Editors)
Elsevier Science Publishers B.V. (North-Holland)
© IMACS, 1988

ANALYSIS OF FLUID EQUATIONS BY GROUP METHODS

W. F. Ames[1] and M. C. Nucci[2,3]

School of Mathematics
Georgia Institute of Technology
Atlanta, Georgia

ABSTRACT

Using the machinery of Lie group analysis several equations
arising in fluid mechanics are studied. In particular, the transonic flow equation, the Navier-Stokes equations, the two-dimensional KdV equation, the pellet-fusion equations, and some wave equations are discussed. In some cases the particular group includes
arbitrary functions of time which permit the transformation of time
dependent equations into the corresponding time independent ones.
Infinitely many time dependent solutions are associated with each
steady solution. Some solutions are constructed and arbitrary
element determinations illustrated.

1. INTRODUCTION

Mathematical models of many real world phenomena are formulated as differential equations. Sophus Lie's (1842-1899) original
aim was to create a theory of integrating ordinary differential
equations similar to the Abelian theory for solving algebraic equations. Lie's research, and its generalization to partial differential equations, is a systematic determination and investigation of
all those continuous transformation groups (called Lie groups) admitted by the system of differential equations under investigation.

One of the main problems of the group analysis is to study
the action of the group, admitted by the given system, on the set
of solutions to the system. The action of the admitted group introduces into the set of solutions an algebraic structure which
may be used to achieve several goals. These include a description

---

[1] Research supported by U.S. Army Grant DAAG-29-84-K-0083.

[2] Permanent address: Dipartimento di Matematica, Universitá di
Perugia, 06100 Perugia, Italy.

[3] Research supported by a NATO-CNR fellowship.

of the general properties of all members of the family of solutions,
the generation of solutions on the basis of known ones, transformation of the system into an equivalent system, as an aid in computation (conversion of boundary value to initial value problems) and
in (nonlinear) eigenvalue computation.

Of course, the admitted group is not usually given with the
system. Determination of the full group, while purely technical,
can be a lengthy complicated task—but an essential one! Another
interesting and practically important problem consists in the group
classifications of differential equations, an idea that is explained
in the sequel. The differential equations of science and engineering, in many cases, contain parameters or functions that are determined experimentally and are not strictly fixed. These are called
arbitrary elements. The group approach enables the determination
of the form of the arbitrary element such that the differential
equation admits a group having definite features or, in general,
the full group.

In this paper we illustrate the full group analysis for the
transonic flow equation including the construction of the full
group, the factoring out of the time dependence and the generation
of an infinite number of time dependent solutions from each steady
state solution.

Following the detailed calculation the paper lists the full
groups for the three-dimensional Navier-Stokes equations, the two-
dimensional KdV equation, the equation of motion of a viscoplastic
medium and a hydrodynamic model of pellet fusion. Various aspects
of group analysis will be discussed using the examples to illustrate the methods.

2. FULL LIE GROUP AND LIE ALGEBRA FOR TRANSONIC FLOW

The Lin-Tsien equation [1]

$$2\phi_{tx} + \phi_x\phi_{xx} - \phi_{yy} = 0 \tag{2.1}$$

where ϕ is a velocity potential, is often used to study dynamic
transonic flow in two space dimensions (x,y). The full group admitted by equation (2.1) is derived and represented by the generators of the Lie algebra, given in Table 1, whose commutator table
is given as Table 2. The development of the full Lie group and
algebra is quite lengthy. Since the detailed description of the
procedure is given in Ames [2], Ovsiannikov [3] and Ibragimov [4],
for example, only an outline of the development is given herein.

TABLE 1   Generators (X_1 to X_6)

$$P_0: \quad X_1 = \frac{4}{3} x \frac{\partial}{\partial x} + y \frac{\partial}{\partial y} + \frac{2}{3} t \frac{\partial}{\partial t} + 2\phi \frac{\partial}{\partial \phi}$$

$$\gamma_0: \quad X_2 = \frac{\partial}{\partial t}$$

$$q(t): \quad X_3 = q'y \frac{\partial}{\partial x} + q \frac{\partial}{\partial y} + \left(2q''xy + \frac{2}{3} q''' y^3\right) \frac{\partial}{\partial \phi}$$

$$w(t): \quad X_4 = w \frac{\partial}{\partial x} + \left(2w'x + 2w''y^2\right) \frac{\partial}{\partial \phi}$$

$$r(t): \quad X_5 = ry \frac{\partial}{\partial \phi}$$

$$s(t): \quad X_6 = s \frac{\partial}{\partial \phi}$$

The commutator ($[X_i, X_j] = X_i X_j - X_j X_i$) table of X_1 to X_6 is given as Table 2.

TABLE 2   Commutator Table

X_i \ X_j	1	2	3	4	5	6
1	0	$-\frac{2}{3} X_2$	X_3 $q \to \frac{2}{3} tq' - q$	X_4 $w \to \frac{2}{3} tw' - \frac{4}{3} w$	X_5 $r \to \frac{2}{3} tr' - r$	X_6 $s \to \frac{2}{3} ts' - 2s$
2		0	X_3 $q \to q'$	X_4 $w \to w'$	X_5 $r \to r'$	X_6 $s \to s'$
3			0	X_5 $r \to 4w''q + 2w'q' - wq''$	X_6 $s \to qr$	0
4				0	0	0
5					0	0
6						0

The most general generator X for the non-extended (original variables only) Lie group

$$X = \alpha(x,y,t,\phi)\,\frac{\partial}{\partial x} + \beta(x,y,t,\phi)\,\frac{\partial}{\partial y} + \gamma(x,y,t,\phi)\,\frac{\partial}{\partial t}$$

$$+ \,\delta(x,y,t,\phi)\,\frac{\partial}{\partial\phi} \tag{2.2}$$

is twice extended (to include first and second derivatives) for the second order equation (2.1) to

$$X_1 = X + \varepsilon\,\frac{\partial}{\partial a} + \omega\,\frac{\partial}{\partial b} + \theta\,\frac{\partial}{\partial c} \tag{2.3}$$

$$X_2 = X_1 + \lambda\,\frac{\partial}{\partial d} + \mu\,\frac{\partial}{\partial e} + \nu\,\frac{\partial}{\partial f} + \xi\,\frac{\partial}{\partial g} + \rho\,\frac{\partial}{\partial h} + \tau\,\frac{\partial}{\partial k} \tag{2.4}$$

where $a = \phi_x$, $b = \phi_y$, $c = \phi_t$, $d = \phi_{xx}$, $e = \phi_{xt}$, $f = \phi_{tt}$, $g = \phi_{ty}$, $h = \phi_{yy}$, and $k = \phi_{xy}$. The coefficients in (2.3) and (2.4) can be derived by implicit differentiation and, therefore, depend upon the variables and coefficients of the lower order extended generators.

To determine the full group of (2.1) the set of generators leaving $F = 2e + ad - h = 0$ invariant must be found. The invariance condition $X_2 F \equiv 0$ becomes

$$\begin{aligned}
0 \equiv\ & [2\delta_{xt} - \delta_{yy}] + a[2\delta_{\phi t} - 2\alpha_{xt} + \delta_{xx} + \alpha_{yy}] \\
& + b[-2\beta_{xt} - 2\delta_{\phi y} + \beta_{yy}] + c[-2\gamma_{xt} + 2\delta_{x\phi} + \gamma_{yy}] \\
& + d[-2\alpha_t + \delta_x] + e[2\delta_\phi - 2\alpha_x - 2\gamma_t] + f[-2\gamma_x] \\
& + g[-2\beta_x + 2\gamma_y] + h[2\beta_y - \delta_\phi] + k[-2\beta_t + 2\alpha_y] \\
& + a^2[-2\alpha_{\phi t} + 2\delta_{\phi x} - \alpha_{xx}] + ab[-2\beta_{\phi t} - \beta_{xx} + 2\alpha_{\phi y}] \\
& + ac[-2\gamma_{\phi t} + 2\delta_{\phi\phi} - 2\alpha_{\phi x} - \gamma_{xx}] + ad[2\delta_\phi - 3\alpha_x] \\
& + ae[-4\alpha_\phi - 2\gamma_x] + af[-2\gamma_\phi] + ag[-2\beta_\phi] + ah[\alpha_\phi] \\
& + ak[-2\beta_x] + b^2[2\beta_{\phi y} - \delta_{\phi\phi}] + bc[-2\beta_{x\phi} + 2\gamma_{\phi y}] \\
& + bd[-\beta_x] + be[-2\beta_\phi] + bg[2\gamma_\phi] + bh[3\beta_\phi] \\
& + bk[2\alpha_\phi] + c^2[-2\gamma_{x\phi}] + cd[-2\alpha_\phi - \gamma_x] + ce[-4\gamma_\phi] \\
& + ch[\gamma_\phi] + ck[-2\beta_\phi] + a^3[-2\alpha_{\phi x} + \delta_{\phi\phi}] + a^2 b[-2\beta_{\phi x}] \\
& + a^2 c[-2\alpha_{\phi\phi} - 2\gamma_{\phi x}] + a^2 d[-4\alpha_\phi] + a^2 e[-2\gamma_\phi] + a^2 k[-2\beta_\phi] \\
& + ab^2[\alpha_{\phi\phi}] + abc[-2\beta_{\phi\phi}] + abd[-2\beta_\phi] + acd[-2\gamma_\phi] \\
& + ac^2[-2\gamma_{\phi\phi}] + b^3[\beta_{\phi\phi}] + b^2 c[\gamma_{\phi\phi}] + a^4[-\alpha_{\phi\phi}] \\
& + a^3 b[-\beta_{\phi\phi}] + a^3 c[-\gamma_{\phi\phi}] \tag{2.5}
\end{aligned}$$

Since all the terms a, b, c, ... can be independently varied, this equation can only vanish identically if the coefficients of all the independent terms vanish. This imposes restrictions (the determining equations) in the form of an overdetermined system of partial differential equations for α, β, γ, and δ. Making use of all the partial differential equations with only one term, there is obtained

$$\alpha = \alpha(x,y,t) \qquad \gamma = \gamma(y,t)$$
$$\beta = \beta(y,t) \qquad \delta = \delta(x,y,t,\phi) \tag{2.6}$$

The remaining restrictions are

$$
\begin{aligned}
&2\delta_{xt} - \delta_{yy} = 0 & \qquad & 2\delta_{\phi x} - \alpha_{xx} = 0 \\
&2\delta_{\phi t} - 2\alpha_{xt} + \delta_{xx} + \alpha_{yy} = 0 & & \delta_{\phi\phi} = 0 \\
&2\delta_{\phi y} - \beta_{yy} = 0 & & 2\delta_{\phi} - 3\alpha_{x} = 0 \\
&2\delta_{\phi x} + \gamma_{yy} = 0 & & \beta_{t} - \alpha_{y} = 0 \\
&2\alpha_{t} - \delta_{x} = 0 & & 2\beta_{y} - \delta_{\phi} = 0 \\
&\delta_{\phi} - \alpha_{x} - \gamma_{t} = 0 & & \gamma_{y} = 0
\end{aligned}
\tag{2.7}
$$

Through further substitutions and reductions, the solution of (2.7) provides the most general generator

$$
\begin{aligned}
X = {} & [4p_0 x/3 + q'(t)y + w(t)]\,\frac{\partial}{\partial x} + [p_0 y + q(t)]\,\frac{\partial}{\partial y} \\
& + [2p_0 t/3 + \gamma_0]\,\frac{\partial}{\partial t} + [2p_0\phi + \tfrac{2}{3}y^3 q''' + 2y^2 w'' \\
& + 2xyq'' + 2xw' + yr(t) + s(t)]\,\frac{\partial}{\partial \phi}
\end{aligned}
\tag{2.8}
$$

for the full group of equation (2.1). Equation (2.8) contains two arbitrary constants p_0 and γ_0 and four arbitrary functions of time, $q(t)$, $w(t)$, $r(t)$, and $s(t)$. Consequently, the Lie algebra for equation (2.1) is said to be infinite dimensional. These arbitrary functions will permit an infinite number of time dependent solutions to be generated from each steady state solution in the next section.

The basic set of generators, obtained by setting all but one of the parameters or functions equal to zero, is given in Table 1.

3. STEADY STATE DECOMPOSITION

Since the generator X_2 of pure time translation exists, it is possible to "factor out" the time dependence and find a subgroup which is the full group for the steady-state transonic flow equation

$$\phi_x \phi_{xx} - \phi_{yy} = 0 \tag{3.1}$$

Then using group analysis, or any other method, we can construct solutions of equation (3.1). From each of these, time dependent solutions can be generated from the factorization. A corresponding result holds for the Navier-Stokes equations, as we shall see, so this result was anticipated.

To obtain the factorization the invariants of the group that are constant in time must be found. This is accomplished by integrating the equation for the invariant I which contains time translation (X_2) and the arbitrary functions of time $(X_3, X_4, X_5,$ and $X_6)$,

$$(X_2 + X_3 + X_4 + X_5 + X_6)I = 0 \tag{3.2}$$

Using the method of characteristics for this first order equation there results

$$x - \tilde{x} = qy - \int (q^2 + w)\, dt = qy - \xi \tag{3.3}$$

$$y - \tilde{y} = \int q\, dt = Q \tag{3.4}$$

$$\begin{aligned}
\phi - \tilde{\phi} = {}& 2q'\tilde{x}\tilde{y} + \frac{2}{3} q''\tilde{y}^3 + 2w\tilde{x} \\
& + \int \{2q'' [q\tilde{y}^2 + 2qQ\tilde{y} - \xi\tilde{y} + qQ^2 - \xi Q] \\
& + \frac{2}{3} q''' [3\tilde{y}^2 Q + 3\tilde{y}Q^2 + Q^3] + 2w'[\tilde{y}q + qQ - \xi] \\
& + 2w'' [2\tilde{y}Q + Q^2] + r\tilde{y} + s\}\, dt \tag{3.5}
\end{aligned}$$

If $\phi(x,y,t)$ satisfies the time dependent equation (2.1), then it follows that "the constants of integration" are the required invariants such that $\tilde{\phi}(\tilde{x},\tilde{y})$ satisfies the steady equation (3.1). Thus, given any solution $\tilde{\phi} = \tilde{\phi}(\tilde{x},\tilde{y})$ of equation (3.1), it follows that time dependent solutions [with four arbitrary functions q, w, r, and s as shown in (3.3)-(3.5)] are constructable by substitution into equations (3.3)-(3.5). Thus,

$$\phi(x,y,t) = \tilde{\phi}(x - q(t)y - \xi, y - Q) \tag{3.6}$$
$$+ \text{ right-hand side of (3.5)}$$

Clearly, since our transformation equations involve four arbitrary functions of time, an infinite number of time dependent solutions are generated from each steady state solution. In the next section some steady solutions will be generated.

4. INVARIANT SOLUTIONS OF THE STEADY STATE EQUATION (3.1)

Since equation (3.1) is an equation in only two variables, and the full group of (3.1) is known, it is possible to generate solutions by group analysis with relative ease. The procedure involves selecting a specific subgroup and finding invariants of that group in terms of the dependent and independent variables. According to a fundamental theorem in group analysis (see Ovsiannikov [3]) the introduction of these invariants into the original equation accomplishes a reduction in the number of independent variables. In the present case, (3.1) reduces to an ordinary differential equation which must be integrated. Several examples will illustrate the procedure.

A. Subgroup of Dilatations and Translations

The subgroup of dilatations and translations for the steady state equation is formed by setting $p_0 = 1$, $\gamma_0 = 0$, $q(t) = \alpha$, $w(t) = \beta$, $r(t) = 0$, and $s(t) = \gamma$, where α, β, and γ are <u>constants</u>, not the functions of equation (2.6). With these, the equation for the invariant I becomes

$$\left(\frac{4}{3}\,\tilde{x} + \beta\right) \frac{\partial I}{\partial \tilde{x}} + (\tilde{y} + \alpha)\,\frac{\partial I}{\partial \tilde{y}} + (2\phi + \gamma)\,\frac{\partial I}{\partial \tilde{\phi}} = 0$$

The integral surfaces of this linear equation are the independent invariant

$$\eta = \frac{\left(\frac{4}{3}\,\tilde{x} + \beta\right)^{3/4}}{\tilde{y} + \alpha} \tag{4.1}$$

and the dependent invariant

$$f(\eta) = \frac{\tilde{\phi} + \gamma}{(\tilde{y} + \alpha)^2} \tag{4.2}$$

or $\tilde{\phi} = (\tilde{y} + \alpha)^2 f(\eta) - \gamma$. Substituting this into equation (3.1), there results the ordinary differential equation for $f(\eta)$,

$$\eta f'f'' - \frac{1}{3}\,(f')^2 - \eta^4 f'' + 2\eta^3 f' - 2\eta^2 f = 0 \tag{4.3}$$

One solution to this equation is $f(\eta) = 9\eta^4/64$. Thus a solution to the steady state equation (3.1) is

$$\tilde{\phi}(\tilde{x},\tilde{y}) = \frac{9}{64}\,\frac{\left(\frac{4}{3}\,\tilde{x} + \beta\right)^3}{(\tilde{y} + \alpha)^2} - \gamma \tag{4.4}$$

An infinite number of time dependent solutions can now be generated by inserting $\tilde{\phi}(\tilde{x},\tilde{y})$ into the transformation (3.3), (3.4), and (3.5) and using arbitrary q, r, s, and w as functions of time. Our previous specification of the arbitrary functions was to get a particular subgroup for the time independent equation.

B. Semi-Translation Subgroup

The semi-translation subgroup is formed by setting $p_0 = 0$, $\gamma_0 = 0$, $q = \alpha$, $w = \beta$, $r = \gamma$, and $s = 0$ with α, β, γ constants. With these, the equation for the invariant becomes

$$\alpha \frac{\partial I}{\partial \tilde{y}} + \beta \frac{\partial I}{\partial \tilde{x}} + \gamma \tilde{y} \frac{\partial I}{\partial \tilde{\phi}} = 0$$

The independent invariant is

$$\eta = \tilde{y} - \frac{\alpha}{\beta} \tilde{x}$$

and the dependent invariant is

$$f(\eta) = \tilde{\phi}(\tilde{x},\tilde{y}) - \frac{\gamma}{2\alpha} \tilde{y}^2$$

The function f satisfies the nonlinear differential equation

$$\frac{1}{2}\left(\frac{\alpha}{\beta}\right)^3 [(f')^2]' + f'' + \frac{\gamma}{\alpha} = 0$$

with two parameters. Two solutions of this equation are

$$f(\eta) = \frac{1}{2\varepsilon}\left[-\eta \pm \frac{2}{3a}(b - a\eta)^{3/2}\right] + D \tag{4.5}$$

where

$$\varepsilon = \frac{1}{2}\left(\frac{\alpha}{\beta}\right)^3 \qquad a = 4\varepsilon \frac{\gamma}{\alpha} \qquad b = 1 - 4\varepsilon c$$

and c and D are arbitrary constants. From both of these solutions an infinity of time dependent solutions can be generated using the transformations (3.3), (3.4), and (3.5).

5. THE NAVIER-STOKES EQUATIONS

The Navier-Stokes equations in Cartesian coordinates

$$u_t + uu_x + vu_y + wu_z = -p_x + \mu\nabla^2 u$$
$$v_t + uv_x + vv_y + wv_z = -p_y + \mu\nabla^2 v$$
$$w_t + uw_x + vw_y + ww_z = -p_z + \mu\nabla^2 w \tag{5.1}$$
$$u_x + v_y + w_z = 0$$

where u, v, and w are the three velocity components, p is pressure

and μ is (constant) viscosity. The derivation of the full group
for the two-dimensional equations was first carried out by Puhnachev [5]. In three dimensions Bitev [6] was apparently first,
followed by the simultaneous work of Lloyd [7] and Boisvert [8,9].
Boisvert realized that there were four arbitrary functions of time
(t) in the group. He demonstrated how these could be exploited in
analysis of the system.

The full group, containing five arbitrary parameters α, β, γ,
λ, and σ and four arbitrary functions $f(t)$, $g(t)$, $h(t)$, and $j(t)$
(in the notation of Table 1) is as follows:

TABLE 3   Generators

$$\alpha: \quad X_1 = \frac{\partial}{\partial t}$$

$$\beta: \quad X_2 = 2t\frac{\partial}{\partial t} + x\frac{\partial}{\partial x} + y\frac{\partial}{\partial y} + z\frac{\partial}{\partial z} - u\frac{\partial}{\partial u}$$
$$- v\frac{\partial}{\partial v} - w\frac{\partial}{\partial w} - 2p\frac{\partial}{\partial p}$$

$$\gamma: \quad X_3 = -y\frac{\partial}{\partial x} + x\frac{\partial}{\partial y} - v\frac{\partial}{\partial u} + u\frac{\partial}{\partial v}$$

$$\lambda: \quad X_4 = -z\frac{\partial}{\partial x} + x\frac{\partial}{\partial z} - w\frac{\partial}{\partial u} + u\frac{\partial}{\partial w}$$

$$\sigma: \quad X_5 = -z\frac{\partial}{\partial y} + y\frac{\partial}{\partial z} - w\frac{\partial}{\partial v} + v\frac{\partial}{\partial w} \tag{5.2}$$

$$X_6(f) = f(t)\frac{\partial}{\partial x} + f'(t)\frac{\partial}{\partial u} - xf''(t)\frac{\partial}{\partial p}$$

$$X_7(g) = g(t)\frac{\partial}{\partial y} + g'(t)\frac{\partial}{\partial v} - yg''(t)\frac{\partial}{\partial p}$$

$$X_8(h) = h(t)\frac{\partial}{\partial z} + h'(t)\frac{\partial}{\partial w} - zh''(t)\frac{\partial}{\partial p}$$

$$X_9(j) = j(t)\frac{\partial}{\partial p}$$

The first five operators generate a finite dimensional Lie algebra
L_5 which is a five-dimensional subalgebra of the infinite-dimen-
sional algebra generated by all nine operators.

To save space we shall indicate several applications of the
subgroups of (5.2) in two space dimensions (x,y) and time t. Eliminating all terms in z and w leaves the operators X_1, X_2, X_3, X_6,
X_7, and X_9 for the two-dimensional system. With $\alpha = 1$ and $\beta = \gamma =$
0 (deleting X_2 and X_3), the Lie operator Q becomes

$$Q = \frac{\partial}{\partial t} + f(t)\,\frac{\partial}{\partial x} + g(t)\,\frac{\partial}{\partial y} + f'(t)\,\frac{\partial}{\partial u} + g'(t)\,\frac{\partial}{\partial v}$$

$$+ \,[j(t) - xf''(t) - yg''(t)]\,\frac{\partial}{\partial p} \qquad (5.3)$$

The first order equation QI = 0 has the characteristics (invariants)

$$\bar{x} = x - F(t) \qquad \bar{y} = y - G(t) \qquad (5.4)$$

where $F' = f$, $G' = g$, while

$$u = \bar{u}(\bar{x},\bar{y}) + f(t)$$

$$v = \bar{v}(\bar{x},\bar{y}) + g(t) \qquad (5.5)$$

$$p = \bar{p}(\bar{x},\bar{y}) - xf'(t) - yg'(t) + k(t)$$

where

$$k(t) = \frac{1}{2}\,f^2 + \frac{1}{2}\,g^2 + \int j(t)\,dt \qquad (5.6)$$

The functions, $\bar{u}$, $\bar{v}$, and $\bar{p}$ satisfy the steady Navier-Stokes equations

$$\bar{u}\bar{u}_{\bar{x}} + \bar{v}\bar{u}_{\bar{y}} = -\bar{p}_{\bar{x}} + \mu[\bar{u}_{\bar{x}\bar{x}} + \bar{u}_{\bar{y}\bar{y}}]$$

$$\bar{u}\bar{v}_{\bar{x}} + \bar{v}\bar{v}_{\bar{y}} = -\bar{p}_{\bar{y}} + \mu[\bar{v}_{\bar{x}\bar{x}} + \bar{v}_{\bar{y}\bar{y}}] \qquad (5.7)$$

$$\bar{u}_{\bar{x}} + \bar{v}_{\bar{y}} = 0$$

As in Section 3 (for the transonic equation), here it is also true that the generator of pure time translation, X_1, exists. Hence it is possible to "factor out" the time dependence. This is a useful result since any solution of the steady state equations (5.7) can be transformed, by means of equations (5.4) and (5.5), into a time dependent solution containing three arbitrary functions of time. A similar result holds in three dimensions.

We now illustrate an application of group analysis for the determination of solutions. Many additional examples are available in the literature (for the Navier-Stokes equations, see Boisvert [8,9])

For convenience we will drop the bars in equations (5.7) and use the stream function form of (5.7), with $v = \psi_x$ and $u = -\psi_y$,

$$\psi_x\psi_{yyy} + \psi_x\psi_{xxy} - \psi_y\psi_{xyy} - \psi_y\psi_{xxx}$$

$$- \,\mu(\psi_{xxxx} + 2\psi_{xxyy} + \psi_{yyyy}) = 0 \qquad (5.8)$$

The rotation group is obtained from (5.2) by setting α and β equal to zero and including only X_3 (in stream function form). The invariants are $\eta = x^2 + y^2$, $\psi = f(\eta)$, and when these are substituted into (5.8) one obtains the <u>linear equation</u>

$$2f'' + 4 f''' + \eta^2 f'''' = 0 \tag{5.9}$$

which is independent of μ! The general solution (first obtained by Hamel [10] using ad hoc methods) is

$$f(\eta) = -c_1 \, \ell n \, \eta + c_2 \eta \, \ell n \, \eta + c_3 \eta + c_4 \tag{5.10}$$

where the c_i are arbitrary constants. The associated time-dependent solutions are

$$u = f(t) + (y - G(t)) [c_1^{-1} - c_5 \, \ell n \, \eta - c_6]$$

$$v = g(t) + (x - f(t)) [-c_1 \eta^{-1} + c_5 \, \ell n \, \eta + c_6]$$

$$p = -xf'(t) - yg'(t) + k(t) - \frac{1}{2} c_1^2 \eta^{-1}$$

$$+ (c_5^2 - c_5 c_6 + \frac{1}{2} c_6^2) \eta - c_1 c_6 \, \ell n \, \eta + (c_5 c_6 - c_5^2) \eta \, \ell n \, \eta$$

$$- \frac{1}{2} c_1 c_5 (\ell n \, \eta)^2 + \frac{1}{2} c_5^2 \eta (\ell n \, \eta)^2 - 4\mu c_5 \, \tan^{-1}\left[\frac{x - F(t)}{y - G(t)}\right]$$

where $\eta = [x - F(t)]^2 + [y - G(t)]^2$.

6. TWO-DIMENSIONAL K-dV EQUATION

One two-dimensional Korteweg-de Vries equation describing certain waves (see Rogers and Shadwick [11]) is

$$u_{xxxx} = -u_{xt} - \alpha u_{yy} - 6u_x^2 - 6uu_{xx} \tag{6.1}$$

We constructed the full group and reported it in [12]. In this section the complete generator is given in the form

$$Q = T \frac{\partial}{\partial t} + X \frac{\partial}{\partial x} + Y \frac{\partial}{\partial y} + U \frac{\partial}{\partial u} \tag{6.2}$$

where

$$T = f(t)$$

$$X = f'(t)x/3 - f''(t)y^2/6\alpha - g'(t)y/2\alpha + h(t)$$

$$Y = 2f'(t)y/3 + g(t) \tag{6.3}$$

$$U = -2f'(t)u/3 + f''(t)x/18 - f'''(t)y^2/36\alpha$$

$$- g''(t)y/12\alpha + h'(t)/6$$

where $f(t)$, $g(t)$, $h(t)$ are arbitrary, so the Lie algebra is infinite dimensional.

The specific choice of the subgroup (the generator $\partial/\partial t$ of pure time translation exists!)

$$T = 1 \qquad X = -g'y/2\alpha + h(t) \qquad Y = g(t)$$
$$U = -g''y/12\alpha + h'/6 \tag{6.4}$$

gives rise to the characteristic variables

$$y = \bar{y} + G(t)$$
$$x = \bar{x} - gy/2\alpha + \Omega(t) \tag{6.5}$$
$$u = \bar{u} - g'y/12\alpha + g^2/24\alpha + h/6$$

where $G' = g$ and $\Omega' = g^2/2\alpha + h$ and $\bar{u} = \bar{u}(\bar{x},\bar{y})$. Under (6.5) equation (6.1) transforms into the time independent equation

$$\bar{u}_{\bar{x}\bar{x}\bar{x}\bar{x}} + 6\bar{u}\bar{u}_{\bar{x}\bar{x}} + 6(\bar{u}_{\bar{x}})^2 + \alpha\bar{u}_{\bar{y}\bar{y}} = 0 \tag{6.6}$$

Each (steady) solution of (6.6) gives rise to a family of time dependent solutions of (6.1) involving two arbitrary functions of time, g and h, when $\bar{u}$, $\bar{x}$, and $\bar{y}$ are replaced by the relations in (6.5).

With the choice of $f(t) = t^3$, $g(t) = h(t) = 0$ in (6.3), the group becomes

$$T = t^3 \qquad X = t^2 x - ty^2/\alpha \qquad Y = 2t^2 y$$
$$U = -2t^2 u + tx/3 - y^2/6\alpha \tag{6.7}$$

Using (6.7) in (6.3) the invariants are found to be

$$\eta = y/t^2 \qquad \xi = xt + y^2/2\alpha t^2$$
$$u = \xi/6 - \eta^2 t^2/12\alpha + F(\eta,\xi)/t^2$$

where F satisfies the equation

$$\alpha F_{\eta\eta} + (3F^2)_{\xi\xi} + F_{\xi\xi\xi\xi} = 0 \tag{6.8}$$

which is the same as (6.6) with η, ξ, and F corresponding to $\bar{y}$, $\bar{x}$, and $\bar{u}$, respectively.

The full group generator for (6.8) is

$$Q = (c_1\xi + c_2)\, \frac{\partial}{\partial\xi} + (2c_1\eta + c_3)\, \frac{\partial}{\partial\eta} - 2c_1 F\, \frac{\partial}{\partial F} \tag{6.9}$$

with the three parameters c_1, c_2, c_3 (the algebra is three-dimensional). Solving $QI = 0$ the invariants are found to be

$$F = \psi(w)/2c_1(2c_1\eta + c_3)$$
$$w = (c_1\xi + c_2)/(2c_1\eta + c_3)^{1/2}$$

where ψ satisfies the ordinary differential equation

$$c_1^3 \psi^{(4)} + 3\psi\psi'' + \alpha c_1 w^2 \psi'' + 3(\psi')^2 + 7c_1 \alpha w \psi' + 8\alpha c_1 \psi = 0 \quad (6.10)$$

From the two solutions

$$\psi_1(w) = -4\alpha c_1 w^2/3 \qquad \psi_2(w) = -4c_1^3 w^{-2}$$

of (6.10) one finds two solutions of (6.1)

$$u(x,y,t) = \frac{x}{6t} - \frac{2\alpha[c_1 xt + c_1 y^2/2\alpha + t^2 c_2]^2}{3t^2[2c_1 y + y^2 c_3]^2}$$

and

$$u(x,y,t) = \frac{x}{6t} - \frac{2c_1^2 t^2}{[c_1 xt + c_1 y^2/2\alpha + t^2 c_2]^2}$$

7. A GAS-DYNAMIC MODEL OF PELLET FUSION

A gas-dynamic model of pellet fusion has been studied by Ervin et al. [13]. The hyperbolic equations governing the process are

$$\frac{\partial \rho}{\partial t} + 2\rho u r^{-1} + \rho \frac{\partial u}{\partial r} + u \frac{\partial \rho}{\partial r} = 0 \tag{7.1}$$

$$\frac{\partial u}{\partial t} + u \frac{\partial u}{\partial r} + RT\rho^{-1} \frac{\partial \rho}{\partial r} + R \frac{\partial T}{\partial r} = 0 \tag{7.2}$$

$$\frac{\partial T}{\partial t} + u \frac{\partial T}{\partial r} + 2(\gamma - 1)Tur^{-1} + (\gamma - 1)T \frac{\partial u}{\partial r} = \beta_0 r^n \tag{7.3}$$

where $\rho(r,t)$ is the density inside the pellet ($\rho g/\mu m^3$), $u(r,t)$ is the velocity inside the pellet ($\mu m/ns$), $T(r,t)$ is the temperature inside the pellet (KeV), r is the radial variable (μm), t is the time (ns), R is the gas constant for lead ($\mu m^2/ns$ KeV), γ is the ratio of specific heats, and r^n is the external energy input (KeV/ns). The boundary/initial conditions are

$$\rho(r,0) = \rho_0(r) \qquad u(r,0) = u_0(r)$$

$$T(r,0) = T_0(r) \qquad u(0,t) = 0$$

The full group admits the dilatation (stretching) subgroup given by

$$\rho = e^{\alpha}\bar{\rho} \qquad u = e^{\beta}\bar{u} \qquad T = e^{\delta}\bar{T} \qquad r = e^{\epsilon}\bar{r} \qquad t = e^{\lambda}\bar{t} \tag{7.4}$$

whose invariants are ($\epsilon \neq 0$)

$$\eta = t/r^{\lambda/\varepsilon} \qquad\qquad g(\eta) = u/r^{\beta/\varepsilon}$$
$$f(\eta) = \rho/r^{\alpha/\varepsilon} \qquad\qquad h(\eta) = T/r^{\delta/\varepsilon} \tag{7.5}$$

The invariance of (7.1), (7.2), and (7.3) under (7.4) requires

$$\beta/\varepsilon = (n + 1)/3$$
$$\delta/\varepsilon = 2(n + 1)/3 \tag{7.6}$$
$$\lambda/\varepsilon = (2 - n)/3$$

with α/ε arbitrary. The functions f, g, and h satisfy the equations

$$f' + (2 + \beta/\varepsilon + \alpha/\varepsilon)fg - (\lambda/\varepsilon)\eta(fg)' = 0 \tag{7.7}$$

$$g' - \frac{\lambda}{\varepsilon}\eta(gg' + R\frac{f'}{f}h + Rh') + \frac{\beta}{\varepsilon}g^2 + \left(\frac{\alpha}{\varepsilon} + \frac{\delta}{\varepsilon}\right)Rh = 0 \tag{7.8}$$

$$h' - \frac{\lambda}{\varepsilon}\eta(gh' + (\gamma - 1)g'h) + \left[\frac{\delta}{\varepsilon} + 2(\gamma - 1)\right.$$
$$\left. + \frac{\beta}{\varepsilon}(\gamma - 1)\right]gh = \beta_0 \tag{7.9}$$

Appropriate boundary/initial conditions also must be invariant under (7.4) and are found to be $\rho(r,0) = r^{\alpha/\varepsilon}f(0) = Ar^a$, forcing $f(0) = A$ and $\alpha/\varepsilon = a$. $u(r,0) = 0$ requires $g(0) = 0$ and $T(r,0) = r^{\delta/\varepsilon}h(0)$ requires $h(0) = B$. Lastly, $u(0,t) = 0$ is satisfied provided $n \geq 2$.

Consider the case when $n = 2$. Now $n = 2$ implies $\beta/\varepsilon = 1$; $\delta/\varepsilon = 2$; $\lambda/\varepsilon = 0$. Equations (7.7)-(7.9) then simplify to

$$f' + \left(3 + \frac{\alpha}{\varepsilon}\right)fg = 0 \tag{7.10}$$

$$g' + g^2 + \left(2 + \frac{\alpha}{\varepsilon}\right)Rh = 0 \tag{7.11}$$

$$h' + (3\gamma - 1)gh = \beta_0 \tag{7.12}$$

with initial conditions $f(0) = A$, $g(0) = 0$, $h(0) = B$.

From (7.10) it follows that

$$f(\eta) = A \exp\left[-\left(3 + \frac{\alpha}{\varepsilon}\right)\int_0^\eta g(\xi)\, d\xi\right]$$

From (7.12)

$$h = -\frac{g' + g^2}{\left(2 + \frac{\alpha}{\varepsilon}\right)R} \tag{7.13}$$

whereupon

$$h' = -\frac{g'' + 2gg'}{\left(2 + \frac{\alpha}{\varepsilon}\right)R} \tag{7.14}$$

Equation (7.11) gives

$$g'(0) = -\left(2 + \frac{\alpha}{\varepsilon}\right) RB$$

Substituting (7.13) and (7.14) into (7.12) yields

$$g'' + (3\gamma + 1)gg' + (3\gamma - 1)g^3 = -\beta_0\left(2 + \frac{\alpha}{\varepsilon}\right)R \qquad (7.15)$$

with initial condition

$$g(0) = 0 \qquad g'(0) = -\left(2 + \frac{\alpha}{\varepsilon}\right)RB \qquad (7.16)$$

The Riccati change of variable

$$s' = \mu gs \qquad (7.17)$$

gives

$$s'' = \mu s(g' + \mu g^2) \qquad (7.18)$$

and

$$s''' = \mu s(g'' + 3\mu g'g + \mu^2 g^3) \qquad (7.19)$$

In particular, for $\gamma = 5/3$ (corresponding to a monoatomic gas) and $\mu = 2$, equation (7.15) becomes

$$s''' + 2\beta_0\left(2 + \frac{\alpha}{\varepsilon}\right)Rs = 0 \qquad (7.20)$$

with initial conditions

$$s(0) = 1 \qquad s'(0) = 0 \qquad s''(0) = -2\left(2 + \frac{\alpha}{\varepsilon}\right)RB \qquad (7.21)$$

The solution of (7.20) and (7.21) is

$$s(\eta) = c_1 e^{-F\eta} + e^{(F/2)\eta}\left(c_2 \cos \frac{\sqrt{3}}{2} F\eta + c_3 \sin \frac{\sqrt{3}}{2} F\eta\right)$$

where

$$F = \left[2\beta_0\left(2 + \frac{\alpha}{\varepsilon}\right)R\right]^{1/3}$$

and c_1, c_2, c_3 are given by

$$\begin{bmatrix} c_1 \\ c_2 \\ c_3 \end{bmatrix} = \frac{1}{3} \begin{bmatrix} 1 & -1 & 1 \\ 2 & 1 & -1 \\ 0 & \sqrt{3} & \sqrt{3} \end{bmatrix} \begin{bmatrix} 1 \\ 0 \\ -\dfrac{B}{\beta_0} F \end{bmatrix}$$

From equation (7.17)

$$g = s'/2s$$

$$= \frac{-F}{2} \; \frac{c_1 e^{-F\eta} + c_4 e^{(F/2)\eta} \sin\left(\frac{\sqrt{3}}{2} F\eta + \phi + \frac{\pi}{3}\right)}{c_1 e^{-F\eta} - c_4 e^{(F/2)\eta} \sin\left(\frac{\sqrt{3}}{2} F\eta + \phi\right)}$$

where

$$c_4 = \sqrt{c_2^2 + c_3^2} \qquad \text{and} \qquad \phi = \tan^{-1} \frac{c_2}{c_3}$$

8. DETERMINATION OF INVARIANT ARBITRARY ELEMENTS

Wave propagation in viscoelastic and viscoplastic transmission lines is governed by the equations (Ames and Suliciu [14])

$$\rho \frac{\partial v}{\partial t} - \frac{\partial \sigma}{\partial x} = 0 \qquad \frac{\partial \varepsilon}{\partial t} - \frac{\partial v}{\partial x} = 0 \qquad \frac{\partial \sigma}{\partial t} - f\left(\varepsilon, \sigma, \frac{\partial \varepsilon}{\partial t}\right) = 0 \qquad (8.1)$$

where $\rho > 0$ is the mass density, v is the particle velocity, σ is the stress, ε is the strain, and f characterizes the nature of the body. In the previous discussion f is the "arbitrary element." Here we show how to obtain the functions f which remain invariant under the dilatation group (another form)

$$x = a^{\alpha}\bar{x} \qquad t = a^{\beta}\bar{t} \qquad v = a^{\gamma}\bar{v} \qquad \sigma = a^{\lambda}\bar{\sigma} \qquad \varepsilon = a^{\theta}\bar{\varepsilon} \qquad (8.2)$$

For the first of equations (8.1) to be invariant under (8.2), the first invariance condition

$$\gamma - \beta = \lambda - \alpha \qquad (8.3)$$

must hold. The corresponding condition for the second equation is

$$\gamma - \alpha = \theta - \beta \qquad (8.4)$$

Upon transforming the third equation, it is found to be invariant if

$$\frac{\partial \bar{\sigma}}{\partial \bar{t}} = a^{\beta - \lambda} f\left(a^{\lambda}\bar{\sigma}, a^{\theta}\bar{\varepsilon}, a^{\theta - \beta} \frac{\partial \bar{\varepsilon}}{\partial \bar{t}}\right) = f\left(\bar{\sigma}, \bar{\varepsilon}, \frac{\partial \bar{\varepsilon}}{\partial \bar{t}}\right)$$

For convenience the bars can be dropped. Upon differentiating both sides of the last equality with respect to a and setting a = 1, the linear equation

$$\lambda\sigma \frac{\partial f}{\partial \sigma} + \theta\varepsilon \frac{\partial f}{\partial \varepsilon} + (\theta - \beta)\mu \frac{\partial f}{\partial \mu} = (\lambda - \beta)f$$

$\mu = \partial\varepsilon/\partial t$, is obtained for f. Using the well known theory of first order equations it is easily seen that

$$f(\varepsilon, \sigma, \mu) = \varepsilon^{(\lambda - \beta)/\theta} h[\sigma/\varepsilon^{\lambda/\theta}, \mu/\varepsilon^{(\theta - \beta)/\theta}] \qquad (8.5)$$

where $h \in C^1$, in both arguments, is arbitrary otherwise.

Next, for any real $a > 0$ consider the action of the spiral (screw) transformation group

$$x = \bar{x} + \alpha a \qquad t = \bar{t} + \beta a \qquad v = e^{\gamma a}\bar{v} \qquad \sigma = e^{\lambda a}\bar{\sigma}$$
$$\varepsilon = e^{\delta a}\bar{\varepsilon} \tag{8.6}$$

on equations (8.1). For the first of those equations to be invariant under (8.6) the first invariance condition

$$\gamma = \lambda \tag{8.7}$$

must hold. The corresponding condition for the second equation is

$$\gamma = \delta \tag{8.8}$$

Upon transforming the third equation it is found to be invariant if

$$f(e^{\lambda a}\bar{\varepsilon}, e^{\lambda a}\bar{\sigma}, e^{\lambda a}\partial\bar{\varepsilon}/\partial\bar{t}) = e^{\lambda a}f(\bar{\varepsilon}, \bar{\sigma}, \partial\bar{\varepsilon}/\partial\bar{t}) \tag{8.9}$$

Again the bars are dropped for convenience. Upon calculating the derivatives with respect to a of both sides of the last equality and setting $a = 0$ (the group identity) the linear equation

$$\varepsilon \frac{\partial f}{\partial \varepsilon} + \sigma \frac{\partial f}{\partial \sigma} + \mu \frac{\partial f}{\partial \mu} = f$$

is obtained for f. Its solution found from first order theory is

$$f(\varepsilon, \sigma, \mu) = \varepsilon k[\sigma/\varepsilon, \mu/\varepsilon]$$

where $k \in C^1$, in both arguments, but is arbitrary otherwise.

REFERENCES

1. K. Oswatitsch (Editor), SYMPOSIUM TRANSONICUM (Achen, 1962), Springer-Verlag, Berlin, 1964.

2. W. F. Ames, NONLINEAR PARTIAL DIFFERENTIAL EQUATIONS IN ENGINEERING, Vol. II, Ch. 2, Academic Press, New York, 1972.

3. L. V. Ovsiannikov, GROUP ANALYSIS OF DIFFERENTIAL EQUATIONS (Russian Edition, NAUKA, 1978), English Edition, Academic Press, New York, 1982.

4. N. H. Ibragimov, Transformation Groups in Mathematical Physics (Russian), NAUKA, Moscow, 1983.

5. V. V. Puhnachev, Group properties of the equations of Navier-Stokes in the plane, JOURNAL OF APPLIED MECHANICS AND TECHNICAL PHYSICS 1 (1960), 83-90 (in Russian).

6.   V. O. Bitev, Group properties of the equations of Navier-
     Stokes, in COLLECTION: NUMERICAL METHODS OF THE MECHANICS OF
     FLUIDS 3, No. 5 (in Russian). Novosibirsk, V. TS. SO, Academy
     of Sciences of USSR, 1972, pp. 13-17.

7.   S. P. Lloyd, The infinitesimal group of the Navier-Stokes
     equations, ACTA MECHANICA 38 (1981), 85-98.

8.   R. E. Boisvert, Group Analysis of the Navier-Stokes Equations,
     Ph.D. Dissertation, Georgia Institute of Technology, Atlanta,
     Georgia, 1982.

9.   R. E. Boisvert, W. F. Ames, and U. N. Srivastava, Group prop-
     erties and new solutions of the Navier-Stokes equations,
     JOURNAL OF ENGINEERING MATHEMATICS 17 (1983), 203-221.

10. G. Hamel, Spiralformige bewegungen zaher flussigkeiten,
     JBER. DEUTSCH. MATH.-VEREIN. 25 (1916), 34-60.

11. C. Rogers and W. F. Shadwick, BACKLUND TRANSFORMATIONS AND
     THEIR APPLICATIONS, Academic Press, New York, 1982.

12. W. F. Ames and M. C. Nucci, Analysis of fluid equations by
     group methods, JOURNAL OF ENGINEERING MATHEMATICS, 1986 (in
     press).

13. V. J. Ervin, W. F. Ames, and E. Adams, Group analysis of the
     pellet fusion process, ARO Report 86-1, 605-618, TRANS. THIRD
     ARMY CONFERENCE ON APPLIED MATHEMATICS AND COMPUTING, 1986.

14. W. F. Ames and I. Suliciu, Some exact solutions for wave prop-
     agation in viscoelastic, viscoplastic and electrical trans-
     mission lines, INTERNATIONAL JOURNAL OF NONLINEAR MECHANICS
     17 (1982), 223-230.

COMPUTATIONAL ACOUSTICS: Wave Propagation
D. Lee, R.L. Sternberg, M.H. Schultz (Editors)
Elsevier Science Publishers B.V. (North-Holland)
© IMACS, 1988

INVARIANT SOLUTIONS OF EQUATIONS
ARISING IN OCEAN ACOUSTICS

P. Childs Richards and W. F. Ames

School of Mathematics
Georgia Institute of Technology
Atlanta, Georgia

1. INTRODUCTION

A widely applicable method for determining analytic solutions
of partial differential equations utilizes the underlying group
structure. The mathematical foundation for determination of the
full group can be found in Ames [1], Bluman and Cole [2], and the
general theory in a Banach space in Ovsiannikov [3]. In this
paper, the method is applied to several equations arising in underwater acoustics.

The system of differential equations that describes the full
group of transformations is determined for the cylindrically symmetric Helmholtz equation (acoustic wave equation) [4],

$$p_{rr} + \frac{1}{r} p_r + p_{zz} + F(r,z)p = 0 \tag{1.1}$$

where subscripts denote differentiation, p is the acoustic pressure variation, $F(r,z) = k_0^2 n^2(r,z)$ is the square of the reference
wave number (k_0) times the index of refraction, n, and (r,z) are
the usual space variables. The dilatation and translation groups
[5] are applied to (1.1) and analytic solutions are found for specific functional forms of $F(r,z)$.

The system that describes the full group of transformations
is determined for the standard parabolic equation [6],

$$u_r = K(r,z)u + \alpha_0 u_{zz} \tag{1.2}$$

where $p(r,z)$ in (1.1) is assumed to be of the form $p(r,z) = H_0^{(1)}(k_0 r)u(r,z)$, $H_0^{(1)}(k_0 r)$ represents the zeroth order Hankel function of the first kind, $k_0 r \gg 1$ (far field approximation), and

$$K(r,z) := \frac{ik_0}{2}(n^2(r,z) - 1) \qquad \alpha_0 := \frac{i}{2k_0} \tag{1.3}$$

The dilatation group is then applied to (1.2) with K and α_0 defined
as in (1.3) and analytic solutions are found for specific functional forms of $K(r,z)$.

2. HELMHOLTZ EQUATION

(a) Full Group Analysis

The infinitesimal generator Q for the transformation group of (1.1) is taken to be

$$Q = \eta(r,z,p)\,\frac{\partial}{\partial p} + \xi(r,z,p)\,\frac{\partial}{\partial r} + \tau(r,z,p)\,\frac{\partial}{\partial z} \qquad (2.1)$$

It is desired to find infinitesimal transformations of the form

$$\begin{aligned}
p' &= p + \varepsilon\eta(r,z,p) + O(\varepsilon^3) \\
r' &= r + \varepsilon\xi(r,z,p) + O(\varepsilon^2) \\
z' &= z + \varepsilon\tau(r,z,p) + O(\varepsilon^2)
\end{aligned} \qquad (2.2)$$

which leave (1.1) invariant. Since the determination of the full group requires lengthy calculations, only an outline will be given. Detailed calculations can be found in [1], [2], and [3].

With $H = p_{rr} + (1/r)p_r + p_{zz} + F(r,z)p = 0$, the invariance condition (with the notation $p_r = q$, $p_z = s$, $p_{rz} = p_{zr} = t$, $p_{zz} = v$, $p_{rr} = w$, and $w = -(1/r)q - v - F(r,z)p$ substituted where appropriate) becomes

$$\begin{aligned}
0 \equiv{} & [\eta_{rr} + \eta_{zz} + \tfrac{1}{r}\eta_r + F\eta] + p[F\cdot(2\xi_r - \eta_p) + \xi F_r + \tau F_z] \\[4pt]
& + q\left[2\eta_{rp} - \xi_{rr} - \xi_{zz} + \frac{\xi_r}{r} - \frac{\xi}{r^2}\right] + s\left[-\tau_{rr} + 2\eta_{zp} - \tau_{zz} - \frac{\tau_r}{r}\right] \\[4pt]
& + q^2\left[\eta_{pp} - 2\xi_{rp} + \frac{2\xi_p}{r}\right] + s^2[\eta_{pp} - 2\tau_{zp}] + t[-2(\tau_r + \xi_z)] \\[4pt]
& + qs[-2(\tau_{rp} + \xi_{zp})] + q^3[-\xi_{pp}] + s^3[-\tau_{pp}] + q^2 s[-\tau_{pp}] \\[4pt]
& + s^2 q[-\xi_{pp}] + v[2(\xi_r - \tau_z)] + tq[-2\tau_p] + ts[-2\xi_p] \\[4pt]
& + vs[-2\tau_p] + vq[2\xi_p] + pq[3F\cdot\xi_p] + ps[F\cdot\tau_p] \qquad (2.3)
\end{aligned}$$

Since (2.3) has to be identically zero for all values of p, q, s, t, q^2, ..., all coefficients of these terms must vanish. From the last six terms it is obvious that

$$\xi = \xi(r,z) \qquad \text{and} \qquad \tau = \tau(r,z) \qquad (2.4)$$

From the coefficient of q^2 it follows that η is linear in p; that is

$$\eta = f(r,z)p + g(r,z) \qquad (2.5)$$

Using (2.3)-(2.5), the following system of partial differential equations is obtained:

$$\xi_r = \tau_z$$

$$\tau_r = -\xi_z$$

$$2f_z - \frac{\tau_r}{r} = 0 \tag{2.6}$$

$$2f_r + \frac{\varepsilon_r}{r} + \frac{\xi}{r^2} = 0$$

$$\xi F_r + \tau F_z = F \cdot (f - 2\xi_r)$$

for η, ξ, τ, and F, where f and g must satisfy Eq. (1.1).

It is evident from (2.6) that the dilatation group ($\xi = c_1 r$ and $\tau = c_1 z$ for c_1 constant) is admitted by the full group. The corresponding finite transformation is obtained from

$$x' = x + \sum_{m=1}^{\infty} \frac{\varepsilon^m Q^{(m)}(x)}{m!}$$

where the generator Q is defined in (2.1) and $Q^{(m)}$ denotes composition of the generator.

In the next two sections, the corresponding finite dilatation group is applied to the Helmholtz equation (1.1) and then the finite translation group is applied to a modified form of (1.1). Exact solutions are obtained for certain cases.

(b) Application of the Dilatation Group

If the index of refraction is restricted to be a function of depth alone, the Helmholtz equation becomes

$$p_{rr} + \frac{1}{r} p_r + p_{zz} + F(z)p = 0 \tag{2.7}$$

The invariants of the dilatation group

$$r = e^{\alpha a}\bar{r} \qquad z = e^{\beta a}\bar{z} \qquad p = e^{\gamma a}\bar{p} \qquad a > 0 \tag{2.8}$$

are well known [1] to be

$$\eta = \frac{r}{z^{\alpha/\beta}} \qquad h(\eta) = \frac{p}{z^{\gamma/\beta}} \tag{2.9}$$

Thus, the invariant solutions are of the form

$$p(r,z) = z^{\gamma/\beta} h(\eta) \tag{2.10}$$

Equation (2.7) is invariant if γ is arbitrary, $\alpha = \beta$, and $e^{2\alpha}F(e^{\beta a}z) = F(z)$ which means F must satisfy

$$zF'(z) = -2F(z)$$

The solution of this equation is

$$F(z) = \frac{k_1}{z^2} \qquad k_1 > 0 \qquad\qquad (2.11)$$

Equation (2.7) is invariant under the dilatation group (2.8) if F has the functional form given in (2.11). It should be noted that due to the translation invariance of z, any F(z) of the form

$$F(z) = \frac{k_1}{(z + k_2)^2} \qquad k_1 > 0,\ k_2 \in \mathbb{R}$$

leaves (2.7) invariant. Thus, in regions where the speed of sound is a linear function of depth, group analysis techniques can be employed.

With F defined as in (2.11), the fundamental invariants of (2.7) under the action of (2.8) are $\eta = r/z$, since $\alpha = \beta$, and $p(r,z) = z^\lambda h(\eta)$, where $\lambda = \gamma/\beta$, $\beta \neq 0$. Since λ is an arbitrary parameter, choose $\lambda \equiv 0$ to give

$$p(r,z) = h(\eta) \qquad\qquad (2.12)$$

Requiring p to satisfy (2.7) gives the ordinary differential equation for h,

$$0 = \eta(1 + \eta^2)h'' + (1 + 2\eta^2)h' + k_1\eta h \qquad k_1 > 0 \qquad (2.13)$$

Exact solutions of (2.13) are possible when $k_1 = -\nu(\nu + 1)$, where $\nu \in (-1,0)$. If $\nu \neq -1/2$, then

$$h(\eta) = c_1 Y_\nu(\sqrt{\eta^2 + 1}) + c_1 Y_{-\nu-1}(\sqrt{\eta^2 + 1})$$

where Y_μ denotes the Legendre function [7] and c_1, c_2 are constants. The acoustic pressure variation in terms of the original variables becomes

$$p(r,z) = c_1 Y_\nu(\sqrt{(r/z)^2 + 1}) + c_2 Y_{-\nu-1}(\sqrt{(r/z)^2 + 1}) \qquad (2.14)$$

If $\nu = 1/2$ then $Y_{-\nu-1}$ is replaced by

$$Y^*_{-1/2} = \lim_{\nu \to -1/2} \frac{Y_\nu - Y_{-\nu-1}}{2\nu + 1}$$

(c)   Application of the Translation Group

The invariants of the dilatation group

$$r = \bar{r} + \alpha a \qquad z = \bar{z} + \beta a \qquad p = \bar{p}$$

are well known [1] to be

$$\eta = r - \frac{\alpha}{\beta} z \qquad h(\eta) = p(r,z)$$

Requiring p to be of the form $p(r,z) = f(r)q(r,z)$ and q to satisfy a partial differential equation of the form $q_{rr} + q_{zz} + G(r,z)q = 0$ gives

$$f(r) = \frac{c}{r^{1/2}} \qquad \text{for constant c} \tag{2.15}$$

With $f(r)$ as defined in (2.15), the equation for q becomes

$$q_{rr} + q_{zz} + \left[\frac{1}{4r^2} + F(r,z)\right]q = 0 \tag{2.16}$$

In order for (2.16) to remain invariant under the translation group

$$r = \bar{r} + \alpha a \qquad z = \bar{z} + \beta a \qquad h(\eta) = q(r,z)$$

F must be of the following form

$$F(r,z) = H\left(r - \frac{\alpha}{\beta} z\right) - \frac{1}{4r^2} \tag{2.17}$$

where H is an arbitrary function of its argument. Requiring q to satisfy (2.16) with F as determined above gives the ODE for h

$$h'' + \lambda H(\eta)h = 0$$

where $\lambda = (1 + (\alpha/\beta)^2)^{-1}$. Exact solutions for some special cases are possible.

Case 1: $H(\eta) = \eta^\gamma$, $\gamma \neq -2$. In this case the exact solution is:

$$h(\eta) = \eta^{1/2} Z_{1/(\gamma+2)}\left[\frac{2\lambda^{1/2}}{(\gamma + 2)} \eta^{1+(\gamma/2)}\right]$$

where

$$Z_\nu(x) = c_1 J_\nu(x) + c_2 J_{-\nu}(x) \tag{2.18}$$

if ν is not an integer. If ν is an integer, then $J_{-\nu}$ is replaced by the Bessel function of the second kind, $Y_\nu(x)$ [7].

In the original variables, the pressure becomes

$$p(r,z) = \left(\frac{r - \frac{\alpha}{\beta} z}{r}\right)^{1/2} Z_{1/(\gamma+2)}\left[\frac{2\lambda^{1/2}}{(\gamma + 2)} \left(r - \frac{\alpha}{\beta} z\right)^{1+(\gamma/2)}\right]$$

The known properties of J_ν, $J_{-\nu}$, and Y_ν permit us to induce properties on p. For example, it is well known that for large values of x,

$$J_\nu(x) \sim \sqrt{\frac{2}{\pi x}} \cos\left(x - \frac{\pi}{4} - \frac{\nu\pi}{4}\right)$$

and

$$Y_\nu \sim \sqrt{\frac{2}{\pi x}} \sin\left(x - \frac{\pi}{4} - \frac{\nu\pi}{2}\right)$$

Case 2: $\lambda H(\eta) = \eta^{-2}$. In this case the exact solution is

$$h(\eta) = \eta^{1/2}\left[c_1 \cos\left(\frac{\sqrt{3}}{2} \ln \eta\right) + c_2 \sin\left(\frac{\sqrt{3}}{2} \ln \eta\right)\right]$$

The acoustic pressure in the original variables is

$$p(r,z) = \left(\frac{r - \frac{\alpha}{\beta} z}{r}\right)^{1/2}\left[c_1 \cos\left(\frac{\sqrt{3}}{2} \ln\left(r - \frac{\alpha}{\beta} z\right)\right)\right.$$
$$\left. + c_2 \sin\left(\frac{\sqrt{3}}{2} \ln\left(r - \frac{\alpha}{\beta} z\right)\right)\right]$$

3. STANDARD PARABOLIC APPROXIMATION

(a) Full Group Analysis

Proceeding in a manner similar to that outlined in Section 2(a), with the infinitesimal generator given by (2.1), the infinitesimal transformations given in (2.2) (with p replaced by u), and $H = u_r - K(r,z)u - \alpha_0 u_{zz}$, yields the following invariance condition for the standard parabolic approximation (1.2) (with the notation $u_r = q$, $u_z = s$, $u_{rz} = u_{zr} = t$, $u_{zz} = v$):

$$0 \equiv [\eta_r - \alpha_0 \eta_{zz} - K\eta] + s[\alpha_0(\tau_{zz} - 2\eta_{zu}) - \tau_r]$$
$$+ q[(\eta_u - \xi_r) + \alpha_0\xi_{zz} - \frac{1}{K}(K_r\xi + K_z\tau)]$$
$$+ s^2[\alpha_0(2\tau_{zu} - \eta_{uu})] + q^2[-\xi_u] + qs[2\alpha_0\xi_{zu} - \tau_u]$$
$$+ s^3[\alpha_0\tau_{uu}] + s^2 q[\alpha_0\xi_{uu}] + v[\alpha_0(2\tau_z - \eta_u)$$
$$+ \frac{\alpha_0}{K}(K_r\xi + K_z\tau)] + t(2\alpha_0\xi_z) + vs[3\alpha_0\tau_u]$$
$$+ vq[\alpha_0\xi_u] + ts[2\alpha_0\xi_u] \qquad\qquad (3.1)$$

where $K(r,z) \neq 0$.

Since all the coefficients of (3.1) must vanish, it is obvious from the last four terms of (3.1) that

$$\xi = \xi(r) \qquad \text{and} \qquad \tau = \tau(r,z) \qquad\qquad (3.2)$$

From the coefficient of s^2 it follows that η is a linear function in u; that is

$$\eta = f(r,z)u + g(r,z) \tag{3.3}$$

Using (3.1)-(3.3), the following relationships between η, ξ, τ, and K are obtained:

$$\tau(r,z) = \frac{1}{2}\xi'(r)z + G(r)$$

$$f(r,z) = -\frac{1}{2\alpha_0}\tau(r,z) + J(z) \tag{3.4}$$

$$\xi K_r + \tau K_z = (f - \xi')K$$

where G(r) and J(z) are arbitrary functions, f and g satisfy equation (1.2).

It is evident from (3.4) that the dilatation group is admitted by the full group. In the next section, the finite dilatation transformation is applied to the standard parabolic equation. Exact solutions are obtained for certain cases.

(b) Application of the Dilatation Group

The first step consists in the determination of those functions $K(r,z)$ which permit (1.2) to remain invariant under the dilatation group

$$r = e^{\alpha a}\bar{r} \qquad z = e^{\beta a}\bar{z} \qquad u = e^{\gamma a}\bar{u} \qquad a > 0 \tag{3.5}$$

Invariants of this group are

$$\eta = \frac{z^{\alpha/\beta}}{r} \qquad \text{and} \qquad f(\eta) = u(r,z) \qquad \alpha/\beta \geq 0$$

Under the action of (3.5), (1.2) becomes

$$e^{(\gamma-\alpha)a}\bar{u}_{\bar{r}} - K(e^{\alpha a}\bar{r}, e^{\beta a}\bar{z})e^{\gamma a}\bar{u} - \alpha_0 e^{(\gamma-2\beta)a}\bar{u}_{\bar{z}\bar{z}} = 0 \tag{3.6}$$

Equation (3.6) is invariant if γ is arbitrary, $\alpha = 2\beta$, and $K(e^{\alpha a}r, e^{\beta a}z)e^{\alpha a} = K(r,z)$, which means K must satisfy

$$\alpha r K_r + 2\alpha K_z = -\alpha K$$

The solution of this equation is

$$K(r,z) = \frac{1}{r}F\left(\frac{z^2}{r}\right) \tag{3.7}$$

where F is an arbitrary function of its argument. If the index of refraction has this form then (1.2) is invariant under the dilatation group and group analysis techniques can be employed.

With K of the form given in (3.7), fundamental invariants $\eta = z^2/r$, $\alpha = 2\beta$, and

$$p(r,z) = f(\eta) \tag{3.8}$$

gives the ordinary differential equation for f:

$$\eta f'' + \frac{(1 + 2\alpha_0)}{2} f' + \frac{F(\eta)}{2} f = 0 \tag{3.9}$$

In some cases, exact solutions are possible. For

$$F(\eta) = 2\left[(bc\eta^{(c-1)})^2 + \frac{a^2 - \nu^2 c^2}{\eta^2} \right]$$

where $a = (1/4)(1 - i/k_0)$, b and ν arbitrary, the solution to (3.9) is

$$f(\eta) = \eta^a Z_\nu(b\eta^\nu)$$

where Z_ν is given by (2.18). In terms of the original variables

$$u(r,z) = \left(\frac{z^2}{r}\right)^a Z_\nu\left(b\left(\frac{z^2}{r}\right)^\nu\right)$$

ACKNOWLEDGMENTS

This work was funded in part by Independent Research and Development Project 6M77, Federal Systems Division, IBM Corporation, and U. S. Army Grant DAAG-29-84-K-0083.

REFERENCES

1. W. F. Ames, NONLINEAR PARTIAL DIFFERENTIAL EQUATIONS IN ENGINEERING, Vol. II, Academic Press, New York, 1972.

2. G. W. Bluman and J. D. Cole, SIMILARITY METHODS FOR DIFFERENTIAL EQUATIONS, Springer-Verlag, New York, 1974.

3. L. V. Ovsiannikov, GROUP ANALYSIS OF DIFFERENTIAL EQUATIONS (Russian edition NAUKA 1978); English Translation edited by W. F. Ames, Academic Press, New York, 1982.

4. D. Lee and J. S. Papadakis, Numerical Solutions of Underwater Acoustic Wave Propagation Problems, Naval Underwater Systems Center Tech. Rept. 5929, February 25, 1979.

5. W. F. Ames, Invariant solutions of the underwater acoustic wave equation, COMPUTER AND MATHEMATICS WITH APPLICATIONS, Vol. II, No. 7/8, pp. 681-686.

6. W. F. Ames and Ding Lee, Current developments in the numerical treatment of ocean acoustic propagation, Proceedings of Workshop on Computational Fluid Dynamics, 1985, Georgia Institute of Technology, in press, North-Holland.

7. M. Abramowitz and I. A. Stegun (Editors), HANDBOOK OF MATHEMATICAL FUNCTIONS, National Bureau of Standards, Applied Math. Series 55, June 1964.

COMPUTATIONAL ACOUSTICS: Wave Propagation
D. Lee, R.L. Sternberg, M.H. Schultz (Editors)
Elsevier Science Publishers B.V. (North-Holland)
© IMACS, 1988

SPINNING THE HELMHOLTZ EQUATION

Roy G. Levers

Ocean Science Division
Admiralty Research Establishment
Portland, Dorset, England, UK

ABSTRACT

In quantum field theory the second-order Klein-Gordon equation may be reduced to a system of coupled first-order equations by the use of Pauli spin matrices. Similar problems arise in attempting to reduce the 2-D Helmholtz equation to a parabolic type equation, in situations where the refractive index varies in both dimensions. At the heart of this problem are the approximations which have to be made to a wavenumber pseudo-differential operator. Unlike previous approaches, we relax the requirements that this operator be a scalar by generalizing Bremmer's coupled forward and backward wave equations. The use of spin matrices yields coupled equations which appear to have rather wide angular validity.

INTRODUCTION

The ability to reduce the elliptic (more generally hyperbolic) wave equation to a simpler "marching-type" parabolic equation for acoustic propagation rests on the accuracy with which a wavenumber-like pseudo-differential operator can be represented. Starting with the so-called "narrow-angle" solutions of Tappert [1] and the "wide-angle" ones of Claerbout [2], there has been continuing intense activity to obtain still wider angle approximations. Much of this has been reviewed by Lee [3]. More recently, Siegmann et al. [4] and Hill [5] have obtained wider-angle solutions to the Claerbout and Tappert approximations, respectively, by "feeding back" second-order range derivatives obtained from first-order equations.

In a completely different approach, Fishman and McCoy [6] derived a number of parabolic approximations in various limits by employing an integral representation of this nonlocal operator.

The present paper uses different techniques to those noted above by generalizing the coupled equations of Bremmer [7]. These link forward and backward waves in an environment without transverse variations in refractive index. Our generalization to the situation where such variations are present is quite formal. It does seem, though, that the methods originally employed by Dirac to take the square root of second-order partial differential equations are appropriate for the Helmholtz equation. Such considerations were addressed in a preliminary manner by myself and Gleaves [8] in 1985. Here we adopt a less restrictive viewpoint, incorporate range backscattering explicitly, and examine briefly the errors in our approximations. It transpires that the greatest limitations on angular validity occur for very low frequencies at thermoclines.

THE HELMHOLTZ AND KLEIN–GORDON EQUATIONS

We denote by $p(r,z)$ the acoustic pressure at range r and depth z due to radiation at a fixed frequency ω from a source at range $r = 0$. Cylindrical spreading may be removed by using the transformation

$$p(r,z) = r^{-1/2} u(r,z) \tag{1}$$

so yielding the far-field ($kr \gg 1$) Helmholtz equation for u

$$u_{rr} + u_{zz} + k^2(r,z) u = 0 \tag{2}$$

with the wavenumber $k(r,z) = \omega/c(r,z)$, c being the local sound speed.

Using the notation

$$D = \partial/\partial r \quad \text{and} \quad d = \partial/\partial z \tag{3}$$

equation (2) may be rewritten as

$$(D^2 + d^2 + k^2) u = 0 \tag{4}$$

or

$$(D^2 + Q^2) u = 0 \tag{5}$$

where

$$Q^2(r,z) = k^2(r,z) + d^2 \tag{6}$$

It would be nice, of course, if equation (5) factorized. Unfortunately, the range dependence of the pseudo-differential operator Q ensures this will not happen due to a lack of commutativity between it and D, i.e.,

$$[D,Q] u = D(Q) u \neq 0 \tag{7}$$

It is relevant to note that the two-dimensional equation (4) is somewhat similar to the four-dimensional Klein-Gordon equation

$$(\Box - m^2)u = 0 \tag{8}$$

for a free particle of mass m in relativistic quantum field theory. This equation was reduced many years ago by Dirac to a system of first-order coupled partial differential equations for the different spin and energy states of the electron. The concept of spin states and the associated so-called Pauli spin matrices will prove to be useful for studying the different "states" of the Helmholtz equation, i.e., the forward and backward waves. One complication of the Helmholtz equation not possessed by the Klein-Gordon equation is the position-dependent nature of the former's "mass" term $ik(r,z)$. The present paper will show in particular how this depth variability places limitations on the angular validity of our derived equations.

PAULI SPIN MATRICES

The Pauli spin matrices σ_i (i = 1, 2, 3) are basic objects of the unitary unimodular group SU(2) satisfying the Lie algebra

$$\sigma_i \sigma_j = \delta_{ij} + i\varepsilon_{ijk}\sigma_k \tag{9}$$

with

$$\begin{aligned}
\varepsilon_{ijk} &= +1 \text{ if } (ijk) \text{ is an even permutation of } (1,2,3) \\
&= -1 \text{ if } (ijk) \text{ is an odd permutation of } (1,2,3) \\
&= 0 \text{ for coincident indices}
\end{aligned}$$

They may be written as

$$\sigma_1 = \begin{pmatrix} 0 & 1 \\ 1 & 0 \end{pmatrix} \qquad \sigma_2 = \begin{pmatrix} 0 & -i \\ i & 0 \end{pmatrix} \qquad \sigma_3 = \begin{pmatrix} 1 & 0 \\ 0 & -1 \end{pmatrix} \tag{10}$$

BREMMER'S COUPLED EQUATIONS

Bremmer [7] considered the coupling of forward (u_+) and backward (u_-) waves in the situation where there was no transverse (depth in our case) variation in refractive index (or equivalently, wave number). Solution of the equation

$$(D^2 + k^2)u = 0 \tag{11}$$

was reduced to the two equations

$$u = u_+ + u_- \tag{12}$$

$$Du = ik(u_+ - u_-) \tag{13}$$

These may simply be rewritten as

$$u_+ = u/2 - iDu/2k$$
$$u_- = u/2 + iDu/2k \tag{14}$$

Equations (11) and (14) yield the coupled equations

$$(D - ik + D(k)/2k)u_+ = (D(k)/2k)u_-$$
$$(D + ik + D(k)/2k)u_- = (D(k)/2k)u_+ \tag{15}$$

The straightforward use of Pauli spin matrices in these equations yields the fairly succinct matrix differential equation

$$(D - ik\sigma_3)U = (D(k)/2k)(\sigma_1 - I)U \tag{16}$$

where U is the two vector

$$U = \begin{pmatrix} u_+ \\ u_- \end{pmatrix} \tag{17}$$

TAPPERT'S COUPLED EQUATIONS

The equations derived by Bremmer have been generalized by Tappert [1] for the case of transverse wavenumber variations. The reduction of equation (5) involves replacing equations (12) and (13) by ones involving the formal use of Q-dependent normalizations, i.e.,

$$u = (Q/k_0)^{-1/2}(u_+ + u_-) \tag{18}$$

$$Du = ik_0(Q/k_0)^{1/2}(u_+ - u_-) \tag{19}$$

where k_0 is a constant reference wavenumber and the refractive index is $n(r,z) = k(r,z)/k_0$. These again yield coupled equations for his "narrow-angle" approximation, which in matrix form may be written as

$$(D - iQ_1\sigma_3)U = (D(\varepsilon)/4)\sigma_1 U \tag{20}$$

with $\varepsilon = n^2 - 1$ and

$$Q_1 = k_0(1 + (\varepsilon + d^2/k_0^2)/2) \tag{21}$$

WIDE ANGLE COUPLED EQUATIONS

Bremmer's equation (13) may be generalized, at least formally, by replacing k by Q to give

$$Du = iQ(u_+ - u_-) \qquad (22)$$

leaving equation (12) unchanged. Q may be formally regarded as a (scalar) horizontal wave operator; it is, in fact, a nonlocal operator. As before these two equations may be replaced by

$$(D - iQ\sigma_3)U = \tfrac{1}{2} D(Q)Q^{-1}(\sigma_1 - I)U \qquad (23)$$

It is this equation, rather than equation (22), which will be utilized in subsequent developments.

POSSIBLE NONSCALAR FORMS FOR Q

Equation (23) suggests the possibility that Q may no longer necessarily be considered to be a scalar operator and could operate in the 2-D vector space spanned by u_+ and u_-.

It would seem rather natural that the square root of the operator Q^2 defined in equation (6) should be linear in both k and d; we therefore define

$$Q = Ak + Bd \qquad (24)$$

A and B are objects which do not depend on r or z but are not necessarily scalars. By substituting equation (24) into equation (6) it is clear that one needs

$$A^2 = B^2 = 1 \qquad (25)$$

and

$$k(AB + BA)d + BAd(k) \equiv 0 \qquad (26)$$

The solution of these equations is aided by the use of the Pauli spin matrices. Let

$$A = \underline{\alpha} \cdot \underline{\sigma} \qquad \text{and} \qquad B = \underline{\beta} \cdot \underline{\sigma} \qquad (27)$$

with

$$\underline{\sigma} = (\sigma_1, \sigma_2, \sigma_3) \qquad (28)$$

and

$$\begin{aligned}
\underline{\alpha} &= (\alpha_1, \alpha_2, \alpha_3) \\
\underline{\beta} &= (\beta_1, \beta_2, \beta_3)
\end{aligned} \qquad (29)$$

where α_i, β_i $(i = 1, 2, 3)$ are scalars. Substituting (27) into (25) and using (9) implies that

$$\underline{\alpha}^2 = \underline{\beta}^2 = 1 \tag{30}$$

Similarly, substituting (27) into (26) and again using (9) yields

$$2\underline{\alpha} \cdot \underline{\beta}d + (\underline{\alpha} \cdot \underline{\beta} - iT)d(k)/k \equiv 0 \tag{31}$$

where

$$T^2 = 1 - (\underline{\alpha} \cdot \underline{\beta})^2 \tag{32}$$

Thus the expression for Q which <u>exactly</u> satisfies expression (6) is

$$Q = \underline{\sigma} \cdot (\underline{\alpha}k + \underline{\beta}d) \tag{33}$$

if equations (30) and (31) are satisfied for $\underline{\alpha}$ and $\underline{\beta}$.

ANGULAR VALIDITY OF COUPLED EQUATIONS

Equation (31) clearly imposes complicated wave function dependent requirements on the vectors $\underline{\alpha}$ and $\underline{\beta}$, which is best avoided. Here we explore the consequences of <u>imposing</u> orthogonality on these vectors; i.e., restrict them to satisfy

$$\underline{\alpha} \cdot \underline{\beta} = 0 \tag{34}$$

This implies that equation (31) cannot be satisfied exactly, but will have an error proportional to $d(k)/k$, which in turn leads to an error in Q^2; i.e.,

$$Q^2 = k^2 + d^2 - id(k) \tag{35}$$

The relative error is therefore

$$d(k)u/(k^2 + d^2)u = d(k)/(k^2 - k_z^2) = d(k)\sec^2\theta/k^2 \tag{36}$$

where k_z is the transverse wavenumber $k \sin \theta$, θ being the angle between k and the horizontal. But $d(k)/k^2 = -d(c)/\omega$ which is clearly very small even at low frequencies on the thermocline. From equation (36) even with $d(c)$ having a magnitude as large as $1.0 \sec^{-1}$, a relative error of 5% occurs at an angle of 56° at 10 Hz, 80° at 100 Hz, and 87° at 1000 Hz.

We now turn our attention to the matrix operator $D(Q)Q^{-1}$ in equation (23) to determine its form when Q is given by (33); note that the meaning of $D(Q)$ is as in equation (7). As $D(Q)Q^{-1} = D(Q)QQ^{-2}$, then

$$D(Q)Q^{-1} = AD(k)(Ak + Bd)/(k^2 + d^2)$$

$$= k^{-1}D(k)(I + ABk^{-1}d)/(1 + k^{-2}d^2) \tag{37}$$

Although this denominator is an integral operator in z, the range-coupling (or equivalently, impedance mismatch) term $D(k)/k$ will generally be small. A good approximation to (37) will then be

$$D(Q)Q^{-1} = k^{-1}D(k)(I(1 - k^{-2}d^2) + ABk^{-1}d) \tag{38}$$

Indeed, $D(k)/k$ will often be a good enough approximation to $D(Q)Q^{-1}$.

ACKNOWLEDGMENT

I would like to acknowledge numerous helpful discussions with David Gleaves of Admiralty Research Establishment, Portland, England, U.K.

REFERENCES

1. Tappert, F. D., WAVE PROPAGATION IN UNDERWATER ACOUSTICS, Lecture Notes in Physics, Vol. 70, Keller and Papadakis, eds., Springer-Verlag, New York, 1977, Chapter 5.

2. Claerbout, J. F., FUNDAMENTALS OF GEOPHYSICAL DATA PROCESSING, McGraw-Hill Book Company, Inc., New York, 1976.

3. Lee, D., The State-of-the-Art Parabolic Equation Approximation as Applied to Underwater Acoustic Propagation, NUSC Technical Document 7247, 1984.

4. Siegmann, W. L., G. A. Kriegsmann, and D. Lee, A wide-angle three-dimensional parabolic wave equation, JOURNAL OF THE ACOUSTICAL SOCIETY OF AMERICA, 78 (1985), 659-664.

5. Hill, R. J., Wider-angle parabolic wave equation, JOURNAL OF THE ACOUSTICAL SOCIETY OF AMERICA, 79 (1986), 1406-1409.

6. Fishman, L., and J. J. McCoy, A new class of propagation models based on a factorization of the Helmholtz equations, GEOPHYSICS JOURNAL OF THE ROYAL ASTRONOMICAL SOCIETY, 80 (1985), 439-461.

7. Bremmer, H., The WKB approximation as the first term of a geometric-optical series, COMMUNICATIONS IN PURE AND APPLIED MATHEMATICS, Vol. 4 (1951), 105-115.

8. Levers, R. G., and D. G. Gleaves, Finding the roots of the acoustic wave equation, PROCEEDINGS OF THE 11TH IMACS WORLD CONGRESS , SYSTEM SIMULATION AND SCIENTIFIC COMPUTATION, Vol. 2 (1985) Oslo, 165-167.

COMPUTATIONAL ACOUSTICS: Wave Propagation
D. Lee, R.L. Sternberg, M.H. Schultz (Editors)
Elsevier Science Publishers B.V. (North-Holland)
© IMACS, 1988

THE ANALOGY BETWEEN NUMERICAL WAVE PROPAGATION AND QUANTUM MECHANICS

Robert Vichnevetsky

Department of Computer Science
Rutgers University
New Brunswick, New Jersey

ABSTRACT

The numerical approximation of hyperbolic equations creates dispersive media in which certain errors may be described with the mathematics of wave propagation theory. We show that there are remarkable analogies between this type of wave propagation and quantum mechanics. In this analogy, numerical wave packets are the analog or elementary physical particles. They can be defined only with the same imprecision than that contained in Heisenberg's uncertainty principle. They are subject to scattering when the mesh size variation contains discontinuities. And the quantum mechanics phenomenon of tunneling also exists in numerical wave propagation. Theoretical descriptions of those phenomena are given, together with the results of numerical experiments which verify them.

1. INTRODUCTION

The numerical approximation of hyperbolic equations creates dispersive media in which certain errors may be described with the mathematics of wave propagation theory. We show that there are remarkable analogies between this type of wave propagation and the motion of elementary particles in quantum mechanics.

This paper consists mostly of a factual description and standard analysis of some of the manifestations of this analogy. Further examination will reveal that it has theoretical foundations that transcend those simple manifestations. The investigation of this question is beyond the present scope and is given separately in [25].

The organization of this paper is as follows: In Sections 2 and 3, the known relations which apply to the propagation of harmonic waves in numerical solutions of hyperbolic equations on uniform and nonuniform spatial grids are reviewed. In Section 4, we show that harmonic wave propagation in conservative numerical schemes has, like mechanical continua, a mass-like invariant and an energy-like invariant. Moreover, the corresponding mathematics are identical to those which describe the motion of a continuum in classical mechanics. In this formulation, the corresponding potential field is proportional to the square of the local mesh size. The resulting equations of motion may then be recast in the standard formulation of Hamiltonian mechanics.

One way in which precise numerical experiments with harmonic solutions may be implemented is with wave packets. But ideal wave packets cannot be created. They can only be approximated, with the same lack of precision than that which is contained in Heisenberg's uncertainty principle of quantum mechanics: it is found that in the relationship between numerical wave propagation and quantum mechanics numerical wave packets are the direct analog of elementary physical particles. This is analyzed in Section 5.

The model used in the analysis of Sections 2 to 5 is a spatial semidiscretization of a simple hyperbolic equation. In Section 6, we show that the same mathematics continue to apply to the discrete time systems which are obtained when time stepping is implemented with a conservative numerical scheme.

As long as mesh size variations are sufficiently slow, wave packets in numerical solutions behave exactly as particles of matter in a correspondingly slowly variable potential field. But partial reflection, or scattering, takes place when mesh size discontinuities are encountered, just as is the case when a beam of monoenergetic physical particles encounters a step change in the potential field. Expressing the amount of reflection requires in both cases that the sinusoidal nature of propagation be invoked, returning to the basic definition of numerical wave packets, and to the wave nature of matter by means of de Broglie's relation. This is analyzed in Section 7.

Evanescent numerical solutions appear when a high frequency sinusoidal condition is applied at the upwind boundary of a computing domain. It is shown in Section 8 that these are the analog of evanescent solutions of Schrodinger's equation inside of a potential wall. This is further verified by the fact that the resulting phenomenon of tunneling in quantum physics also exists with numerical solutions, as described and verified by experiments in Section 9.

2. WAVE PROPAGATION IN A DISCRETE COMPUTING DOMAIN

The propagation of small perturbations (typically error terms) in numerical solutions of hyperbolic equations, may be analyzed by considering the simple model equation:

$$\frac{\partial U}{\partial t} + c \frac{\partial U}{\partial x} = 0 \qquad U = U(x,t) \qquad c = \text{constant} > 0 \qquad (2.1)$$

and its approximation with the central differences semidiscretization:

$$\frac{du_n}{dt} = -c\left(\frac{u_{n+1} - u_n}{\delta_n + \delta_{n+1}}\right) \equiv A_n \cdot u_n \qquad n = 0, \pm 1, \pm 2, \ldots \qquad (2.2)$$

on a nonuniform division of the x axis. The mesh points are denoted:

$$\cdots < x_{n-1} < x_n < x_{n+1} < \cdots \qquad (2.3)$$

and $\delta_n = x_n - x_{n-1}$ are the mesh increments, $u_n(t) \simeq U(x_n,t)$ are the discrete approximations of U at the mesh points.

This semidiscretization creates a nonhomogeneous dispersive medium which can be described with standard tools of wave propagation theory: when δ is constant, then (2.2) admits sinusoidal waves of the form:

$$u_n = e^{i(\omega t + \xi x_n)} \qquad (2.4)$$

where the frequency ω and the wave number ξ are linked by the dispersion relation

$$\omega = -\frac{c}{\delta} \sin(\xi\delta) \qquad (2.5)$$

This may be extended to the case where δ is a function of position by invoking standard results in linear wave propagation theory. We shall assume (ignoring for the moment the points of mesh size discontinuity), that δ_n varies continuously and slowly with respect to n, and that, accordingly, the discrete sequence $\{\delta_n\}$ may be embedded in a continuous function $\delta(x)$ which varies slowly with respect to x. With nonconstant δ, sinusoidal waves which generalize (2.4) are of the form [12,23,27]:

$$u(x,t) = \alpha(x,t)e^{i(\omega t + \varphi(x))} \qquad (2.6)$$

Here, $\varphi(x)$ is the phase of u, which defines a local wave number by the relation:

$$\xi(x) = \frac{d\varphi}{dx} \tag{2.7}$$

and $\alpha(x,t)$ is an amplitude function, assumed to vary slowly with respect to individual oscillations of the wave described by $e^{i\varphi(x)}$

The frequency ω is a constant; waves of the form (2.6) are monochromatic waves. The wavenumber ξ is not constant, but varies with x. Its relationship to ω is still the dispersion relation, which is now x dependent:

$$\omega = - \frac{c}{\delta(x)} \sin(\xi(x)\delta(x)) \tag{2.8}$$

Actual computations are discrete in time as well as in space. However, when time marching is achieved with a conservative integration scheme (such as, for instance, the Crank Nicolson method), then all the results derived in the semidiscrete case continue to hold, modulo minor changes in definition and notation which do not affect the nature of the theoretical results. A formal proof shall be given in Section 6 below.

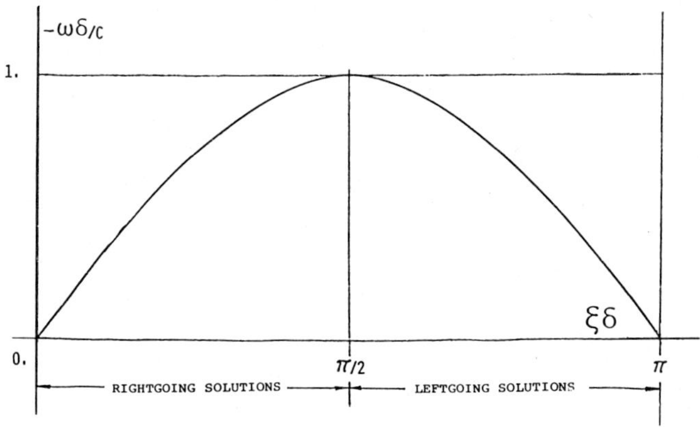

Fig. 2.1. Dispersion relation.

3. CONSERVATION

It is easily shown that solutions of the Cauchy problem for the semidiscretization (2.2) conserve

$$m \equiv \sum_{-\infty}^{\infty} \frac{(\delta_n + \delta_{n+1})}{2} u_n^2 \tag{3.1}$$

(when it is finite) with respect to time. A simple proof consists in taking the derivative of this expression and then using Eq. (2.2).

The scalar expressed by (3.1) is the square of the usual ℓ_2 norm
of $\{u_n\}$ and asking that it be finite is equivalent to asking that
u belong to a Hilbert space; (3.1) is usually called the energy
of u in numerical analysis. In the present context, however, we
shall refer to this quantity as m, the "mass" of u. The motivation for this name will be found later on, when analogies with
classical and quantum mechanics will be brought to light.

Conservation laws of this kind are, of course, genuine properties of hyperbolic systems. Indeed, (3.1) is identical to the
expression obtained when the integral

$$\int_{-\infty}^{\infty} |U(x,t)|^2 \, dx \tag{3.2}$$

is approximated by trapezoidal quadrature on the grid (2.3). And
(3.2) is also an invariant with respect to time when U is a solution of the Cauchy problem for the original equation (2.1).

In the particular case of monochromatic solutions of the form
(2.6) on a slowly variable grid, the mass of u may be expressed
(approximately and under the same assumptions than those stated
in relation with (2.6)) as:

$$m = \int |u(x,t)|^2 \, dx = \int |\alpha(x,t)|^2 \, dx \tag{3.3}$$

where $|u|^2$ is the corresponding mass distribution or density func-
tion at the time t. It is known from wave propagation theory (see,
for example, Lighthill, 1978, or Whitham, 1974) that the propaga-
tion of $|u|^2$ takes place locally at the group velocity $G(\omega,x)$, de-
fined by

$$G = -\frac{\partial \omega}{\partial \xi} \tag{3.4}$$

where $\omega(\xi)$ is the dispersion relation. Whence, with (2.5) (Fig.
3.1):

$$G(\omega,x) = c \, \cos(\xi(x)\delta(x)) \tag{3.5}$$

$$= \pm c \sqrt{1 - \left[\frac{\omega\delta(x)}{c}\right]^2} \tag{3.5}$$

Important to note is that to each ω there corresponds a rightgoing solution (G > 0) and a leftgoing solution (G < 0), both
with equal absolute values of the group velocity.

The invariance of (3.3) and the fact that $|u|^2$ propagates at
the group velocity may be combined to produce the following

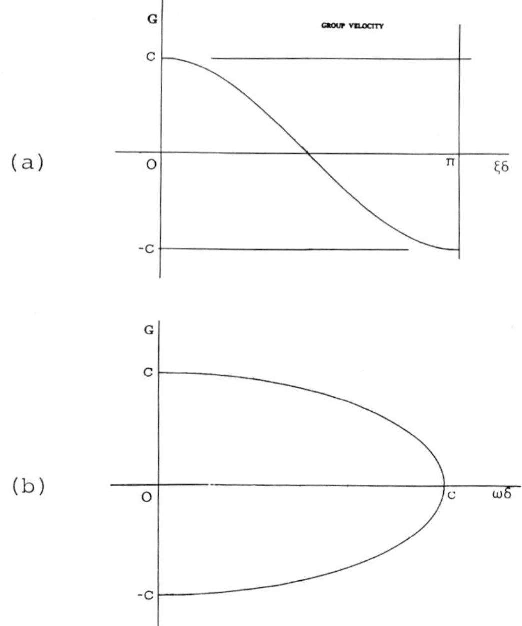

Fig. 3.1. (a) Group velocity versus δ and wave number; (b) group velocity versus δ and frequency.

conservation law (see, for instance, Lighthill, 1978, Sec. 4.5, Eq. 115):

$$\frac{\partial |u|^2}{\partial t} + \frac{\partial}{\partial x} (G|u|^2) = 0 \qquad (3.6)$$

An equivalent expression is derived by Giles and Thompkins [7], using an asymptotic technique in which an expression reducible to (3.6) appears as the first term of the series expansion of the node equation for a sinusoidal solution on a slowly variable grid.

4. HAMILTONIAN FORMULATION

As a first indication that there may be a close analogy between numerical wave propagation and quantum mechanics, we show in this section that monochromatic waves in an irregular grid conserve not only what we have called "mass," but also a second (energy-like) quantity. With two invariants, the corresponding mathematics become identical to the equations of motion of a continuum in classical mechanics, and they may be recast in the standard formulation of Hamilton's canonical equations. Consider monochromatic solutions, i.e., solutions of the form (2.6), for

which the mass is defined by (3.3). For such solutions we also
define:

$$\text{Momentum:} \quad P(t) = \int_{-\infty}^{\infty} G|u|^2 \, dx \qquad (4.1)$$

$$\text{Kinetic energy:} \quad T(t) = \int_{-\infty}^{\infty} \frac{G^2}{2} |u|^2 \, dx \qquad (4.2)$$

$$\text{Potential energy:} \quad V(t) = \int_{-\infty}^{\infty} \frac{(\omega\delta)^2}{2} |u|^2 \, dx \qquad (4.3)$$

We then find with (3.6) that the total energy, defined as

$$T + V = \int_{-\infty}^{\infty} \frac{c^2}{2} |u|^2 \, dx \qquad (4.4)$$

is constant. To monochromatic waves may thus be associated, in
addition to mass, a kinetic energy and a potential energy whose
sum is an invariant. This is identical in form to the equations
describing the dynamics of classical mechanical systems, as may
be verified by recasting the preceding results in the formulation
of Hamilton's canonical equations. To that effect, we rewrite
the above as:

$$P = \int P|u|^2 \, dx \qquad P = G(\omega, x) \qquad (4.1a)$$

$$T = \int T|u|^2 \, dx \qquad T = \frac{G^2(\omega, x)}{2} \qquad (4.2a)$$

$$V = \int V|u|^2 \, dx \qquad V = \frac{(\omega\delta(x))^2}{2} \qquad (4.3a)$$

where P, T, and V are the specific momentum, kinetic energy, and
potential energy, respectively. To these we add:

$$Q \equiv \frac{1}{m} \int x|u|^2 \, dx = \frac{1}{m} \int Q|u|^2 \, dx \qquad (Q = x) \qquad (4.5)$$

(Q = position of the center of gravity of m) and rewrite (4.4) as

$$H = T + V = \int H|u|^2 \, dx \qquad (4.6)$$

where H, the specific Hamiltonian, is to be expressed as a function
of P and Q:

$$H(P, Q) = \frac{P^2}{2} + \frac{(\omega\delta(Q))^2}{2} \qquad (4.7)$$

We then find, invoking (3.6) and integration by parts:

$$m \frac{dQ}{dt} = \int Q \frac{\partial |u|^2}{\partial t} \, dx = \int G|u|^2 \, dx \tag{4.8a}$$

$$\frac{dP}{dt} = \int P \frac{\partial |u|^2}{dt} \, dx = \int G \frac{\partial G}{\partial x} |u|^2 \, dx \tag{4.9a}$$

which may also be rewritten as:

$$m \frac{dQ}{dt} = \int \frac{\partial H}{\partial P} |u|^2 \, dx \tag{4.8}$$

$$\frac{dP}{dt} = - \int \frac{\partial H}{\partial Q} |u|^2 \, dx \tag{4.9}$$

Moreover, we find:

$$\frac{dH}{dt} = \int H \frac{\partial |u|^2}{\partial t} \, dx = \int G \left[\frac{\partial H}{\partial Q} \frac{\partial Q}{\partial x} + \frac{\partial H}{\partial P} \frac{\partial P}{\partial x} \right] |u|^2 \, dx = 0 \tag{4.10}$$

These are indeed identical to Hamilton's canonical equations for
the dynamics of a distributed mass system in classical mechanics
(see, e.g., Goldstein, 1950).

5. NUMERICAL EXPERIMENTS AND WAVE PACKETS

Instead of the general monochromatic waves used in the preceding section, one may recast the analysis in terms of "wave
packets." Wave packets are one of the convenient devices with
which numerical experiments may be conducted. Their definition
is given in classical wave mechanics somewhat heuristically as
"a sinusoidal function of finite length, comprising only a finite
number of wavelengths" (see Brillouin, 1946).

But, as will be found, ideal wave packets cannot be created.
They can only be approximated, with the same lack of precision
than that which is contained in Heisenberg's uncertainty principle of quantum mechanics. In fact, numerical wave packets and
the elementary particles of quantum mechanics are very similar
entities; both behave globally as particles of finite mass and
energy, described by the equations of classical mechanics; both
have internal sinusoidal oscillations (given for physical particles by de Broglie's relation). And for both, those internal
sinusoidal oscillations manifest themselves when sharp discontinuities in the potential field are encountered.

A usual form for the expression of wave packets (to be used,
for instance, for the generation of initial data) is that of a
sinusoidal function modulated by a Gaussian envelope of standard
deviation σ:

$$u = e^{-(1/2)[(x-x_0)/\sigma]^2} e^{i\xi_0 x} \tag{5.1}$$

Its Fourier transform also has a Gaussian envelope, of standard deviation $1/\sigma$:

$$\hat{u}(\xi) = \sqrt{2\pi}\ \sigma e^{-(1/2)[(\xi-\xi_0)\sigma]} e^{-i\xi x_0} \tag{5.2}$$

It may be recognized that (5.1) is only the approximation of an ideal wave packet. Indeed, both (5.1) and (5.2) have infinite support. By contrast, we would like an ideal wave packet to be of finite support in x and of vanishingly small wavenumbers bandwidth in ξ. But this turns out to be unfeasible and (5.1) is the best one can do; while not of finite support, the mass of (5.1) is nevertheless concentrated to within a negligible remainder in a finite region near x_0 in physical space and that of (5.2) near ξ_0 in Fourier space (for example, the mass of (5.1) which is contained outside of $(x_0 - 5\sigma,\ x_0 + 5\sigma)$ is of the order of 10^{-7} times the total mass). The product of the approximate width of the two corresponding distribution functions is of order one; this is known as the bandwidth theorem in Fourier analysis. Loosely stated, it says that this product is bounded from below by a number of order one for any function u(x). Moreover, this minimum is reached by Gaussians (and only by Gaussians) of the form (5.1). While one cannot create ideal wave packets of finite support, (5.1) is the best approximation thereof one may get. When σ is reasonably large with respect to the wavelength $\lambda_0 = 2\pi/\xi_0$, then (5.1) approximates a wave packet of wavenumber ξ_0.

This difficulty of constructing ideal numerical wave packets is no different from that encountered in quantum mechanics where elementary particles and their associated wave train are subjected to Heisenberg's uncertainty principle. This principle and the bandwidth theorem of Fourier analysis are essentially expressions of the same mathematics. In fact, our preceding discussion about wave packets is remarkably similar to discussions found in quantum mechanics texts in relation to the analytic expression of the wave equivalent of a particle [13,14].

The form taken by the canonical equations becomes particularly interesting when the analysis is restricted to wave packets. Their definition is equivalent to assuming that when u is a single wave packet, then G and δ may be approximated by constants in the integrals of Sections 5 and 6 which thus become (to within negligible remainders):

Mass:

$$m = \int |u(x,t)|^2 \, dx \tag{5.3}$$

Position:

$$Q = \frac{1}{m} \int x|u(x,t)|^2 \, dx \tag{5.4}$$

Momentum:

$$P = mG(\omega,Q) \tag{5.5}$$

Kinetic energy:

$$T = \frac{m}{2} G^2(\omega,Q) = \frac{P^2}{2m} \tag{5.6}$$

Potential energy:

$$V = \frac{m}{2} (\omega\delta(Q))^2 \tag{5.7}$$

Hamiltonian:

$$H = \frac{P^2}{2m} + \frac{m}{2} (\omega\delta(Q))^2 \tag{5.8}$$

And equations (4.8), (4.9) become simply:

$$\frac{\partial Q}{\partial t} = G(\omega,Q) = \frac{\partial H}{\partial P} \tag{5.9}$$

$$\frac{\partial P}{\partial t} = -\frac{m}{2} \frac{\partial}{\partial Q} (\omega\delta(Q))^2 = -\frac{\partial H}{\partial Q} \tag{5.10}$$

which are identical to Hamilton's canonical equations for a single
physical particle moving in a potential field. That is, the dynamics of the wave packet are identical to those of a classical
matter particle of mass m given by (5.3), and position Q given by
(5.4). As long as the effective support of (5.1) remains in some
sense small with respect to the length scale over which variations
of δ are significant, the motion of the "particle" may be described
by its group velocity alone, and the sinusoidal oscillations of u
inside of the wave packet's envelope are irrelevant.

Examples which illustrate the motion of numerical wave packets
in a particle-like manner are shown in Figs. 6.1 and 6.2 below.

6. FULLY DISCRETE SCHEMES

It will now be shown that the mass and energy conservation
principles which have been found to hold for the semidiscrete computational model used so far continue to do so when the system
(2.2) is integrated in time with a conservative discrete time
stepping scheme. As a consequence, the satisfaction of basic

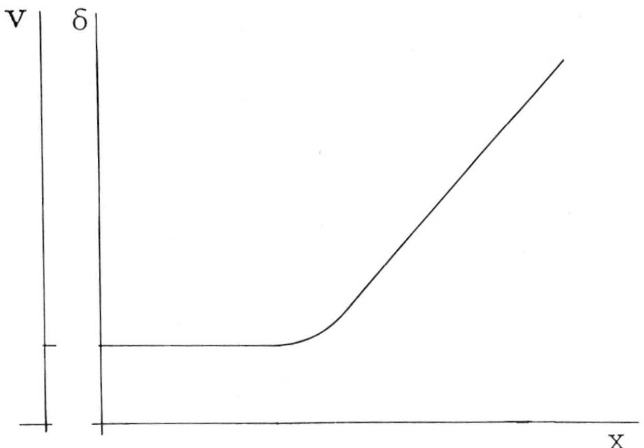

Fig. 6.1. Potential versus distance for the numerical experiment
shown in Fig. 6.2 (obtained by a corresponding variation in the
mesh size δ with position).

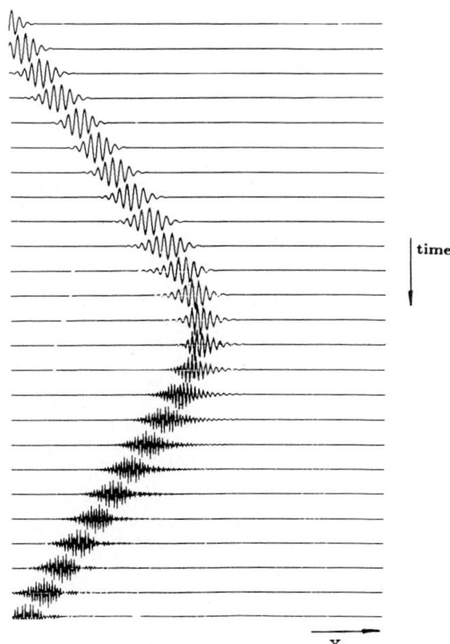

Fig. 6.2. Reflection of a wave packet (the analog of a particle)
by the potential gradient of Fig. 6.1.

equations of mechanics, including the Hamiltonian formulation,
which have been found to hold for monochromatic solutions and
wave packets in the semidiscrete case will, save for a minor
change in definitions, also hold for solutions of the fully discrete calculations.

As an example of a conservative time stepping scheme, we consider the familiar Crank-Nicolson method which may be expressed as:

$$\frac{u_n^{j+1} - u_n^j}{\Delta t} = \frac{1}{2} \mathbf{A}_n (u_n^j + u_n^{j+1}) \tag{6.1}$$

where

$$u_n^j \simeq U(x_n, j\Delta t) \tag{6.2}$$

This may also be rewritten in operator notations as

$$\mathbf{M}u_n^j = \mathbf{A}_n u_n^j \tag{6.3}$$

where

$$\mathbf{M} = \frac{2}{\Delta t}\left(\frac{\mathbf{Z} - 1}{\mathbf{Z} + 1}\right) \tag{6.4}$$

and $\mathbf{Z}$ is the time-shift operator defined by the identity

$$\mathbf{Z}u_n^j \equiv u_n^{j+1} \tag{6.5}$$

Sinusoidal solutions on a uniform spatial grid are still admitted
by the fully discrete equations (6.3). This may be verified by
simple substitution of

$$u_n^j = e^{i(\omega j\Delta t + \xi x_n)} \tag{6.6}$$

for u. The corresponding dispersion relation is thus found:

$$\mu(\omega) = - \frac{c}{\delta} \sin(\xi\delta) \tag{6.7}$$

where $i\mu(\omega)$ is the spectral function or Fourier symbol of the
operator $\mathbf{M}$ defined as:

$$i\mu(\omega) \equiv \frac{\mathbf{M}e^{i\omega j\Delta t}}{e^{i\omega j\Delta t}} \tag{6.8}$$

When $\mu(\omega)$ is a real function of ω then, by the dispersion relation, solutions which are sinusoidal in space are also sinusoidal
in time, i.e., their amplitude remains constant. Since a general

solution of finite mass may be expressed in Fourier integral form
as the sum of such space sinusoidal solutions, this implies, via
Parseval's relation, the conservation of m with respect to time.
Time stepping methods for which this is the case are the methods
that we have previously called conservative.

For the Crank-Nicolson method we find:

$$\mu(\omega) = \frac{2}{\Delta t} \tan \frac{\omega \Delta t}{2} \tag{6.9}$$

which is real. This is one of the expressions of the conservative
property of the Crank-Nicolson method which was referred to earlier.
A more general proof of the conservative property of this method
(which applies also to nonuniform grids) consists in multiplying
(6.1) by

$$(u_n^{j+1} + u_n^{j})(\delta_n + \delta_{n+1})/2 \tag{6.10}$$

and summing over all n in $(-\infty, \infty)$. This results, for solutions of
the Cauchy problem which are in a Hilbert space, in:

$$m^{j+1} - m^{j} = 0 \tag{6.11}$$

Other (stable) time stepping methods may result in a complex $\mu(\omega)$,
introducing exponential damping with respect to time; these are
called dissipative methods. It is, however, the case that for
small values of Δt, dissipative methods become almost conservative. Indeed, the consistency condition requires that

$$\mu(\omega) = \omega + \text{higher order terms} \tag{6.12}$$

and those higher order terms become negligible when Δt is small.

The group velocity of a fully discrete scheme is [from (3.4)
and (6.7)]:

$$G = c \cos(\xi \delta) \frac{d\omega}{d\mu} = \pm c \sqrt{1 - \left(\frac{\mu(\omega)\delta}{c}\right)^2} \frac{d\omega}{d\mu} \tag{6.13}$$

For the Crank-Nicolson method:

$$G = \pm c \sqrt{1 - \left[\frac{2}{R} \tan\left(\frac{\omega \Delta t}{2}\right)\right]^2 \cos^2\left(\frac{\omega \Delta t}{2}\right)} \tag{6.14}$$

where

$$R = \frac{c \Delta t}{\delta} \tag{6.15}$$

is the Courant number.

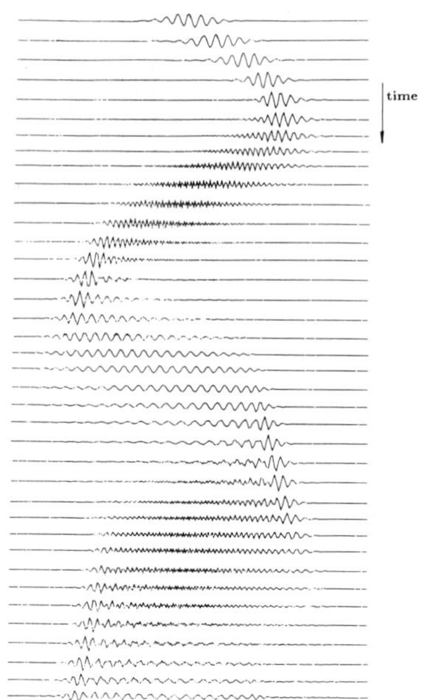

Fig. 6.3. Multiple reflection of a wave packet in a "grid well."
The mesh size δ is smallest at the center of the computing domain,
and increases geometrically in both directions toward the boundaries (from [23]). Note the gradual "dispersion" of the wave
packet.

It may be observed that other than in the replacement of d/dt
by **M** and the replacement of ω by $\mu(\omega)$, all the derivations of Sections 4 and 5 continue to be strictly valid. In particular, for
the fully discrete system (6.3) where **M** describes any conservative
time stepping method, and a solution u which is a single wave
group, we obtain the discrete-time form of Hamilton's canonical
equations (with the appropriate expression for the group velocity):

$$\mathbf{M} \cdot Q^j = G(\omega, Q^j) = \frac{\partial H}{\partial P^j} \tag{6.16}$$

$$\mathbf{M} \cdot P^j = - \frac{m}{2} \frac{\partial}{\partial Q^j} (\omega h(Q^j))^2 = - \frac{\partial H}{\partial Q^j} \tag{6.17}$$

and

$$\mathbf{M} \cdot H(P^j, Q^j) = 0 \tag{6.18}$$

It should be noted that we have implicitly defined the existence
of dynamical systems which satisfy strictly the discrete-time
form (6.16)-(6.17) of the equations of classical mechanics.

Numerical Experiments

The numerical experiment shown in Fig. 6.2 illustrates the motion of a numerical wave packet in a nonconstant potential field: A rightgoing wave packet (or "particle") arrives from the left (where δ = constant) into a region where the mesh size (or potential) undergoes a smooth, gradual increase. The particle climbs the hill up to the point where its total energy equals the potential energy (i.e., where its kinetic energy equals zero), then returns to the left with a symmetrical trajectory.

What this example illustrates is a process of internal reflection. From the computational viewpoint this process is entirely parasitic, due to the numerical approximation, since nothing of the sort is present in solutions of the original equation (2.1). There are applications (for example in the implementation of large aerodynamic codes [10]), where the same wave group undergoes more than one internal reflection and thereby remains trapped inside of the computing domain (cf. Giles and Thompkins [7], Vichnevetsky [23]). An illustration of this is given by the numerical experiment whose results are shown in Fig. 6.3.

7. MESH DISCONTINUITIES AND SCATTERING

The solid particle model of a wave packet's propagation assumes a slow, continuous variation of the mesh properties and breaks down when this condition ceases to hold. This is the case, for example, when abrupt variations of the mesh size are present in the computing domain. What then happens is partial reflection or "scattering" at the point of discontinuity, as will now be illustrated.

Consider two piecewise uniform spatial grids which interface in x = 0 [Fig. 7.1(b)]

$$\delta = \begin{cases} \delta_L & \text{when } x < 0 \\ \delta_R > \delta_L & \text{when } x > 0 \end{cases} \tag{7.1}$$

creating a mesh discontinuity. Consider now a wave packet which arrives from the left with a specific total energy satisfying:

$$\frac{c^2}{2} = H > \frac{1}{2} (\mu(\omega)\delta_R)^2 \tag{7.2}$$

i.e., the specific total energy exceeds the potential energy in x > 0. If the mesh size variation were smooth, then the theory

     R. Vichnevetsky

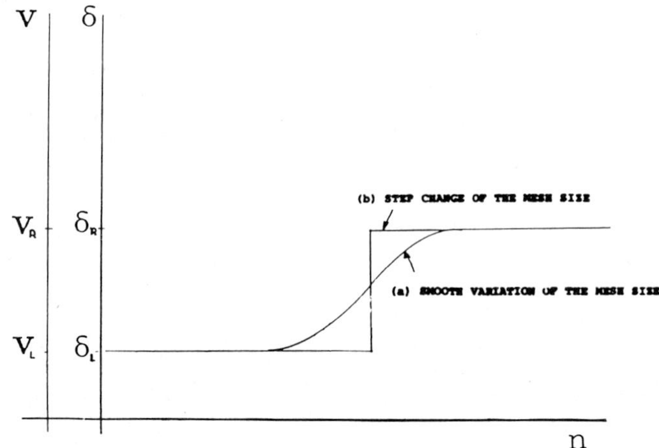

Fig. 7.1. Mesh size and potential variation for the numerical experiments of Figs. 7.2 and 7.3. (a) Smooth variation of the mesh size; (b) abrupt variation of the mesh size.

of the preceding sections would apply and this wave packet would pass from $x < 0$ to $x > 0$ as a whole, without any reflection. That this indeed is the case is illustrated in Fig. 7.2.

But things are different when the mesh size variation is in the form of a step such as (7.1). This situation has been well documented in the standard numerical analysis literature, and we

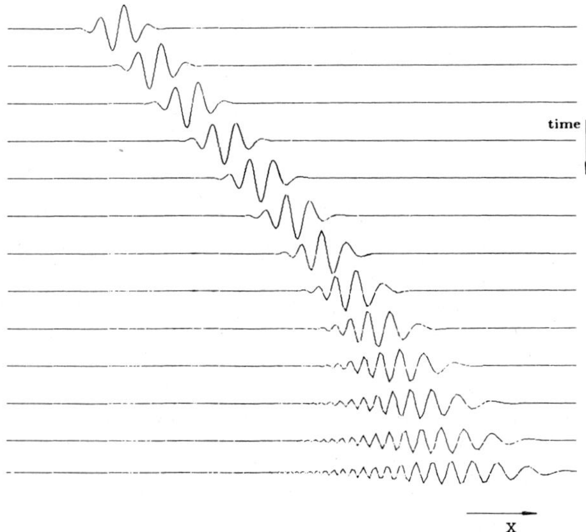

Fig. 7.2. A wave packet passes without reflection across the smooth mesh size (or potential) change shown in Fig. 7.1(a).

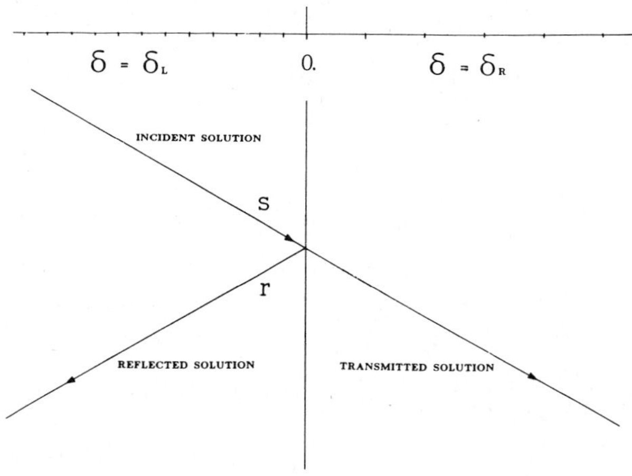

Figure 7.3

shall briefly recall the relevant results: describing analytically the process of reflection requires that the sinusoidal nature
of the wave packet be invoked. Specifically, use is made of discrete-time Fourier transforms, defined as

$$\bar{v}(\omega) \equiv \Delta t \sum_{j=-\infty}^{\infty} v^j e^{-ij\omega\Delta t} \tag{7.3}$$

(and which become standard Fourier transforms in the semidiscrete
case, i.e., when $\Delta t \to 0$).

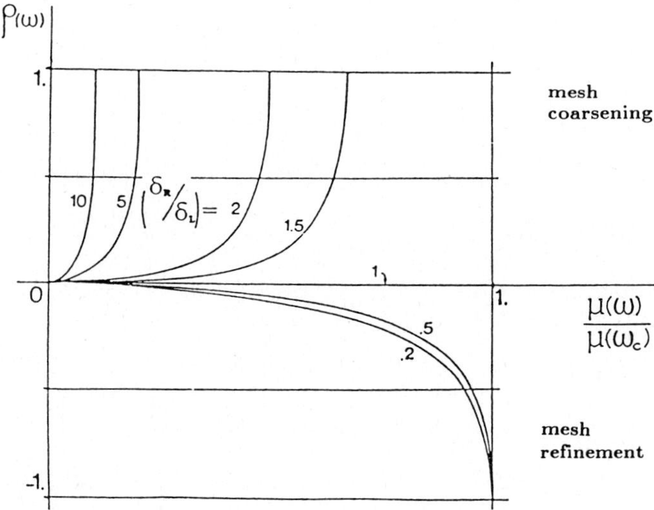

Fig. 7.4. Amplitude reflection ratio at a point of mesh size
discontinuity.

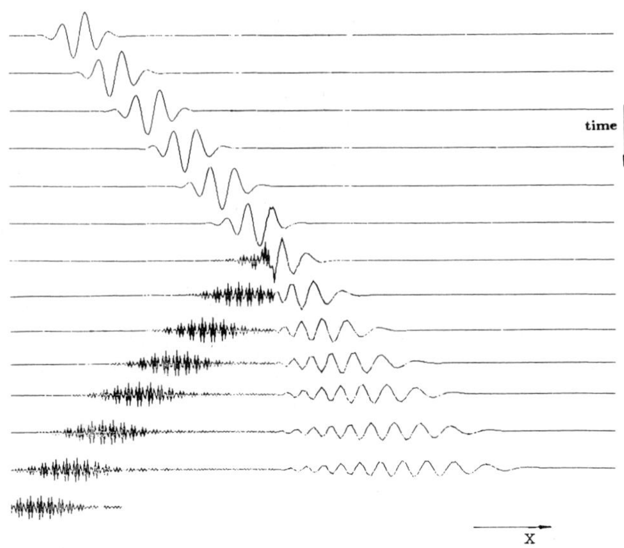

Fig. 7.5. Same experiment as that shown in Fig. 7.2, but with the
mesh size versus position variation shown in Fig. 7.1(b). Although
the total mesh size change is the same as in the case shown in Fig.
7.2, the fact that the variation is abrupt creates scattering.
(Time integration with the Crank-Nicolson method, $\Delta t = 0.25$.)

The amount of scattering may be expressed by an amplitude reflection ratio

$$\rho \equiv \frac{\bar{r}_0(\omega)}{\bar{s}_0(\omega)} = \frac{\sqrt{1 - \left(\frac{\mu\delta_L}{c}\right)^2} - \sqrt{1 - \left(\frac{\mu\delta_R}{c}\right)^2}}{\sqrt{1 - \left(\frac{\mu\delta_L}{c}\right)^2} + \sqrt{1 - \left(\frac{\mu\delta_R}{c}\right)^2}} = \frac{G_L - G_R}{G_L + G_R} \qquad (7.4)$$

where $\bar{s}_0(\omega)$ and $\bar{r}_0(\omega)$ are the Fourier transforms of the incident
and reflected solution in $x = 0$, and G_L and G_R are the group
velocities corresponding to ω and to the mesh sizes δ_L and δ_R,
respectively (Figs. 7.3 and 7.4) [18].

Results of numerical experiments which illustrate this case
case are given in Figs. 7.5 and 7.6.

As was mentioned earlier, that relationships between Fourier
transforms have been used to derive (7.4) implies that the wave
(or sinusoidal) nature of the solutions has been taken into account. By contrast, the analysis of the situations illustrated
by Figs. 6.1, 6.2, and 7.2, where scattering was absent, ignored this information, since only group velocities were invoked.

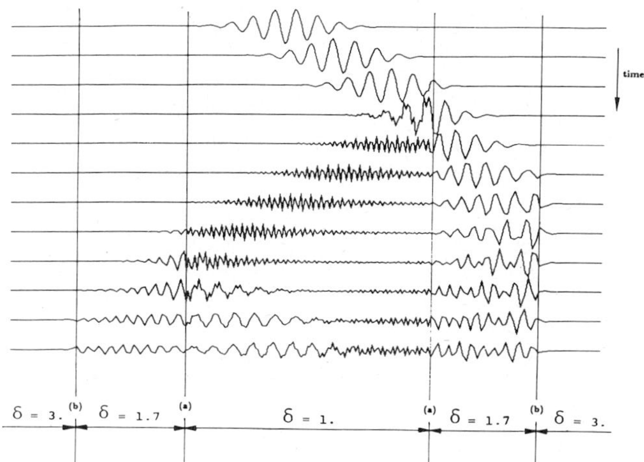

Fig. 7.6. Same experiment as that of Fig. 6.3, but the continuous variation in the mesh size (or potential) has been replaced by a step-like variation, creating internal scattering where there was none.

This duality in behavior illustrated by the difference between Figs. 7.2 and 7.5 has a direct analog in the particle/wave duality of matter in quantum mechanics: When a beam of monoenergetic particles passes across a smooth, continuous change in the potential field, then all the particles stay with the beam. But when the potential change is in the form of a step, then a certain number of particles are reflected (or scattered). A reflection ratio similar to (the square of) (7.4) gives the probability for a single particle in the beam to be reflected back. Here, too, the derivation of this ratio invokes the wave nature of matter (via de Broglie's relation), and mathematics identical to those used here. The reflection ratio of both is identically related to the potential function (cf. [13], p. 80 et seq.). By contrast, classical mechanics would predict (erroneously) that all particles cross the potential step, i.e., the absence of scattering.

There remains an apparent difference between the two phenomena. In the numerical propagation model, a single "particle" is broken down in two pieces of which one is reflected. In quantum mechanics, by contrast, one considers many particles (in a beam) of which a certain proportion is scattered. But expressed in terms of mass ratio, the numbers are the same. Moreover, one may repeat the numerical experiment with a continuous monochromatic

wave generated by a sinusoidal condition at the upwind boundary;
the situation then becomes mathematically undistinguishable from
the description of a beam or "swarm" of particles with continuous
distributions in de Broglie's theory (see, e.g., [1], Chap. XXIX).

8.   TOTAL REFLECTION AND EVANESCENT SOLUTIONS

The absolute value of the amplitude reflection ratio (7.4)
corresponding to

$$|\mu(\omega)| > \frac{c}{\delta_R}$$

or

$$|\omega| > \omega_c = \frac{2}{\Delta t} \text{ atan } \frac{c\Delta t}{2\delta_R} \qquad (8.1)$$

(where ω_c is called the cut-off frequency) becomes:

$$|\rho(\omega)| = \left| \frac{\sqrt{1 - \left(\frac{\mu\delta_L}{c}\right)^2} - i\sqrt{\left(\frac{\mu\delta_R}{c}\right)^2 - 1}}{\sqrt{1 - \left(\frac{\mu\delta_L}{c}\right)^2} + i\sqrt{\left(\frac{\mu\delta_R}{c}\right)^2 - 1}} \right| = 1 \qquad (8.2)$$

That $|\rho| = 1$ means that there is total reflection. But the corre-
sponding solution in x > 0 is not identically zero. Indeed, to
Fourier components of frequency $|\omega| > \omega_c$ there corresponds in x >
0 solutions which satisfy [26]:

$$\frac{\bar{u}_{n+1}(\omega)}{\bar{u}_n(\omega)} = -i\left[\left(\frac{\mu\delta_R}{c}\right) - \sqrt{\left(\frac{\mu\delta_R}{c}\right)^2 - 1}\right]$$

$$= e^{-i(\pi/2)} e^{-\text{Arg ch}(\mu\delta_R/c)} \qquad (8.3)$$

These are evanescent solutions. They are harmonic in time, but
decay exponentially in space at a rate which increases with $|\omega|$ -
ω_c. The mass and energy flows associated with those solutions
are zero (as is the group velocity).

Consider now the case where u is a single wave packet of frequency ω satisfying (8.1). This wave packet will be entirely reflected toward x < 0. But while it is contact with the point of
mesh size discontinuity, it will generate a transient evanescent
solution to the right of that point, decaying exponentially with
distance (by "transient," it is meant that this solution exists
in x > 0 as long as reflection is in progress, but that it dis-
appears thereafter). There is again a direct analogy with quantum

mechanics; when a beam of monoenergetic particles is directed
against an abrupt potential step whose height exceeds their energy,
then all the particles will bounce back, but some travel a finite
distance beyond the point of potential discontinuity before doing
so. The probability of the presence of particles in x > 0 is then
described by an evanescent solution of Schrodinger's equation which
also decays exponentially with distance and is mathematically
equivalent to (8.2).

One of the manifestations of these evanescent solutions is in
the process of tunneling, which shall be described next.

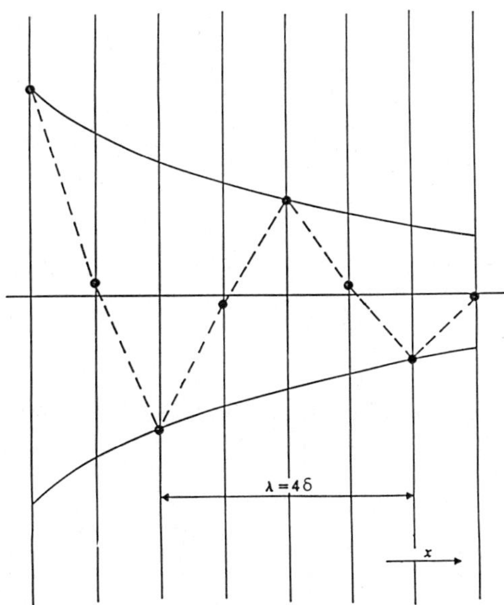

Fig. 8.1. Evanescent solution in x > 0. There is amplitude decay with distance as described by equation (8.3).

9. TUNNELING

Consider a computing domain which is discretized with a uniform fine grid everywhere, except in a wall of finite thickness L
where the mesh is coarser (Fig. 9.1):

$$\delta = \begin{cases} \delta_0 & \text{when } x < 0 \text{ and } x > L \\ \delta_1 > \delta_0 & \text{when } 0 < x < L \end{cases}$$

Consider now a wave packet in x < 0 with a specific energy
which is less than the potential barrier created by the coarse
grid wall:

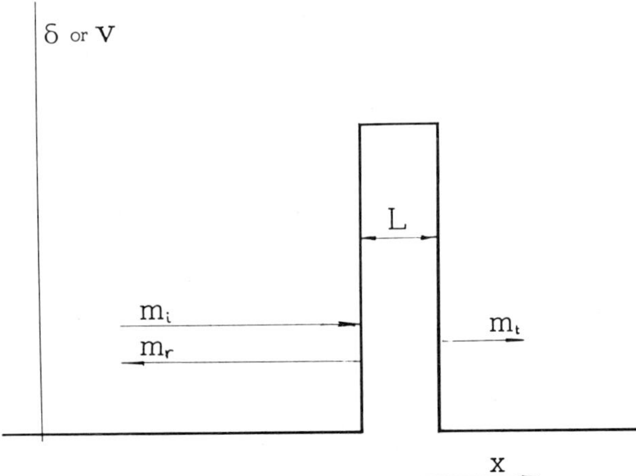

Fig. 9.1. Potential wall of finite thickness created by a coarsen-
ing of the mesh. The mass m_i of a numerical wave packet incident
from the left is not entirely reflected (as m_r) but partially tunnels across the potential wall.

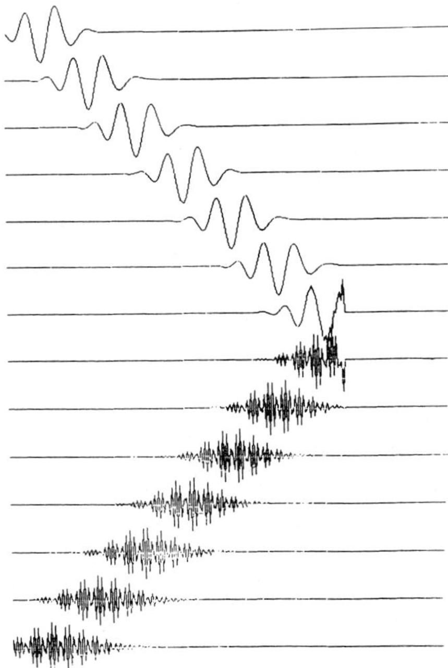

Fig. 9.2. Reflection of a numerical wave packet (or "particle") by
the potential wall illustrated in Fig. 9.1. If the coarse mesh did
extend to infinity, then the reflected mass would be strictly equal
to the incident mass. But with a coarse grid over a region of finite (small) thickness, some of the mass tunnels across the wall
(the amplitude of the evanescent solution that exists to the right
of the reflection point is not zero, but is too small to be visible
on this figure).

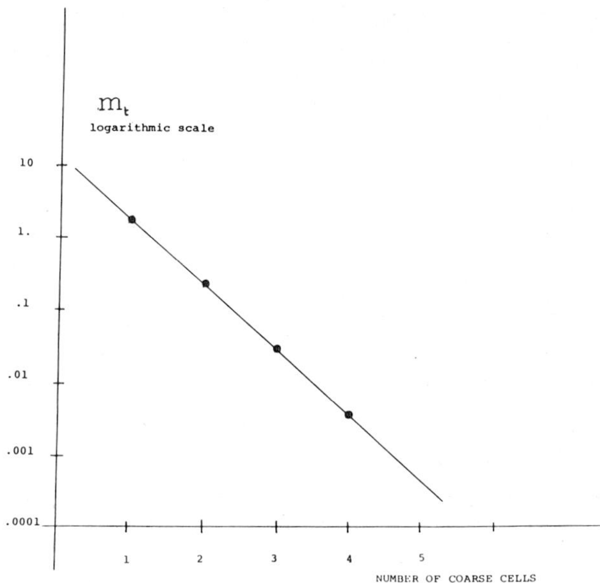

Fig. 9.3. Mass having tunneled (m_t) measured in the numerical experiment described in Figs. 9.1 and 9.2. That the experimental points lie on a straight line of this semi-log graph expresses the exponential decay of m_t with L.

$$\frac{1}{2} \, (\mu(\omega)\delta_0)^2 < H = \frac{c^2}{2} < \frac{1}{2} \, (\mu(\omega)\delta_1)^2 \tag{9.2}$$

According to the particle model, total reflection would occur and the wave group would return entirely toward the left. But this is not what happens.

Before analyzing this in more detail, we recall the phenomenon of tunneling in physics: When a beam of monoenergetic elementary particles collides with a potential wall of finite thickness and of a height which exceeds the particles' energy, then classical mechanics would predict that all the particles will be reflected by the potential wall. But by quantum mechanics theory, the probability of the presence of a particle inside of the wall is not zero. As was mentioned earlier, it is given by an evanescent solution of a corresponding Schrodinger's equation, decaying exponentially with distance from the incident surface. There is a finite probability that particles will reach the other end of the wall, exit and not return, having "tunneled" across the potential barrier.

The same applies to numerical solutions. With L finite, it is observed experimentally that a wave group which satisfies (9.2) is not entirely reflected, but that it partially "tunnels" through the coarse mesh; some of its mass is found to the right of the wall after reflection. The analytic explanation is identical to that which applies to elementary particles; during the process of total reflection of the numerical wave packet against the potential wall, a transient evanescent solution (i.e., a solution of the general form (8.3) but with an amplitude which varies slowly with time) is generated inside the wall. When this wall is of infinite thickness, then all the mass contained in that transient evanescent solution eventually returns with the reflected wave packet, as described in Section 8. But with a wall of finite thickness, some of the mass contained in the tail of the evanescent solution reaches the other end of the wall and escapes by tunneling. Given in Figs. 9.1-9.3 are the results of a numerical experiment which verifies the numerical tunneling of a single wave packet.

As in the preceding example, the remaining difference between numerical and quantum mechanics tunneling is that the former deals with a single wave packet, a piece of which tunnels across the coarse mesh, while the latter deals with many particles in a beam, which each have a finite probability to tunnel across the potential wall. Here, too, the numerical experiment may be repeated with a sustained monochromatic wave which is the analog of a beam of monoenergetic physical particles [25].

10. CLOSING COMMENTS

We have illustrated in this paper the close analogy that is found to exist between numerical wave propagation and quantum mechanics. The obvious question to ask is where this stems from.

The first part of this analogy, that with classical mechanics, has a source which is easily traced; equation (2.2) is not a first order difference equation but is of second order, as are the equations of mechanics. Indeed, this equation may, modulo a relabeling of variables, be cast in the form of a staggered central difference approximation of the (second order) wave equation written in the form of a first order system [5,26].

The analogy with quantum mechanics is most vividly illustrated in experiments with wave packets. What those experiments bring to the fore is the fact that the particle/wave duality of matter is also found to apply to those numerical wave packets and that the corresponding special phenomena, in particular scattering

and those involving evanescent solutions of Schrodinger's equation
(such as in tunneling) are present in both cases. But it may be
argued that the computing domains of numerical analysis are no
different in that respect from many other physical dispersive
media, that they indeed closely resemble several known physical
periodic structures in which wave propagation has been analyzed
in great detail [cf. Brillouin's "Wave propagation in periodic
structures" (1946)] and that the analogy must also apply to wave
propagation in those and other dispersive media. While this must
indeed be the case, one finds little reference to this fact in
the relevant literature. An explanation of this may be found in
that numerical wave propagation has attributes not found elsewhere. There is, of course, the fact that numerical systems are
discrete in time. But this is of no real consequence. What is
of consequence is that in numerical wave propagation (and only in
numerical wave propagation) does one have the possibility of creating dispersive media (periodic structures) with well defined
space dependent properties, of conducting experiments with precisely formulated initial conditions (in particular, Gaussian wave
packets) and of measuring with great accuracy every characteristic
of the subsequent time history. Those special phenomena which are
only theoretical entities in physical systems where they can hard-
ly be observed become real entities in numerical computing domains.

As may be expected, a more detailed analysis of the analogy
between quantum mechanics and numerical wave propagation reveals
common foundations that go well beyond the similarity of observations reported in this paper; this commonality stems of course
from the fact that both classes of phenomena are concerned with
harmonic wave propagation in which certain variables may change
only in discrete steps or quanta (energy in the former case, spatial position in the latter). The investigation of this question
is somewhat beyond the scope of the present paper, and is reported
separately in [25].

REFER   REFERENCES

1. d'Abro, A., THE RISE OF THE NEW PHYSICS, Dover Publications,
      1952.

2. Born, M., and K. Huang, DYNAMICAL THEORY OF CRYSTAL LATTICES,
      Oxford University Press, 1954.

3. Brillouin, L., WAVE PROPAGATION IN PERIODIC STRUCTURES, McGraw Hill, New York, 1946.

4. Browning, G., H. D. Kreiss, and J. Oliger, Mesh refinement, MATHEMATICS OF COMPUTATION 27 (1973), 29-39.

5. Chin, R. C. Y., and G. W. Hedstrom, Scattering of waves from a staggered difference scheme on a variable grid, PROCEEDINGS OF THE 10TH IMACS WORLD CONGRESS, Montreal, 1982.

6. Cohen-Tadoudji, C., B. Diu, and F. Laloe, QUANTUM MECHANICS, J. Wiley and Sons, New York, 1977.

7. Giles, M. B., and W. T. Thompkins, Jr., Propagation and stability of wavelike solutions of finite difference equations with variable coefficients, JOURNAL OF COMPUTATIONAL PHYSICS 58 (1985), 349-360.

8. Goldstein, H., CLASSICAL MECHANICS, Addison Wesley, New York, 1950.

9. Hedstrom, G. W., Models of difference schemes for $Ut + Ux = 0$ by partial differential equations, MATHEMATICS OF COMPUTATION 29, No. 132 (1975), 969-977.

10. Jameson, A., W. Schmidt, and E. Turkel, Numberical solution of the Euler equations by finite-volume methods using Runge-Kutta time stepping schemes, AMERICAN INSTITUTE OF AERONAUTICS AND ASTRONAUTICS JOURNAL (1981), Paper 81-1259.

11. Kreiss, H., and J. Oliger, Methods for the approximate solution of time dependent problems, GARP Publication Series No. 10, World Meteorological Organization, Geneva, 1973.

12. Lighthill, J., WAVES IN FLUIDS, Cambridge University Press, 1978.

13. Messiah, A., QUANTUM MECHANICS, J. Wiley and Sons, New York, 1959.

14. Pain, H. J., THE PHYSICS OF VIBRATIONS AND WAVES, J. Wiley and Sons, New York, 1968 (3d edition, 1983).

15. Papoulis, A., THE FOURIER INTEGRAL AND ITS APPLICATIONS, McGraw Hill, New York, 1962.

16. Trefethen, L. N., Group velocity of finite difference schemes, SIAM REVIEW 23 (1982), 113-136.

17. Vichnevetsky, R., Energy and group velocity in semidiscretizations of hyperbolic equations, MATHEMATICS OF COMPUTERS IN SIMULATION 23 (1981), North Holland, 333-343.

18. Vichnevetsky, R., Propagation through numerical mesh refinement for hyperbolic equations, MATHEMATICS OF COMPUTERS IN SIMULATION 23 (1981), North Holland, 344-353.

19. Vichnevetsky, R., Group velocity and reflection phenomena in numerical approximations of hyperbolic equations, JOURNAL OF THE FRANKLIN INSTITUTE (1983),

20. Vichnevetsky, R., Propagation and spurious reflection in finite element approximations of hyperbolic equations, COMPUTERS AND MATHEMATICS WITH APPLICATIONS 11 (1985), 733-746.

21. Vichnevetsky, R., Partical like properties of numerical wave propagation in irregular grids, Report MAE 1714, Princeton University, Mechanical and Aerospace Eng. Dept., 1985.

22. Vichnevetsky, R., Invariance theorems concerning reflection at numerical boundaries, JOURNAL OF COMPUTATIONAL PHYSICS 63 (1986), 268-282.

23. Vichnevetsky, R., Wave propagation and reflection in irregular grids for hyperbolic equations, APPLIED NUMERICAL MATHEMATICS, vol. 2, no. 1-2 (1987), North Holland.

24. Vichnevetsky, R., Wave propagation analysis of different schemes for hyperbolic equations: A review, INTERNATIONAL JOURNAL FOR COMPUTER METHODS IN FLUIDS, J. Wiley and Sons, New York, 1987.

25. Vichnevetsky, R., Quantum phenomena in numerical wave propagation, in ADVANCES IN COMPUTER METHODS FOR PARTIAL DIFFERENTIAL EQUATIONS VI, IMACS, New Brunswick, N.J., 1987.

26. Vichnevetsky, R., and J. B. Bowles, FOURIER ANALYSIS OF NUMERICAL APPROXIMATIONS OF HYPERBOLIC EQUATIONS, Society for Industrial and Applied Mathematics, Philadelphia, 1982.

27. Whitham, G. B., LINEAR AND NONLINEAR WAVES, J. Wiley and Sons, New York, 1974.

COMPUTATIONAL ACOUSTICS: Wave Propagation
D. Lee, R.L. Sternberg, M.H. Schultz (Editors)
Elsevier Science Publishers B.V. (North-Holland)
© IMACS, 1988

VARIATIONAL METHODS TO IMPROVE COMPUTATIONAL ALGORITHMS
FOR ACOUSTIC RADIATION AND SCATTERING

Allan D. Pierce

School of Mechanical Engineering
Georgia Institute of Technology
Atlanta, Georgia

ABSTRACT

Kirchhoff-Helmholtz integral corollaries of the wave equation
express acoustic pressure at either interior or exterior points in
terms of pressure and its normal derivative (acceleration multiplied by mass density) over any closed surface. These lead to a
variety of integral or differential-integral equation formulations
of acoustic radiation and scattering involving rigid, perfectly
soft, or elastic bodies of non-standard shapes. The dependent
function is a field quantity distributed over the surface. Computational algorithms based on such formulations are widely used,
but are still not wholly satisfactory. The present paper develops
arguments to the effect that variational formulations ensuing from
the corollaries offer possibilities of improving upon the existing
algorithms or of developing new superior algorithms, and shows
that it is often possible to express a variational principle as a
stationary functional whose value is a quantity of direct physical
interest, such that mediocre estimates of the dependent function
yield accurate estimates of that quantity. The present work can
be viewed as an extension (although restricted to acoustics) of
work done by Schwinger, Levine, and others in the 1940s and 1950s,
whereby appropriate stationary expressions for total radiated
power and target strength are constructed using a technique introduced in more recent times by Gerjuoy, Rau, and Spruch. In some
instances, the so-constructed stationary expression will require
estimation of functions ancillary to the actual problem of interest, but it is demonstrated that the construction can be carried
through such that these ancillary quantities are the solutions to
closely related physical problems.

I. INTRODUCTION

Variational formulations of physical problems are often convenient departure points for introducing approximations; they help

to maintain a degree of self-consistency that might otherwise be absent. Also, they offer some assurance that, when one seeks to approximate a solution by a function of restricted (but initially incompletely specified) type with, for example, various adjustable parameters, the variational formulation will automatically select, from among all such functions of such a type, one which in some sense is optimal.

A disconcerting feature of many variational formulations, however, is that one often does not necessarily know at the outset the actual sense in which such an approximate solution is optimal. Nevertheless, it is a strong philosophical tenet among many theoretical physicists, engineers, and applied mathematicians that, all other things being seemingly equal and provided the variational formulation was constructed in a natural manner, without any artificial mathematical contrivances, the approximation yielded by the variational formulation is vastly to be preferred.

There is a class of variational principles, however, whose appeal does not rest on aesthetic considerations alone. These are those of the generic form $\delta J = 0$ where the stationary quantity J is a measurable physical quantity J. Rayleigh's principle [1] for vibrations, in which J is the square of the natural frequency, is perhaps the best known of such variational principles. Others are Fermat's principle of stationary travel time for optical [2] and acoustical [3] ray paths and the principle of minimum potential energy [4] in statics. Ideally, one would prefer principles that enable one to "keep score" on the goodness of alternative approximations. For example, when applying Rayleigh's principle to determine the lowest natural frequency and the fundamental mode's shape, one takes the "best" choice as being that approximate trial function which, when inserted into the variational expression, yields the lowest estimate for ω^2, and the latter is indisputably an upper bound to the actual lowest ω^2. True minimum or maximum principles are relatively scarce, and are probably nonexistent for wave propagation problems, except in the limit of extremely low frequencies. Nevertheless, stationarity principles do retain some features that could be of considerable interest to one seeking improved algorithms for physical problems.

One might, for example, place high priority on the accurate estimate of the physical quantity J which corresponds to the variational expression J. The latter could be a functional of some other physical quantity $\psi(\eta)$ that depends on some position parameter

η. One might have some insight into the nature of the unknown function $\psi(\eta)$ and use this insight and the variational principle to refine the quantitative estimate of the function. The resulting approximation to the actual $\psi(\eta)$ might be characterized as being "good but not great," or simply as being "mediocre." However, this mediocre estimate for $\psi(\eta)$, when inserted into the variational expression, should result in a highly accurate estimate of the physical quantity J. The reason for this is that the error in J should be of second order when the error in $\psi(\eta)$ is taken as first order.

Variational principles with the feature just described were developed for a number of wave propagation applications during the 1940s and 1950s by Schwinger, Levine, Lippmann, and others [5-9]; a good summary of such can be found in the second volume of Morse and Feshbach's [10] 1953 treatise; a number of papers [11-15] appeared that were in this general spirit and specifically directed toward acoustical applications. However, the enthusiasm within the acoustical community for such methods waned considerably during the subsequent two decades [16], presumably because of the computational efforts involved and the formidable sophistication of the analytical techniques that had been used in prior implementations of such variational formulations. Another factor was undoubtedly the widespread belief that anything accomplishable using a variational formulation could be equally well accomplished using Galerkin's method [17]. [Debating this point would unduly lengthen the present paper.]

In regard to acoustic radiation and scattering by bodies of nonstandard shape, the variational principles commonly presented in the literature up until relatively recently had integrals over singular integrands, and the procedure for direct numerical evaluation seemed undefined [18]. In retrospect, however, this was easily circumventable [19] using techniques applied by Maue [20] and Stallybrass [21] in earlier formulations of integral equations for similar problems; variational principles not involving singularity difficulties were independently developed by Hamdi [22] and by the present author [23], and were used in a number of applications by Ginsberg, Wu, and others [24-26]. However, these were not associated at the outset with the stationarity of a desirable physical quantity, so the question of the sense in which the variational solution is "optimal" remained unanswered.

Relatively recently, the author's attention was drawn to a
1983 paper (which is actually the culmination of a long series of
papers) by Gerjuoy, Rau, and Spruch [27] that provides the key to
the answer to the question just posed and which, moreover, gives
a general technique for constructing variational principles that
allow accurate estimation of any physical quantity of interest.
The present paper shows how this technique can be applied to the
formulation of novel and practical variational principles for
acoustic radiation and scattering.

II. KIRCHOFF-HELMHOLTZ COROLLARIES FOR BOUNDARY VALUES

Before proceeding with the construction of the variational
principles discussed in the present paper, it is convenient to
first define a number of operators that apply for problems of
acoustic radiation and scattering.

With any given closed surface S one may in general associate
a number of different types of acoustic radiation and scattering
problems. Although one may be interested in solving only one such
problem, some knowledge of related problems involving the same
surface may be helpful in producing a good approximation for the
problem of interest. In general, one distinguishes _interior_ and
exterior problems [28]. For the former, the complex amplitude of
the acoustic pressure p_{int} satisfies the scalar Helmholtz equation
throughout the volume V enclosed by S. For exterior problems, the
complex amplitude p_{ext} satisfies the Helmholtz equation throughout
the region external to S. In many cases such exterior problems
can be idealized such that the external region is unbounded in all
directions; even when there are external boundaries, the solution
for an unbounded external environment may be a valuable building
block in the construction (using ray concepts or the method of
images) of the solution for the bounded external region. The
present discussion assumes, for simplicity, that the external region is unbounded.

Exterior problems can be classified as _radiation_ or _scatter-
ing_ problems. For radiation problems, the complex pressure amplitude p_{ext} satisfies the Sommerfeld radiation condition at great
distances from the source. For scattering problems, p_{ext} is regarded as a sum of an incident part $p_{ext,inc}$ and a scattered part
$p_{ext,sc}$, where the former is taken as given throughout the external region, while the latter satisfies the Sommerfeld radiation
condition. Consequently, general statements that one might make

for radiated fields apply equally well to scattered fields. For
brevity, the expression p_{ext} is understood throughout the present
section to refer to either the radiated field for a radiation problem or the scattered field for a scattering problem.

Following well-known procedures [29], one can derive <u>Kirch-
hoff-Helmholtz integral corollaries</u> to the wave equation for both
the interior problem and the exterior problem. These can be written in such a form that they relate the appropriate acoustic pressure p on the surface to a quantity f which is the negative of
the outward normal component of the gradient of p at the surface.
For the interior problem one has

$$M\{\vec{x}, p_{int}, f_{int}\} = \begin{cases} -p_{int}(\vec{x}) & \text{if } \vec{x} \text{ inside } V \\ 0 & \text{if } \vec{x} \text{ outside } V \end{cases} \tag{1}$$

while for the exterior problem one has

$$M\{\vec{x}, p_{ext}, f_{ext}\} = \begin{cases} p_{ext}(\vec{x}) & \text{if } \vec{x} \text{ outside } V \\ 0 & \text{if } \vec{x} \text{ inside } V \end{cases} \tag{2}$$

Both Eqs. (1) and (2) use the abbreviation

$$M\{\vec{x}, p, f\} = \frac{1}{4\pi} \iint [f(\vec{\xi}) G(\vec{x}|\vec{\xi}) + p(\vec{\xi}) \vec{n}(\vec{\xi}) \nabla_\xi G(\vec{x}|\vec{\xi})] \, dS_\xi \tag{3}$$

where

$$G(\vec{x}|\vec{\xi}) = \frac{e^{ikR}}{R} \tag{4}$$

is the so-called "free space" Green's function with

$$R = |\vec{x} - \vec{\xi}| \tag{5}$$

denoting the distance between "source" and "receiver" points. In
the integrand of Eq. (3), the point $\vec{\xi}$ (after evaluation of any
requisite normal derivatives) is understood to range over the surface S, with the point $\vec{x}$ held fixed during the integration. The
unit outward normal vector $\vec{n}(\vec{\xi})$ points out of the enclosed volume
V at the surface point $\vec{\xi}$. The scalar quantity $f(\vec{\xi})$ can be re-
garded as $-i\omega\rho v(\vec{\xi})$ where $v(\vec{\xi})$ is the normal component $\vec{n}(\vec{\xi}) \cdot \vec{v}(\vec{\xi})$
of the complex fluid velocity vector amplitude $\vec{v}(\vec{\xi})$; in accord
with Euler's equation of motion for a fluid, one has

$$-i\omega\rho\vec{v}(\vec{\xi}) = -\nabla_\xi p(\vec{\xi}) \tag{6}$$

In Eq. (3) one should note that the integral M is a function
of the point $\vec{x}$, but a functional (function of a function) of the
function arguments p and f.

From either Eq. (1) or Eq. (2), one can derive two types of relations (distinguished by subscripts I and II) between the surface values of p and f. One such type of relation results [28] when the off-surface point $\vec{x}$ is allowed to approach an arbitrary but fixed surface point $\vec{\eta}$. For one of the terms in the integral defining the quantity M, the limit as $\vec{x}$ approaches $\vec{\eta}$ of the integral is not the same as the integral over the limit of the integrand as $\vec{x}$ approaches $\vec{\eta}$. Instead, one has [30],

$$\lim_{\varepsilon \to 0} \iint p(\vec{\xi})\vec{n}(\vec{\xi})\nabla_\xi G(\vec{x}_\varepsilon|\vec{\xi})\,dS_\xi = \pm 2\pi p(\vec{\eta}) + \iint p(\vec{\xi})\vec{n}(\vec{\xi}) \times \nabla_\xi G(\vec{\eta}|\vec{\xi})\,dS_\xi \tag{7}$$

where

$$\vec{x}_\varepsilon = \vec{\eta} + \varepsilon\vec{n}(\vec{\eta}) \tag{8}$$

The plus sign in Eq. (7) applies if the limit is taken with ε kept positive during the process, while the minus sign applies if it is kept negative during the process. With this subtlety taken into account, one obtains

$$p_{int}(\vec{\eta}) + L_I\{\vec{\eta},p_{int}\} = -H_I\{\vec{\eta},f_{int}\} \tag{9}$$

$$p_{ext}(\vec{\eta}) - L_I\{\vec{\eta},p_{ext}\} = H_I\{\vec{\eta},f_{ext}\} \tag{10}$$

where

$$L_I\{\vec{\eta},p\} = \frac{1}{2\pi} \iint p(\vec{\xi})\vec{n}(\vec{\xi}) \cdot \nabla_\xi G(\vec{\eta}|\vec{\xi})\,dS_\xi \tag{11}$$

$$H_I\{\vec{\eta},f\} = \frac{1}{2\pi} \iint f(\vec{\xi})G(\vec{\eta}|\vec{\xi})\,dS_\xi \tag{12}$$

such that the symbols L_I and H_I can be regarded as linear operators which operate on the surface values of p and f, respectively, with the resultant in each case being a function of the position of the surface point $\vec{\eta}$.

The second type of surface relationship is obtained by taking the gradient of both sides of Eq. (1) or Eq. (2), subsequently setting $\vec{x}$ to $\vec{\eta} + \varepsilon\vec{n}(\vec{\eta})$, where $\vec{\eta}$ is an arbitrary point on the surface, taking the dot product with $\vec{n}(\vec{\eta})$, then taking the limit as ε goes to zero. To express the limit of one of the integrals as the integral of a limit, it is appropriate to make use of a relation previously derived by Maue [20] and by Stallybrass [21], holding for $\varepsilon \neq 0$, to the effect that

$$(\vec{n}(\vec{\eta})\cdot\nabla_x)(\vec{n}(\vec{\xi})\cdot\nabla_\xi)G(\vec{x}|\vec{\xi}) = k^2\vec{n}(\vec{\eta})\cdot\vec{n}(\vec{\xi})G(\vec{x}|\vec{\xi})$$
$$- (\vec{n}(\vec{\xi})\times\nabla_x)\cdot(\vec{n}(\vec{\xi})\times\nabla_\xi)G(\vec{x}|\vec{\xi}) \tag{13}$$

(The derivation makes use of the symmetry of the Green's function and of the fact that the Green's function satisfies the Helmholtz equation for $\vec{x} \neq \vec{\xi}$.) Moreover, a version of Stokes' theorem allows a type of integration by parts, by which the operator $\vec{n}(\vec{\xi}) \times \nabla_\xi$ is transferred, with the usual accompanying change in sign, from $G(\vec{x}|\vec{\xi})$ to $p(\xi)$. A relation analogous to Eq. (7) is also used to transform one of the terms. In such a manner, one obtains the relations

$$L_{II}\{\vec{n}, p_{int}\} = f_{int}(\vec{n}) - H_{II}\{\vec{n}, f_{int}\} \tag{14}$$

$$-L_{II}\{\vec{n}, p_{ext}\} = f_{ext}(\vec{n}) + H_{II}\{\vec{n}, f_{ext}\} \tag{15}$$

where

$$L_{II}\{\vec{n}, p\} = (\vec{n}(\vec{n}) \times \nabla_\eta)\,\frac{1}{2\pi} \iint (\vec{n}(\vec{\xi}) \times \nabla_\xi p(\vec{\xi}))G(\vec{n}|\vec{\xi})\ dS_\xi$$

$$+ \frac{k^2}{2\pi} \iint \vec{n}(\vec{n}) \cdot \vec{n}(\vec{\xi}) p(\vec{\xi}) G(\vec{n}|\vec{\xi})\ dS_\xi \tag{16}$$

and

$$H_{II}\{\vec{n}, f\} = \frac{1}{2\pi} \iint f(\vec{\xi})\,\vec{n}(\vec{n}) \cdot \nabla_\eta G(\vec{n}|\vec{\xi})\ dS_\xi \tag{17}$$

In regard to Eq. (16), one should note that the operator $\vec{n}(\vec{n}) \times \nabla_\eta$ involves only derivatives tangential to the surface, so that the integral on which it acts need only be evaluated at surface points $\vec{n}$.

III. ADJOINT OPERATORS

Having now identified characteristic surface operators L_I, L_{II}, H_I, H_{II}, it is appropriate for the derivations given further below involving variational expressions to identify their corresponding adjoint operators. The notation of the present paper distinguishes such by the superscript †; their definitions are such that if $\Phi(\vec{n})$ and $\Psi(\vec{n})$ are functions defined over the surface S, then, for example,

$$\iint \Psi(\vec{n})\,L_I\{\vec{n}, \Phi\}dS_\eta = \iint \Phi(\vec{n})\,L_I^\dagger\{\vec{n}, \Phi\}dS_\eta \tag{18}$$

Using such a definition of an adjoint operator, one identifies with reference to Eqs. (11), (12), (16), and (17) that

$$L_I^\dagger\{\vec{n}, p\} = \frac{1}{2\pi} \iint p(\vec{\xi})n(\vec{n}) \cdot \nabla_\eta G(\vec{\xi}|\vec{n})dS_\xi = H_{II}\{\vec{n}, p\} \tag{19}$$

$$H_I^\dagger\{\vec{n},f\} = \frac{1}{2\pi} \iint f(\vec{\xi})G(\vec{\xi}|\vec{n})\,dS_\xi = H_I\{\vec{n},f\} \tag{20}$$

$$L_{II}^\dagger\{\vec{n},p\} = (\vec{n}(\vec{n}) \times \nabla_\eta)\cdot\frac{1}{2\pi} \iint (\vec{n}(\vec{\xi}) \times \nabla_\xi p(\vec{\xi}))G(\vec{\xi}|\vec{n})\,dS_\xi$$

$$+ \frac{k^2}{2\pi} \iint \vec{n}(\vec{n})\cdot\vec{n}(\vec{\xi})p(\vec{\xi})G(\vec{\xi}|\vec{n})\,dS_\xi$$

$$= L_{II}\{\vec{n},p\} \tag{21}$$

$$H_{II}^\dagger\{\vec{n},f\} = \frac{1}{2\pi} \iint f(\vec{\xi})\vec{n}(\vec{\xi}) \cdot \nabla_\xi G(\vec{\xi}|\vec{n})\,dS_\xi = L_I\{\vec{n},f\} \tag{22}$$

Thus the operators H_I and L_{II} are self-adjoint, while L_I and H_{II} are adjoints of each other (an adjoint pair).

IV.   STATIONARY EXPRESSION FOR RADIATED POWER

A quantity often of interest is the net radiated power for a radiation problem or the net scattered power for a scattering problem. With the nomenclature and conventions introduced in the previous sections, this can be expressed as 1/2 of the real part of the surface integral of $p_{ext}(\vec{n})v^*_{ext}(\vec{n})$ or as

$$\text{Power} = \frac{1}{4i\omega\rho} \iint [p_{ext}f^*_{ext} - p^*_{ext}f_{ext}]\,dS_\eta \tag{23}$$

In what follows the method of Gerjuoy, Rau, and Spruch [27] is adopted to derive variational expressions that will yield accurate estimates of this quantity when one or the other of the quantities f_{ext} or p_{ext} is specified on the surface S. For scattering problems, a priori knowledge of f_{ext} on the surface would be appropriate for the problem of scattering by a perfectly rigid body, while a priori knowledge of p_{ext} on the surface would be appropriate for scattering by a perfectly soft body (where sum of incident and scattered pressure vanishes on S).

To construct an appropriate stationary expression which yields the radiated power when $f_{ext}(\vec{n})$ is known at the outset, one sets

$$J = \iint \tilde{p}_{ext}(\vec{n})f^*(\vec{n})\,dS_\eta$$

$$+ \iint \tilde{\Phi}_I(\vec{n})[\tilde{p}_{ext}(\vec{n}) - L_I\{\vec{n},\tilde{p}_{ext}\} - H_I\{\vec{n},f\}]\,dS_\eta$$

$$+ \iint \tilde{\Phi}_{II}(\vec{n})[L_{II}\{\vec{n},\tilde{p}_{ext}\} + f(\vec{n}) + H_{II}\{\vec{n},f\}]\,dS_\eta \tag{24}$$

where $\tilde{p}_{ext}(\vec{n})$ is a trial function for $p_{ext}(\vec{n})$, while $\tilde{\Phi}_I$ and $\tilde{\Phi}_{II}$ are regarded as trial functions for variables $\Phi_I(\vec{n})$ and $\Phi_{II}(\vec{n})$; the untilded Φ's will eventually be defined as the exact (although perhaps a priori unknown) solutions of related problems. Note that the subscript "ext" has been dropped from f_{ext}; since $f(\vec{\eta})$ is here a prescribed boundary value, it is not necessarily associated with the exterior region.

If $\tilde{p}_{ext}$ is identically p_{ext}, then the second and terms on the right side of Eq. (24) vanish by virtue of Eqs. (10) and (15), and the radiated power is

$$\text{Power} = \frac{1}{2\omega\rho} \, \text{Im}\{J\} \tag{25}$$

What is desired is that the error in Eq. (25) be at most only of second order in the variations of p_{ext}, Φ_I, and Φ_{II} when slightly erroneous functions $\tilde{p}_{ext}$, $\tilde{\Phi}_I$, and $\tilde{\Phi}_{II}$ are used in Eq. (24) to calculate J. This will be so if the expression on the right side of (24) is stationary under independent variations of these three quantities. This requirement, with application of the techniques of the calculus of variations, yields the three equations

$$p_{ext}(\vec{\eta}) - L_I\{\vec{\eta},p_{ext}\} - H_I\{\vec{\eta},f\} = 0 \tag{26a}$$

$$L_{II}\{\vec{\eta},p_{ext}\} + f(\vec{\eta}) + H_{II}\{\vec{\eta},f\} = 0 \tag{26b}$$

$$f^*(\vec{\eta}) + \Phi_I(\vec{\eta}) - L_I^\dagger\{\vec{\eta},\Phi_I\} + L_{II}^\dagger\{\vec{\eta},\Phi_{II}\} = 0 \tag{26c}$$

The first two of the above three equations are the same as Eqs. (10) and (15); the development of Section II shows that they are mathematically consistent (not contradictory). This implies that one is at liberty to choose either Φ_I or Φ_{II} to be zero, or to impose any independent relation among these two quantities.

Ideally, one would want the unknown functions that are used in the estimate of the radiated power to have tangible physical interpretations, for such would facilitate their estimation. To this purpose, one rewrites Eq. (26c) using the adjoint relations (19) and (20) as

$$f^*(\vec{\eta}) + \Phi_I(\vec{\eta}) - H_{II}\{\vec{\eta},\Phi_I\} + L_{II}\{\vec{\eta},\Phi_{II}\} = 0 \tag{27}$$

Then, after a comparison with Eqs. (14) and (15), a simple choice becomes apparent:

$$\Phi_I(\vec{\eta}) = -\frac{1}{2} f^*(\vec{\eta}) \tag{28a}$$

$$\Phi_{II}(\vec{\eta}) = \frac{1}{2} p_{ext,aux}(\vec{\eta}) \tag{28b}$$

where $p_{ext,aux}(\vec{x})$ is the solution of an auxiliary exterior problem with the surface value of f specified as

$$f_{aux}(\vec{\eta}) = f^*(\vec{\eta}) \tag{28c}$$

Given the choice described by Eqs. (28), the expression for the stationary quantity J reduces (omitting the tildes and the subscript "ext" for brevity) after some manipulations to

$$
\begin{aligned}
J = \frac{1}{2} \iint & [p(\vec{\eta})f_{aux}(\vec{\eta}) + p_{aux}(\vec{\eta})f(\vec{\eta})]dS_{\eta} \\
+ \frac{1}{4\pi} \iiiint & [p(\vec{\eta})f_{aux}(\vec{\xi}) + p_{aux}(\vec{\eta})f(\vec{\xi})]\vec{n}(\vec{\eta}) \cdot \nabla_{\eta}G(\vec{\eta}|\vec{\xi})dS_{\xi}dS_{\eta} \\
- \frac{1}{4\pi} \iiiint & (\vec{n}(\vec{\eta}) \times \nabla_{\eta}p_{aux}(\vec{\eta})) \cdot (\vec{n}(\vec{\xi}) \times \nabla_{\xi}p(\vec{\xi}))G(\vec{\eta}|\vec{\xi})dS_{\xi}dS_{\eta} \\
+ \frac{k^2}{4\pi} \iiiint & \vec{n}(\vec{\eta}) \cdot \vec{n}(\vec{\xi})p_{aux}(\vec{\eta})p(\vec{\xi})G(\vec{\eta}|\vec{\xi})dS_{\xi}dS_{\eta} \\
+ \frac{1}{4\pi} \iiiint & f_{aux}(\vec{\eta})f(\vec{\xi})G(\vec{\eta}|\vec{\xi})dS_{\xi}dS_{\eta}
\end{aligned} \tag{29}
$$

In the variation principle $\delta J = 0$ that corresponds to Eq. (29), the functions p and p_{aux} are to be varied independently, so the corresponding Euler-Lagrange equations are

$$f_{aux}(\vec{\eta}) + H_{II}\{\vec{\eta},f_{aux}\} + L_{II}\{\vec{\eta},p_{aux}\} = 0 \tag{30a}$$

$$f(\vec{\eta}) + H_{II}\{\vec{\eta},f\} + L_{II}\{\vec{\eta},p\} = 0 \tag{30b}$$

both of which are of the form of Eq. (15). Note that one does not recover Eq. (10), even though it was used during the construction.

If the acoustic pressure $p_{ext}(\vec{\eta})$ [rather than $f_{ext}(\vec{\eta})$] is specified on the surface S at the outset and $f_{ext}(\vec{\eta})$ is unknown, an analogous stationary expression for the radiated power can be similarly constructed. Such a development yields

$$\text{Power} = \frac{1}{2\omega\rho} \text{Im}\{N\} \tag{31}$$

$$
\begin{aligned}
N = - \frac{1}{2} \iint & [p(\vec{\eta})f_{aux}(\vec{\eta}) + p_{aux}(\vec{\eta})f(\vec{\eta})]\ dS_{\eta} \\
+ \frac{1}{4\pi} \iiiint & [p(\vec{\eta})f_{aux}(\vec{\xi}) + p_{aux}(\vec{\eta})f(\vec{\xi})]\vec{n}(\vec{\eta}) \\
& \times \nabla_{\eta}G(\vec{\eta}|\vec{\xi})dS_{\xi}dS_{\eta} \\
- \frac{1}{4\pi} \iiiint & (\vec{n}(\vec{\eta}) \times \nabla_{\eta}p_{aux}(\vec{\eta})) \cdot (\vec{n}(\vec{\xi}) \times \nabla_{\xi}p(\vec{\xi}))G(\vec{\eta}|\vec{\xi})dS_{\xi}dS_{\eta} \\
+ \frac{k^2}{4\pi} \iiiint & \vec{n}(\vec{\eta}) \cdot \vec{n}(\vec{\xi})p_{aux}(\vec{\eta})p(\vec{\xi})G(\vec{\eta}|\vec{\xi})dS_{\xi}dS_{\eta} \\
+ \frac{1}{4\pi} \iiiint & f_{aux}(\vec{\eta})f(\vec{\xi})G(\vec{\eta}|\vec{\xi})dS_{\xi}dS_{\eta}
\end{aligned} \tag{32}
$$

where

$$p_{aux}(\vec{\eta}) = p^*(\vec{\eta}) \tag{33}$$

and N is stationary to first order when either $f(\vec{n})$ or f_{aux} are varied.

Note that the sole difference between the form of Eq. (33) and that of Eq. (29) is the sign of the first term. The Euler-Lagrange equations ensuing from $\delta N = 0$ are considerably different, however, these being

$$p_{aux}(\vec{n}) - L_I\{\vec{n}, p_{aux}\} - H_I\{\vec{n}, f_{aux}\} = 0 \tag{34a}$$

$$p(\vec{n}) - L_I\{\vec{n}, p\} - H_I\{\vec{n}, f\} = 0 \tag{34b}$$

and are of the form of Eq. (10) rather than of Eq. (15).

Returning to the expression (29) which yields a variational expression for the acoustic pressure $p(\vec{n})$, one may note that a considerable simplification results if each point on the surface is vibrating either exactly in phase or exactly 180° out of phase with other points. Then $f^*(\vec{n})$ is $Kf(\vec{n})$, where K is a complex constant of magnitude unity. It follows from the linearity of Eqs. (30) that $p_{aux} = Kp$, so one need not vary p and p_{aux} separately when using the variational principle. Instead, one can insert Kp for p_{aux} at the outset into Eq. (29). Since one can always choose the time origin such that K is unity, no generality is lost if one simply sets p_{aux} to p and f_{aux} to f, with the stipulation that $f(\vec{n})$ is real (positive or negative) at each point on the surface. Doing such yields

$$J = \iint p(\vec{n}) f(\vec{n}) dS_\eta + \frac{1}{2\pi} \iiiint p(\vec{n}) f(\vec{\xi}) \vec{n}(\vec{n}) \cdot \nabla_\eta G(\vec{n}|\vec{\xi}) dS_\xi dS_\eta$$

$$- \frac{1}{4\pi} \iiiint (\vec{n}(\vec{n}) \times \nabla_\eta p(\vec{n})) \cdot (\vec{n}(\vec{\xi}) \times \nabla_\xi p(\vec{\xi})) G(\vec{n}|\vec{\xi}) dS_\xi dS_\eta$$

$$+ \frac{k^2}{4\pi} \iiiint \vec{n}(\vec{n}) \cdot \vec{n}(\vec{\xi}) p(\vec{n}) p(\vec{\xi}) G(\vec{n}|\vec{\xi}) dS_\xi dS_\eta$$

$$+ \frac{1}{4\pi} \iiiint f(\vec{n}) f(\vec{\xi}) G(\vec{n}|\vec{\xi}) dS_\xi dS_\eta \tag{35}$$

The corresponding Euler-Lagrange equation derived from $\delta J = 0$ is Eq. (15).

Insofar as one desires a variational principle for the acoustic pressure (for the exterior problem) on the surface, then that corresponding to Eq. (35) will suffice, regardless of whether or not $f(\vec{n})$ is real; the introduction of the acoustic pressure p_{aux} for the auxiliary exterior problem into the expression for J is necessary only if one desires substantially more accuracy in the radiated power estimate than in the estimate of surface acoustic pressure. A simplified variational principle, equivalent to that of Eq. (35), has been previously derived by other methods independently by Hamdi [22] and the present author [23].

V. STATIONARY EXPRESSION FOR TARGET STRENGTH

Another example of interest to underwater acoustics is the calculation of target strength TS, which is a type of logarithmic measure of the apparent area of a body in backscatter, the definition [29] being

$$TS = 10 \ln\left(\frac{\sigma_{bs}}{\pi R_o^2}\right) \tag{36}$$

where σ_{bs} is the backscattering cross-section and R_o is a reference length normally taken as 1 m. The backscattering cross-section is 4π times the differential cross-section in the backward direction, the latter being the limiting value of square of distance times intensity back toward the source, divided by the incident intensity at the scattering object. A value for the far field backscattered intensity can be formally calculated from Eqs. (2), (3), and (4) by approximating the R in the exponent of the Green's function as $r - e_{bs} \cdot \xi$, where e_{bs} is the unit vector from the target back toward the receiver. With the pressure amplitude of the incident wave normalized to unity, one has

$$\sigma_{bs} = \frac{1}{4\pi} \left|A_{exact}\right|^2 \tag{37}$$

where

$$A_{exact} = \iint e^{ik\cdot\xi}[f_{ext,sc}(\vec{\xi}) + i\vec{k}\cdot\vec{n}(\vec{\xi})p_{ext,sc}(\vec{\xi})]dS_\xi \tag{38}$$

is a surface integral that characterizes the amplitude of the scattered wave. The wave vector $\vec{k}$ that appears here is understood to be that of the incident wave, which is opposite in direction to $\vec{e}_{bs}$.

For simplicity, attention is restricted here to when the scattering object is perfectly rigid, so that the normal component of the scattered part of the fluid velocity at the surface S is opposite to that of the incident wave. The incident wave is a plane wave, so its velocity is $\vec{k}/\omega\rho$ times the acoustic pressure. Consequently, one has

$$f_{ext,sc}(\vec{\xi}) = i\vec{k}\cdot\vec{n}(\vec{\xi})e^{i\vec{k}\cdot\vec{\xi}} \tag{39}$$

In what follows, we abbreviate this quantity simply as $f(\vec{\xi})$ and $p_{ext,sc}$ simply as $p(\vec{\xi})$.

It is evident that a stationary expression for target strength will result if one has a stationary expression for A, so one can follow the procedure outlined in Section IV and set

$$
\begin{aligned}
A = &\iint e^{ik\eta}[f(\vec{\eta}) + ik\vec{n}(\vec{\eta})\tilde{p}(\vec{\eta})]dS_\eta \\
&+ \iint \tilde{\Psi}_I(\vec{\eta})[\tilde{p}(\vec{\eta}) - L_I\{\vec{\eta},\tilde{p}\} - H_I\{\vec{\eta},f\}]dS \\
&+ \iint \tilde{\Psi}_{II}(\vec{\eta})[L_{II}\{\vec{\eta},\tilde{p}\} + f(\vec{\eta}) + H_{II}\{\vec{\eta},f\}]dS_\eta
\end{aligned}
\tag{40}
$$

Varying Ψ_I and Ψ_{II} once again yields Eqs. (26a) and (26b), while varying p yields

$$
i\vec{k}\cdot\vec{n}(\vec{\eta})e^{i\vec{k}\cdot\vec{\eta}} + \Psi_I(\vec{\eta}) - L_I^\dagger(\vec{\eta},\Psi_I) + L_{II}(\vec{\eta},\Psi_{II}) = 0
\tag{41}
$$

or

$$
f(\vec{\eta}) + \Psi_I(\vec{\eta}) - H_{II}(\vec{\eta},\Psi_I) + L_{II}(\vec{\eta},\Psi_{II}) = 0
\tag{42}
$$

Comparison of the latter with Eq. (15) leads to the choice of

$$
\psi_I(\vec{\eta}) = -\frac{1}{2} f(\vec{\eta})
\tag{43}
$$

such that Ψ_{II} can consequently be identified as

$$
\Psi_{II}(\vec{\eta}) = \frac{1}{2} p(\vec{\eta})
\tag{44}
$$

Subsequent substitutions of (43) and (44) into the expression (40) reduce the stationary expression for A to

$$
\begin{aligned}
A = &\iint f(\vec{\eta})[e^{i\vec{k}\cdot\vec{\eta}} + p(\vec{\eta})]dS_\eta \\
&+ \frac{1}{2\pi} \iiiint p(\vec{\eta})f(\vec{\xi})\vec{n}(\vec{\eta})\cdot\nabla_\eta G(\vec{\eta}|\vec{\xi})dS_\xi dS_\eta \\
&- \frac{1}{4\pi} \iiiint (\vec{n}(\vec{\eta})\times\nabla_\eta p(\vec{\eta}))\cdot(n(\vec{\xi})\times\nabla_\xi p(\vec{\xi}))G(\vec{\eta}|\vec{\xi})dS_\xi dS_\eta \\
&+ \frac{k^2}{4\pi} \iiiint \vec{n}(\vec{\eta})\cdot\vec{n}(\vec{\xi})p(\vec{\eta})p(\vec{\xi})G(\vec{\eta}|\vec{\xi})\ dS_\xi dS_\eta \\
&+ \frac{1}{4\pi} \iiiint f(\vec{\eta})f(\vec{\xi})G(\vec{\eta}|\vec{\xi})dS_\xi dS_\eta
\end{aligned}
\tag{45}
$$

Here the quantity $f(\vec{\eta})$ is understood to be given by Eq. (39), while $p(\vec{\eta})$ is the scattered part of the acoustic pressure at the rigid body's surface, when the incident acoustic pressure wave has unit amplitude. The above expression is stationary to small variations in the dependent variable p.

Comparison of Eq. (45) with Eq. (35) indicates that the variational principles, $\delta J = 0$ and $\delta A = 0$, are equivalent if the f's appearing in the two expressions are the same.

VI. NUMERICAL EXAMPLE

To demonstrate the stationary expressions constructed along the lines described in this paper can yield accurate estimates with relatively minimal effort, the power radiated by an unbaffled rigid disk in transverse sinusoidal motion is considered here. The disk is taken as infinitesimally thin, of radius a, and oriented such that its faces are parallel to the x-y plane with its center nominally at the origin. The disk is moving backward and forward in the z direction with velocity $v_c \sin(\omega t)$, where v_c is a constant.

For this example, one can use Eq. (35) for the calculation of the radiated power because the velocity on the surface is everywhere in phase. The quantity f is $\omega \rho v_c$ on the $+z$ side and $-\omega \rho v_c$ on the $-z$ side. This antisymmetry of f in z requires that p also be antisymmetric, so considerable simplification results in the expression (35). With appropriate multiplications of each integral by a constant (0, 2, or 4), one need only carry out integrations over the front side of the disk. Some subtlety is involved, however, with the term

$$+ \frac{1}{2\pi} \iiint \left[p(\vec{\eta}) f(\vec{\xi}) n(\vec{\eta}) \cdot \nabla_\eta G(\vec{\eta}\,|\,\vec{\xi}) \, dS_\xi dS_\eta \right.$$

In the limit as the disk thickness goes to zero, there is no contribution when $\vec{\eta}$ and $\vec{\xi}$ are on the same side of the disk, but when they are on opposite sides, considerations such as those that enter into the derivation of Eq. (7) apply, with the net result that the above term reduces to

$$\iint p(\vec{\eta}) f(\vec{\eta}) \, dS_\eta$$

which effectively doubles the first term in the right side of (35). Taking full advantage of the symmetry subsequently yields

$$J = 4 \iint p(\vec{\eta}) f(\vec{\eta}) \, dA_\eta$$
$$- \frac{1}{\pi} \iiiint (\vec{e}_z \times \nabla_\eta p(\vec{\eta})) \cdot (\vec{e}_z \times \nabla_\xi p(\vec{\xi})) G(\vec{\eta}\,|\,\vec{\xi}) \, dA_\xi dA_\eta$$
$$+ \frac{k^2}{\pi} \iiiint p(\vec{\eta}) p(\vec{\xi}) G(\vec{\eta}\,|\,\vec{\xi}) \, dA_\xi dA_\eta \qquad (46)$$

where here the area integrals (no longer surface integrals) extend over only the front surface of the disk.

Calculations based on this expression were carried out for several values of ka with trial functions consisting of a linear combination of N basis functions, the sum having the generic form

$$p(r) = \sum_{n=1}^{N} C_n [1 - (r/a)^2]^{1/2} (r/a)^{2(n-1)} \tag{47}$$

This was done for N = 1, 2, 3, 4, 5, and 6. For each such N the coefficients C_n were determined by a Rayleigh-Ritz procedure: Eq. (47) was inserted into (46), the requisite integrals were evaluated and the coefficients C_n were determined from the system of linear algebraic equations resulting from setting each of the derivatives $\partial J/\partial C_n$ to zero. These so-derived coefficients were then substituted back into the expression for J to derive the corresponding N-term approximation for radiated power via Eq. (25). These coefficients were also inserted back into Eq. (47) with r set to 0 to determine the N-term approximation for the acoustic pressure amplitude $|p(0)|$ at the center of the disk.

The contention of the present paper's theoretical development is that the calculated values for the radiated power should converge much more rapidly with increasing N than do the values for acoustic pressure at the center of the disk. This of course was the case. The following numbers, normalized to the values for N = 6, are representative:

Calculations for ka = 5:

    N = 1: |p(0)| = 2.306093   Power = 1.051240
    N = 2: |p(0)| = 1.948102   Power = 0.951187
    N = 3: |p(0)| = 1.137665   Power = 0.998393
    N = 4: |p(0)| = 1.009615   Power = 0.999957
    N = 5: |p(0)| = 0.999587   Power = 1.000000

Thus, for N = 1, the radiated power estimate is only 5% in error, even though the pressure at the center of the disk is off by more than a factor of 2. At N = 4 the error in power is approximately 4×10^{-6}, while the error in acoustic pressure at disk center is approximately 4×10^{-4}, the latter being nearly 100 times as large as the former.

VII. CONCLUDING REMARKS

The variational principles exhibited in the present paper
were selected primarily for convenience of presentation. In particular, there seems to be no reason why the internal structure
(e.g., its elastic properties) of the body cannot be taken into
account. This has been demonstrated recently by Ginsberg [26]
for the example of sound radiation from a vibrating point-driven
circular elastic plate. The extent to which computational algorithms based on variational formulations can compete with other
techniques for acoustic radiation and scattering calculations remains a subject for further investigation.

ACKNOWLEDGMENTS

The author thanks W. Möhring for bringing the cited work by
Gerjuoy, Rau, and Spruch to his attention. He would also like to
thank Peter H. Rogers, Michael Stallybrass, and Jerry H. Ginsberg
for helpful discussions. The calculations reported in the latter
part of the paper were carried out by X.-F. Wu. The work reported
here was supported by the Office of Naval Research.

REFERENCES

1. Rayleigh, J. W. S., Some general theorems relating to vibra-
    tions, PROCEEDINGS OF THE LONDON MATHEMATICAL SOCIETY, 4
    (1873), 357-368.

2. Born, M. and E. Wolf, PRINCIPLES OF OPTICS, 6th Ed., Pergamon
    Press, Oxford, 1980, pp. 128-132.

3. Ugincius, P., Ray acoustics and Fermat's principle in a mov-
    ing inhomogeneous medium, JOURNAL OF THE ACOUSTICAL SOCIETY
    OF AMERICA, 51 (1972), 1759-1763.

4. Love, A. E. H., A TREATISE ON THE MATHEMATICAL THEORY OF ELAS-
    TICITY, 4th Ed., published 1927, Dover Publications, New York,
    1944, pp. 171-172.

5. Schwinger, J., and D. S. Saxon, DISCONTINUITIES IN WAVEGUIDES.
    Notes on lectures (pre-1945) by Julian Schwinger, Gordon and
    Breach, New York, 1968, pp. 25-56.

6. Levine, H., and J. Schwinger, On the theory of diffraction
    by an aperture in an infinite plane screen. I, PHYSICAL
    REVIEW, 74(8), Oct. 15, 1948; II, PHYSICAL REVIEW, 75(9),
    May 1, 1949, 1423-1432.

7. Lippmann, B. A., and J. Schwinger, Variational principles
    for scattering processes, I, PHYSICAL REVIEW, 79(3), Aug. 1,
    1950, 469-480

8. Kodis, R. D., An introduction to variational methods in elec-
    tromagnetic scattering, JOURNAL OF THE SOCIETY FOR INDUSTRIAL
    AND APPLIED MATHEMATICS, 2(2), June, 1954, 89-112.

9. Sleator, F. B., A variational solution to the problem of
    scalar scattering by a prolate spheroid, JOURNAL OF MATHE-
    MATICS AND PHYSICS, 39 (1960), 105-120.

10. Morse, P. M., and H. Feshbach, METHODS OF THEORETICAL PHYSICS,
    Vol. II, Mc-Graw-Hill, New York, 1953, Chapter 9.

11. Levine, H., Variational principles in acoustic diffraction
    theory, JOURNAL OF THE ACOUSTICAL SOCIETY OF AMERICA, 22(1)
    (1950), 48-55.

12. Levitas, A., and M. Lax, Scattering and absorption by an
    acoustic strip, JOURNAL OF THE ACOUSTICAL SOCIETY OF AMERICA,
    23(3) (1951), 316-322.

13. Miles, J. W., On acoustic diffraction through an aperture in
    a plane screen, ACUSTICA, 2 (1952), 287-291.

14. Miles, J. W., On acoustic diffraction cross sections for
    oblique incidence, JOURNAL OF THE ACOUSTICAL SOCIETY OF AMER-
    ICA, 24(3) (1952), 324.

15. Morse, P. M., Transmission of sound through a circular mem-
    brane in a plane wall, JOURNAL OF THE ACOUSTICAL SOCIETY OF
    AMERICA, 40(2) (1966), 354-366.

16. Note, for example, the small number of papers in the special
    session on variational techniques in acoustics at the Spring
    1970 meeting of the Acoustical Society of America [JOURNAL
    OF THE ACOUSTICAL SOCIETY OF AMERICA, 48(1) (July 1970),
    Part 1, 82-83].

17. Jones, D. S., A critique of the variational method in scat-
    tering problems, IRE TRANSACTIONS ON ANTENNAS AND PROPAGA-
    TION, AP-4(3) (1956), 297-301.

18. Baker, B. B., and E. T. Copson, THE MATHEMATICAL THEORY OF
    HUYGENS' PRINCIPLE, 2nd Ed., Oxford, 1950, see the footnote
    on page 186.

19. Bouwkamp, C. J., DIFFRACTION THEORY. A CRITIQUE OF SOME RE-
     CENT DEVELOPMENTS, Research Report No. EM-50, New York Univ-
     ersity, Mathematics Research Group, April 1953, pp. 23-36.

20. Maue, A.-W., Zur Formulierung eines allgemeinen Beugungsprob-
     lems durch eine Integralgleichung, ZEITSCHRIFT FUR PHYSIK,
     126 (1949), 601-618.

21. Stallybrass, M. P., On a pointwise variational principle for
     the approximate solution of linear boundary value problems,
     JOURNAL OF MATHEMATICS AND MECHANICS, 16(11) (1967), 1247-
     1286.

22. Hamdi, M. A., Une formulation variationnelle par équations
     intégrales pour la résolution de l'équation de Helmholtz avec
     des conditions aux limites mixtes, COMPTES RENDUS DE D'
     ACADEMIE SCIENCES DE PARIS, 292(II) (Jan. 5, 1981), 17-20.

23. Pierce, A. D., and X.-F. Wu, Variational method for predic-
     tion of acoustic radiation from vibrating bodies, JOURNAL OF
     THE ACOUSTICAL SOCIETY OF AMERICA, SUPPLEMENT 1, 74, S107
     (Fall 1983).

24. Ginsberg, J. H., A. D. Pierce, X.-F. Wu, and J. S. DiMarco,
     Rayleigh-Ritz analysis and finite element descriptions de-
     rived from variational principles of diffraction by a circu-
     lar disk, JOURNAL OF THE ACOUSTICAL SOCIETY OF AMERICA, SUP-
     PLEMENT 1, 77, S60 (Spring 1985).

25. Wu, X.-F., A. D. Pierce, and J. F. Ginsberg, Application of
     a variational principle to acoustic radiation from a vibra-
     ting finite cylinder, JOURNAL OF THE ACOUSTICAL SOCIETY OF
     AMERICA SUPPLEMENT, 1 79 S35 (Spring 1986).

26. Ginsberg, J. H., and A. D. Pierce, Variational principle for
     harmonic fluid-structure interaction applied to an unbaffled
     elastic disk, JOURNAL OF THE ACOUSTICAL SOCIETY OF AMERICA
     SUPPLEMENT 1 79 S91 (Spring 1986).

27. Gerjuoy, E., A. R. P. Rau, and L. Spruch, A unified formula-
     tion of the construction of variational principles, REVIEWS
     OF MODERN PHYSICS, 55(3) (July 1983), 725-774.

28. Copley, L. G., Fundamental results concerning integral repre-
     sentations in acoustic radiation, JOURNAL OF THE ACOUSTICAL
     SOCIETY OF AMERICA, 44(1) (Aug. 1968), 28-32.

29. Pierce, A. D., ACOUSTICS: AN INTRODUCTION TO ITS PHYSICAL
     PRINCIPLES AND APPLICATIONS, Mc-Graw-Hill, New York, 1981,
     pp. 180-182, 428-429.

30. Kellogg, O. D., FOUNDATIONS OF POTENTIAL THEORY, originally
     published 1929, Dover, New York, 1953, pp. 160-172.

COMPUTATIONAL ACOUSTICS: Wave Propagation
D. Lee, R.L. Sternberg, M.H. Schultz (Editors)
Elsevier Science Publishers B.V. (North-Holland)
© IMACS, 1988

ACOUSTICAL RADIATION IN THE PRESENCE OF TEMPERATURE GRADIENTS: A WAVE ENVELOPE FINITE ELEMENT APPROACH

R. J. Astley

Department of Mechanical Engineering
University of Canterbury
New Zealand

ABSTRACT

The calculation of acoustical fields in an unbounded compressible fluid within which is embedded a finite region of significant temperature variation is the general problem addressed in this paper. The method proposed is demonstrated by application to a specific example, that of sound propagating along a heated pipe and radiating from an unflanged exhaust. In this case significant distortion of the radiated sound field occurs as a consequence of diffraction through the axial and radial thermal gradients in the vicinity of the outlet. Analysis and results are presented for the axially symmetric case.

1. INTRODUCTION

The numerical simulation of acoustical radiation into an infinite domain poses some severe problems even in the absence of temperature gradients. The physical requirement that the sound field should be entirely radiative in the far field must be simulated by an appropriate numerical boundary condition. The obvious treatment in such cases involves the use of a Sommerfield radiation condition but is appropriate only if the computational boundary is sufficiently distant from radiating or reflecting surfaces. In practice it may be prohibitively expensive to carry a conventional numerical scheme into the far field in this way. The problem is exacerbated if the characteristic acoustical wavelength is relatively small compared with the geometrical lengthscale of the problem although for extremely short wavelengths it may be resolved by the application of ray acoustical theory [1,2].

In the absence of temperature or flow gradients many of the difficulties associated with the problem may be resolved by the use of boundary integral representations which intrinsically incorporate the correct radiation conditions [3,4]. When temperature or

flow gradients are present, however, the source functions required
for such an approach are no longer readily available; although for
the particular case where flow gradients are present in low Mach
number flows a transformation may be used to formulate the problem
in boundary integral form [5]. In the more general case, however,
it appears unavoidable that a full numerical representation must
be used at least in those regions where significant temperature or
velocity gradients occur.

Considerable effort has been expended on the development of
numerical schemes for such problems—much of it motivated by aero-
acoustic application. Linear acoustical theory is usually quite
adequate for such applications and "steady state" solutions may be
obtained by considering the sound field as a superposition of time
harmonic components. Conventional finite difference or finite element methodology [6] may then be applied—the latter [7,8,9] generally finding greater acceptance than the former in view of its
ability to more easily accommodate arbitrarily shaped bounding
surfaces. Transient, finite difference methods have also been developed [10] however and demonstrated to be effective for bounded
in-duct sound fields [11].

Whichever numerical scheme is employed in such regions, the
problem still remains of extending such representations into the
far field and of coupling them to an appropriate radiative boundary condition. To do so using a conventional numerical representation [7], requiring many grid or node points for each wavelength
variation in the solution, is seldom practicable except for low
frequency, long wavelength ('long' compared with geometric length-
scales) disturbances. Two alternative procedures have been proposed. Both have been demonstrated to be effective for the case
where the initial inhomogeneities result from velocity gradients
in the near field. The first is that of the "hybrid" finite ele-
ment/boundary element approach whereby a conventional finite element scheme in the near field is coupled to a boundary element
representation in the uniformly flowing outer region. This approach
requires an iterative matching of impedance at an intermediate surface and has been applied with success to the turbofan radiation
problem [12]. Solution times are high, however, and the requirement to use an iterative solution appears to be a less than opti-
mally efficient procedure for what is fundamentally a linear
problem.

A second approach which has been employed by the present author involves the use of a conventional finite element representation within the near field and a wave envelope finite element representation in the intermediate and far field. Both representations are compatibly matched at an intermediate surface. The wave envelope elements incorporate some aspects of ray acoustical behavior and can accommodate several wavelength variations of the solution within a single element. They remain valid in the high frequency limit. Their first application was for ducted sound fields [13] and in this context they constituted a finite element analog of the finite difference wave envelope approach already demonstrated to be effective for such problems [14,15]. The method was subsequently extended to describe simultaneously the in-duct propagation of sound in a turbofan nacelle and its subsequent radiation into the far field [16]. The effects of mean flow velocity gradients were included in later analyses of this type [17,18] as were the effects of acoustical source distributions within the flow [19] (the latter in an attempt to model noise generation by propfan as well as turbofan engines).

In the current paper a further extension of the method is presented, that of its application to sound propagation in the presence of temperature gradients. A specific problem of practical interest which falls into this category is that of sound propagating from a heated pipe and diffracting into a cooler region. Although the method to be presented in this paper may be applied to a more general class of problem, the analysis will be presented in detail only for this specific problem. The reader is referred elsewhere [20] for a review of the problem itself and past attempts, theoretical and experimental, to obtain solutions.

2. GEOMETRY AND GOVERNING EQUATIONS

The geometry of the problem to be solved is illustrated in Figure 1. The computational domain is shown in the upper half of the figure, the temperature field in the lower half. Axial symmetry exists about the z axis. The computational domain is divided into an inner region R_1 and an outer region R_2 bounded by surfaces of revolution Γ_a, Γ_b, and Γ_c. Γ_a denotes the plane of excitation within the pipe. Γ_b defines the inner and outer pipe surfaces and includes also the surface of a conical baffle behind the plane of the duct outlet. (It has been found in previous studies [16] that the presence of a computational boundary of this type has little

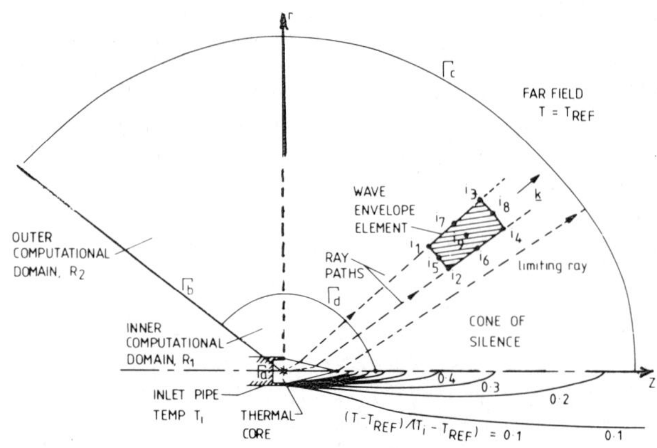

Fig. 1.   Inlet Geometry

effect on the acoustical field forward of the outlet.) The contour Γ_c defines a far field boundary at a large distance from the
outlet. An additional contour Γ_d, interior to the computational
domain, represents the interface between the inner and outer
regions.

The temperature field, $T(r,z)$, within the computational region
is taken to be that of a turbulent jet at temperature T_i exhausting
into an ambient gas of reference temperature T_r (denoted by T_{ref} in
Fig. 1). A typical temperature field of this type displays the
temperature contours shown in the lower half of the figure. These
are characterized [20] by a thermal core in the vicinity of the
pipe outlet followed by a mixing regior and a fully developed 'similarity' region as $z/a \to \infty$ (a denotes the inner radius of the pipe).
It is the propagation of sound from the generating plane Γ_a within
the pipe through this region of severe thermal gradient and into
the far field which is the object of the current analysis.

Velocity gradients in the mean flow will be neglected in the
present analysis; their effect on the sound field is likely to be
small compared with that of temperature gradients unless the Mach
number of the jet is of a magnitude comparable to the ratio T_i/T_r.
The linearized continuity, momentum, and convected entropy equations [21] then yield the acoustical pressure equation

$$\rho_o \underline{\nabla} \cdot ((1/\rho_o)\underline{\nabla}p^1) - (1/c_o^2)\ \partial^2 p^1/\partial t^2 = 0 \tag{1}$$

where $\rho_o(r,z)$, $c_o(r,z)$, and $p^1(r,z,t)$ are the mean density, local
adiabatic sound speed, and acoustical pressure perturbation, respectively. By neglecting variations of steady state pressure

within the thermal jet and assuming a perfect gas with a constant
ratio of specific heats, the mean local density and sound speed
may be specified as functions of temperature only and written

$$\rho_o/\rho_r = (T/T_r)^{-1} \quad \text{and} \quad c_o/c_r = \sqrt{T/T_r}$$

where ρ_r and c_r are ambient reference values. If time harmonic
solutions of the form $p^1(r,z,t) = p(r,z) \exp(i\omega t)$ are then sought,
equation (1) becomes

$$\underline{\nabla} \cdot (\alpha(r,z)\underline{\nabla}p) + k_r^2 p = 0 \tag{2}$$

where $\alpha(r,z) = T(r,z)/T_r$ and $k_r = \omega/c_r$. This is a modified form
of the Helmholtz equation which clearly reduces to the standard
form for the case $T(r,z) \equiv T_r$ (i.e., $\alpha = 1$). A solution to equation (2) is now sought which satisfies the appropriate boundary
conditions on Γ_a, Γ_b, and Γ_c. These are:

$$\underline{\nabla}p \cdot \hat{\underline{n}} = 0 \quad \text{on } \Gamma_b \quad \text{and} \quad \underline{\nabla}p \cdot \hat{\underline{n}} = i(k_r/\sqrt{\alpha})p \quad \text{on } \Gamma_c \tag{3}$$

where $\hat{\underline{n}}$ is an outward normal from R_1 or R_2. The first represents
a condition of zero normal particle velocity at a solid boundary
and the second a Sommerfield radiation condition which uses the
local thermally adjusted wavenumber $k_r/\sqrt{\alpha}$ at the far field boundary. The boundary condition at the generating plane Γ_a is somewhat
more complicated and must be written in modal form:

$$p = \sum_{i=1}^{\infty} \{A_i^+ f_i^+(r) + A_i^- f_i^-(r)\} \quad \text{on } \Gamma_a$$

$$\underline{\nabla}p \cdot \hat{\underline{n}} = - \sum_{i=1}^{\infty} \{A_i^+ \lambda_i^+ f_i^+(r) + A_i^- \lambda_i^- f(r)\} \quad \text{on } \Gamma_a \tag{4}$$

where $A_i^{\pm}$, $\lambda_i^{\pm}$, and $f_i^{\pm}(r)$ are coefficients, axial wavenumbers, and
radial eigenfunctions for incident (propagating in the positive z
direction) and reflected (propagating in the negative z direction)
duct eigenmodes. The coefficients A_i^+ may be assumed to be known
incident modal amplitudes. The coefficients A_i^- which represent
reflections at open end of the pipe must be determined by the
solution.

3. THE RESIDUAL SCHEME

The acoustical pressure is now approximated by a trial function $\tilde{p}$. The trial function is formed as an expansion of known

basis functions ($\psi_i(r,z)$, $i = 1, \ldots, n$) and unknown coefficients (a_i, $i = 1, \ldots, n$) giving

$$\tilde{p} = \sum_{i=1}^{n} a_i \psi_i \tag{5}$$

An independent set of weighting functions W_i ($i = 1, \ldots, n$) is also selected and a Galerkin procedure applied to the minimization of field and boundary residuals resulting from substitution of the trial function into equations (2) and boundary conditions (3) and (4). The analysis which follows is then similar to that described in Ref. 16, Sec. 2.3, except that the variable coefficient $\alpha(r,z)$ occurs in equations (2) and (4) but is not present in the equations of Ref. 16. With this modification, however, the analysis proceeds in an almost identical manner and results in a final statement of the discretized problem as

$$\begin{bmatrix} D^- & B^{-T} \\ \hline B^- & C \end{bmatrix} \begin{Bmatrix} A^- \\ \hline a \end{Bmatrix} = - \begin{bmatrix} D^+ \\ \hline B^+ \end{bmatrix} \{A^+\} \tag{6}$$

where the partitioned submatrices $B^\pm$, $D^\pm$, and C are given by:

$$B^\pm_{ij} = \int_{\Gamma_a} \alpha W_j (i\lambda^\pm_i f^\pm_i) r\,dr \qquad i = 1, \ldots, m; \ j = 1, \ldots, n \tag{7}$$

$$C_{ij} = \int_{R_1+R_2} (-\alpha \underline{\nabla} W_j \cdot \underline{\nabla} \psi_i + k_r^2 W_j \psi_i) r\,dr\,dz$$

$$\qquad - \int_{\Gamma_c} ik\sqrt{\alpha} W_j \psi_i r\,ds \qquad i,j = 1, \ldots, n \tag{8}$$

and

$$D^\pm_{ij} = -\int_{\Gamma_a} \alpha i\lambda^-_i f^-_i f^\pm_i r\,dr \qquad i,j = 1, \ldots, m \tag{9}$$

The vectors $\{A^\pm\}$ and $\{a\}$ contain respectively the modal coefficients $A^\pm_i$ and the unknown trial function coefficients a_i of expression (5). Clearly, equation (6) may now be solved for a given set of incident modal coefficients $\{A^+\}$ to yield the reflected modes within the duct and the trial function coefficients.

4. SELECTION OF BASIS AND WEIGHTING FUNCTIONS

The residual scheme of the preceding section is quite general in that it leaves open the choice of both the basis functions ψ_i and the weighting functions W_i. Different functions are chosen in

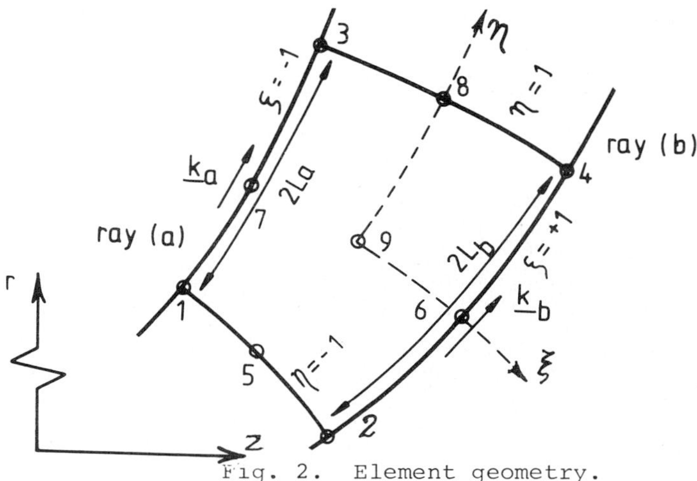

Fig. 2.   Element geometry.

the inner and outer regions. In the inner region a straightforward Galerkin finite element method [22] is employed. The region
is divided into elements—in this case nine-noded Lagrangian isoparametric elements of the type indicated in Fig. 2— and the basis
functions defined by nodal shape functions within each element.
The weight functions are equated with the basis functions. Basis
and weight functions selected in this way have the property of being equal to unity at node i (in the case of W_i and ψ_i) and zero
at all other nodes. The coefficients a_i therefore became nodal
values of the trial function $\tilde{p}$. A quadratic interpolation in ξ
and η (local curvilinear coordinates as shown in Fig. 2) is implicit in such a model and it is consequently anticipated that several such elements will be required in any direction to adequately
resolve a single wavelength variation in the solution. A typical
inner mesh of this type is shown in Fig. 3A. Also shown is a
representation of the characteristic wavelength for a typical solution to be presented in the following section.

      In the outer region a different finite element model is em-
ployed. First, the elements are set up so that the lines η = constant (see Fig. 2) lie along idealized ray paths emanating from a
source in the mouth of the inlet. The assumption here is that at
large distances from the radiating pipe the acoustical field will
behave locally as if it were generated by a complex source of the
origin. These rays have of course been refracted in their journey
through the thermal jet. If ray acoustical theory were applicable
all acoustical energy would travel along such rays. No such restrictive assumption is made in the present analysis. It will be

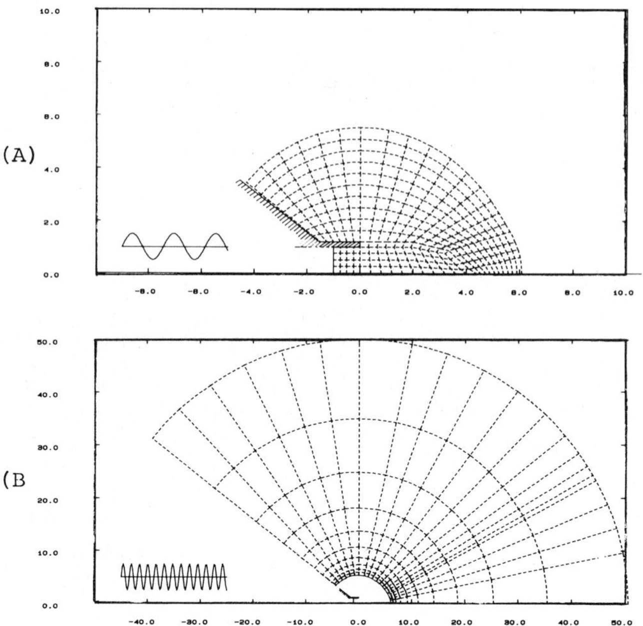

Fig. 3. (A) Inner mesh (conventional elements). (B) Outer mesh
(wave envelope elements).

assumed, however, that the local wavelike behavior of the solution
will resemble that of a wave packet travelling along such rays.
Within each 'wave envelope' element, therefore, a local nondimensional wavenumber $k'(\xi)$ is defined as a linear interpolation of
the nondimensional wavenumbers $k_a L_a$ and $k_b L_b$ on ray paths (a) and
(b), where k_a and k_b are defined arbitrarily as wavenumbers at mid-
side nodes 7 and 6, as indicated in Fig. 2. This yields the expression

$$k'(\xi) = ((1 - \xi)k_a L_a + (1 + \xi)k_b L_b)/2 \tag{10}$$

The shape functions, S_1, S_2, ..., S_9, say, within such an element
are now defined by

$$S_i(\xi,\eta) = N_i(\xi,\eta) \exp(-ik'(\xi)(\eta - \eta_i)) \qquad i = 1, \ldots, 9 \tag{11}$$

where $N_i(\xi,\eta)$ is the shape function for a conventional element of
the same topology. The basis functions ψ_i are now defined implicitly as the appropriate shape functions within each element. The
weight functions are then defined as the complex conjugates of the
basis functions. This definition has the advantage of leading to

the cancellations of the harmonic factors, i.e., factors of the
form $\exp(\pm ik'(\xi)\eta)$, in all products of ψ_i and W_j and hence in the
integrands of the expressions for the components of the matrix [C]
[see equation (8)]. Conventional Gaussian integration may thus be
used within each element in the evaluation of these integrals.
The above definition of ψ_i retains the property that $\psi_i \equiv 1$ at
node i and thus preserves the significance of the coefficients a_i
as nodal values of acoustic pressure.

The advantage of the above choices of ψ_i and W_i over the conventional definitions is to be found in their ability to represent
many wavelength variations of the solution within a single element
so long as the local wavelike behavior of that solution is approximately that predicted by ray acoustical theory. The difference
between the actual solution and such a wave packet is accommodated
within each element by the nonharmonic portion of the shape function S_i. Since only an approximation to the wavelike behavior
within each element is required, it is quite appropriate to use
approximate, rather than exact, ray paths in setting up such elements. In the results to follow, for example, the ray paths in
the outer region are generated, not from the exact temperature
field but from a somewhat gross idealization which approximates it
by a thermal core of temperature T_i surrounded by gas at ambient
temperature T_r. Each ray is then bent by a single refraction at
the edge of the core rather than by a continuous process through
the mixing region. Several idealized rays of this type are indicated in Fig. 1. A typical outer mesh formed using such rays and
extending to approximately 50 duct radii is shown in Fig. 3B. Also
shown in Fig. 3B is a representation of a typical characteristic
wavelength which is clearly seen to be small compared with the dimensions of the elements. A problem arises in constructing elements within the region of silence formed by the cone of limiting
rays, shown in Fig. 1. These travel initially along the axis of
symmetry and are refracted outwards at the apex of the thermal
core. An additional imaginary source was placed at the apex of
the core and used to generate rays in this region. A similar de-
vice was used to generate the elements behind the plane of the
pipe outlet.

5. RESULTS AND DISCUSSION

The numerical scheme outlined in the preceding sections was
applied to a particular configuration [23] for which measured data

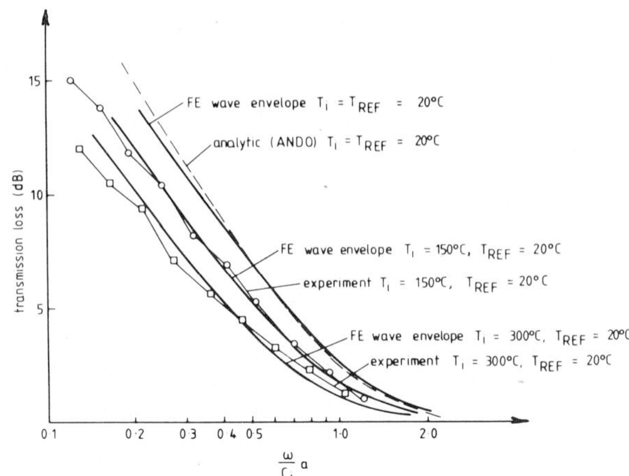

Fig. 4. Comparison of measured, computed, and analytic transmission loss for a heated unflanged pipe of radius a.

is available. This comprised an unflanged pipe of internal diameter 85 mm and thickness 10 mm with values of T_i in the range 20°C to 300°C and T_r = 20°C. The flow velocity in the tube was extremely low (less than 1 m/sec). Computed results are presented for the three cases T_i = 20°C, 150°C, and 300°C. The first equates to the case without thermal gradients and has been solved by analytic means [24].

Measurements of the temperature profile along the axis of symmetry for the above case [23] indicate a thermal core of length L where L is of the order of three to four times the radius of the duct. For the case T_i = 300°C the value is close to three and for T_i = 150°C it is closer to four. These values were used in establishing the temperature field for the computed solutions in each case. A comparison of measured, analytic, and computed values of the transmission loss in the pipe is presented in Fig. 4 (the transmission loss is readily calculated from the computed modal coefficients in the pipe). The nondimensional frequency parameter against which the transmission losses are plotted is that used in Ref. 23 and is formed using the inlet sound speed c_i. An incident plane mode is assumed in the duct (all other modes are cut off within the range of frequencies shown). The agreement between the computed and measured values for T_i = 150°C and T_i = 300°C is clearly good as is the agreement between the computed solution and Ando's analytic result [24] for the case T_i = T_r = 20°C. Further comparisons, not presented here, between measured, analytic, and

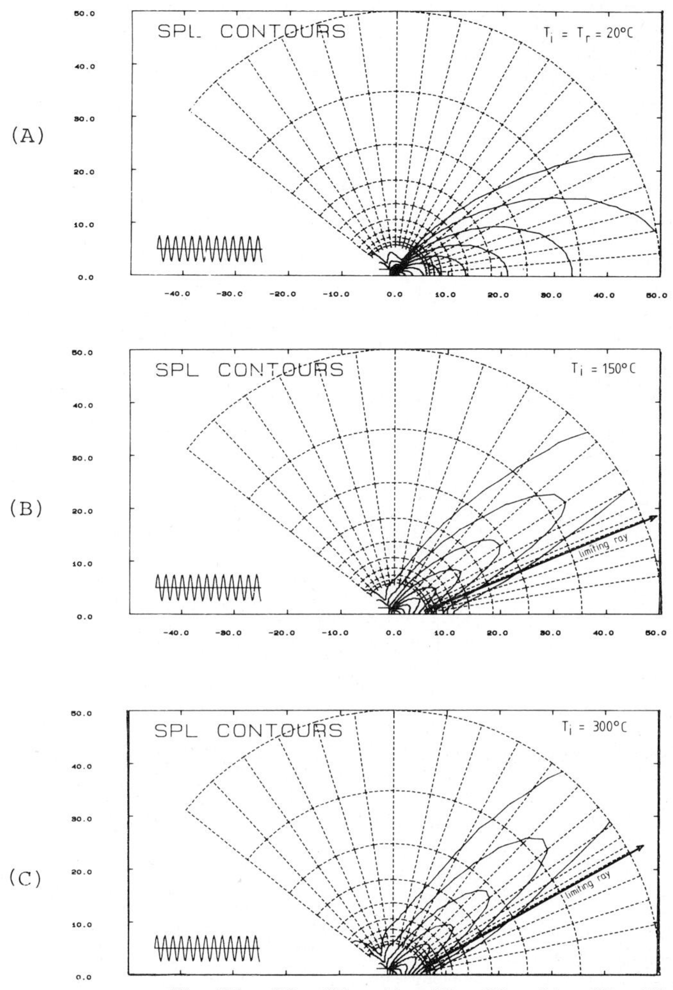

Fig. 5. Computed radiation patterns for a heated unflanged pipe of radius a. Plane mode incident in the pipe, ka = 4.0. (A) $T_i = T_r = 20°C$; (B) $T_i = 150°C$, $T_r = 20°C$; (C) $T_i = 300°C$, $T_r = 20°C$.

computed values of reactance and resistance at the pipe exit also show good agreement.

The extent to which the far field is modified by the temperature gradients in the vicinity of the pipe is illustrated by plots of computed SPL contours, at 4 dB intervals, over the entire computational region for the three cases $T_i = 20°C$, 150°C, and 300°C. These are shown in Figs. 5A, B, and C. The limiting ray in Figs. 5B and 5C is indicated by a heavy solid line. The frequency is

well above the range used in Fig. 4 and is, in fact, slightly
above cut-on of first radial mode in the pipe. The plane wavelength of the solution in the far field is indicated as in Fig. 3B
by a wavelike insert. The computational region extends to approximately fifty pipe radii from the pipe. This is clearly a computa-
tionally-demanding configuration requiring an enormous number of
mesh points if a conventional numerical approach were adopted.
For the results presented in Fig. 5, 1980 degrees of freedom gave
solutions which were effectively invariant to further mesh refinement. The resulting solutions illustrate the dramatic way in which
the radiation lobe is refracted away from the center line as the
temperature is increased. This is, in fact, very much as might
have been anticipated from ray acoustical considerations and it is
noticeable that the limiting ray does indeed delineate a sharp
change in SPL, although it does not, of course, define a forward
region of total silence as predicted in the high frequency limit.
No measured values of far field SPL are currently available to
compare with these predicted radiation patterns. It is hoped that
some may be obtained to confirm these results.

6. CONCLUSIONS

To conclude, the comparison of computed results with measured
and analytic values and the qualitative features of the computed
far field indicate that the numerical scheme is producing useful
results both in the near and far field. Further measurements in
the far field would be useful in confirming this assertion. The
facility to model many wavelength variations within a single wave
envelope element renders the present scheme computationally practicable in situations where conventional numerical approaches would
require an unacceptable number of node or mesh points. The extension to fully three-dimensional situations although more costly
than the present axisymmetric model is a practicable proposition.
The number of degrees of freedom which such a model would involve
if wave envelope elements were employed would certainly not exceed
that of many large 3-D solid and structural analyses.

REFERENCES

1. Lighthill, M. J., WAVES IN FLUIDS, Cambridge University Press,
    1978, Sections 1.11 and 4.5.

2. Boyd, W. K., and A. J. Kempton, Ray theory predictions of the
    noise radiated from aeroengines, AIAA Paper 84-2322, AIAA/NASA

9th Aeroacoustics Conference, Williamsburg, 1984.

3. Meyer, W. L., W. A. Bell, B. T. Zinn, and M. P. Stallybrass, Boundary integral solutions of three dimensional acoustic radiation problems, JOURNAL OF SOUND AND VIBRATION, 59(2) (1978), 245-262.

4. Meyer, W. L., W. A. Bell, M. P. Stallybrass, and B. T. Zinn, Prediction of the sound field radiated from axisymmetric surfaces, JOURNAL OF THE ACOUSTICAL SOCIETY OF AMERICA, 65(3) (1979), 631-638.

5. Astley, R. J., and J. G. Bain, A three dimensional boundary element scheme for acoustical radiation in low mach number flows, JOURNAL OF SOUND AND VIBRATION, 109(3) (1986), 445-465.

6. Baumeister, K. J., Numerical techniques in linear duct acoustics - A status report, JOURNAL OF ENGINEERING FOR INDUSTRY, 103 (1981), 270-

7. Sigman, R. K., R. K. Majjigi, and B. T. Zinn, Determination of turbofan inlet acoustics using finite elements, AIAA JOURNAL, 16 (1978), 1139-45.

8. Kagawa, Y., T. Yamabuchi, T. Yoshikawa, S. Ooie, N. Kyouno, and T. Shindou, Finite element approach to acoustic transmission-radiation systems and application to horn and silencer design, JOURNAL OF SOUND AND VIBRATION, 69(2) (1980), 207-228.

9. Astley, R. J., AND W. Eversman, Acoustic transmission in lined ducts with flow; Part 2: The finite element method, JOURNAL OF SOUND AND VIBRATION, 74 (1981), 103-121.

10. Baumeister, K. J., Time dependent difference theory for noise propagation in a two-dimensional duct, AIAA JOURNAL, 18 (1980), 1470-76.

11. Cabelli, A., Duct acoustics - A time dependent difference approach for steady state solutions, JOURNAL OF SOUND AND VIBRATION, 85(4) (1982), 423-434.

12. Horowitz, S. J., R. K. Sigman, and B. T. Zinn, An iterative finite element-integral technique for predicting sound radiation from turbofan inlets in steady flight, AIAA JOURNAL, AIAA Paper 82-0124 (1982).

13. Astley, R. J., AND W. Eversman, A note on the utility of a wave envelope approach in finite element duct transmission

studies, JOURNAL OF SOUND AND VIBRATION, 76 (1981), 595-601.

14. Baumeister, K. J., Finite-difference theory for sound propagation in a lined duct with uniform flow using the wave envelope concept, NASA TP 1001 (1977).

15. Kaiser, J. E., and A. H. Nayfeh, A wave-envelope technique for wave propagation in non-uniform ducts, AIAA JOURNAL, 15 (1977), 533-537.

16. Astley, R. J., Wave envelope and infinite elements for acoustical radiation, INTERNATIONAL JOURNAL FOR NUMERICAL METHODS IN FLUIDS, 3 (1983), 507-526.

17. Parrett, A. V., and W. Eversman, Applications of finite element and wave envelope element approximations to turbofan engine noise radiation including flight effects, AIAA JOURNAL, AIAA Paper 84-2333 (1984).

18. Astley, R. J., A finite element wave envelope formulation for acoustical radiation in moving flows, JOURNAL OF SOUND AND VIBRATION, 103(4) (1985), 471-485.

19. Eversman, W., and J. E. Steck, Finite element modelling of acoustic singularities with application to near and far field propeller noise, AIAA JOURNAL, AIAA Paper 84-2286 (1984).

20. Rajaratnam, N., TURBULENT JETS, Elsiever Publishing Company, 1976.

21. Kapur, A., A. Cummings, and P. Mungur, Sound propagation in a combustion can with axial temperature and density gradients, JOURNAL OF SOUND AND VIBRATION, 25(1) (1972), 129-138.

22. Norrie, D. H., and G. DeVries, AN INTRODUCTION TO FINITE ELEMENT ANALYSIS, Academic Press, 1978.

23. Cummings, A., High temperature effects on the radiation impedance of an unflanged duct exit, JOURNAL OF SOUND AND VIBRATION, 52(2) (1977), 299-304.

24. Ando, Y., On the sound radiation from semi-infinite circular pipe of certain wall thickness, ACUSTICA, 22 (1969), 219-225.

COMPUTATIONAL ACOUSTICS: Wave Propagation
D. Lee, R.L. Sternberg, M.H. Schultz (Editors)
Elsevier Science Publishers B.V. (North-Holland)
© IMACS, 1988

RAPID EVALUATION OF RADIATION FIELDS
IN UNDERWATER ACOUSTICS

I. B. Bernstein and V. Rokhlin

Department of Computer Science
Yale University
New Haven, Connecticut

ABSTRACT

A method is proposed for the rapid, accurate numerical solution of the Helmholtz equation for the case of long range ocean acoustics. It is based on using the rays of geometric acoustics to generate a coordinate system which is then employed to transform both the dependent and independent variables in the Helmholtz equation. The acoustic field near a source can be accurately generated numerically by solving an integral equation equivalent to the original Helmholtz equation and outgoing conditions. The solution of the new equation near the acoustic source should then have no spatial variation on the scale of the local wave length, although some structure may be expected to develop gradually with increasing range due to diffraction. In general, there can be more than one solution of the new problem associated with a given point in the original system, corresponding to crossing of two or more of the quasi-plane waves of geometric acoustics. The acoustic fields on each of the associated Riemann sheets must be linearly superposed. The sheets are joined at caustics where a local solution of the original Helmholtz equation must be generated to join solutions on the several sheets. The resulting equations lend themselves to rapid, efficient numerical solution using modern numerical algorithms.

1.   INTRODUCTION

A problem of considerable interest is the numerical computation of acoustic fields in the ocean when the wave length λ is much less than the minimum scale length ℓ associated with temperature and salinity variation. In many cases, ℓ is comparable to the top to bottom height h, but much less than the ranging distance R. For harmonic waves the simplest approximate method for solving the governing Helmholtz equation $\nabla^2 \rho + k^2 \rho = 0$ is geometric acoustics where one writes $\rho = \mathrm{Re}(ae^{i\psi})$, chooses $(\nabla\psi)^2 = k^2$, and solves

approximately for the amplitude a. This provides an asymptotical-
ly correct solution in the limit $\lambda/\ell = 0$, but which for finite λ/ℓ
accumulates error as R increases. The parabolic approximation to
the Helmholtz equation has been proposed as a remedy, but simple
arguments suggest that it too fails as R increases. The direct
numerical solution of the Helmholtz equation is intractable, even
in two dimensions and certainly in three, because of the necessity
of choosing a grid size which is a fraction of λ, since a huge number of wavelengths are required when R is large. We wish to describe here a method which serves to convert the problem, by a simultaneous change of dependent and independent variables, to an
elliptic equation, from the solution of which the bulk of the fine
structure on the scale of the wavelength λ has been removed.

The method is suggested by the nature of the geometric acoustic solutions. For concreteness consider a two-dimensional situation where the geometric acoustics field radiated by a localized
source is ducted between two caustics, so that at each point within the geometric acoustics field there are two rays. Thus in order
to get the total field one must linearly syperpose the contribu-
tions of the two rays, viz., $\rho = \mathrm{Re}(a_1 e^{i\psi_1} + a_2 e^{i\psi_2})$. In more general circumstances there may be more than two rays passing through
the point under consideration, and the superposition is correspondingly more complicated. Geometric acoustics fails in the neighborhood of caustics, surfaces which are envelopes of the systems of
rays. A local solution of the Helmholtz equation in the neighborhood of a caustic surface provides the information required to
link the geometric acoustics solutions incident on and emergent
from the caustic surface. As the range increases the errors in
the geometric acoustics solution accumulate, but one does not expect an abrupt appearance of fine structure.

The observation that there are two or more quasi-plane waves
$ae^{i\psi}$ contributing to the solution suggests that one view the problem in terms of Rieman sheets, on each of which one system of rays
$\chi(x,y) = \mathrm{const.}$ (which are tangent to $\nabla\psi$), and the orthogonal trajectories thereunto (the wave fronts $\psi = \mathrm{const.}$) provide a perspic-
uous coordinate system. When a solution of the form $ae^{i\psi}$ is sought,
the unapproximated elliptic partial differential equation for a
has a solution which is a slowly varying function. In the geomet-
ric acoustics limit, the result is that $a^2 |\nabla\chi|/|\nabla\psi|$ is constant
along a ray, that is, it depends on χ alone. The counterpart of
the parabolic approximation when applied to the transformed

equation should yield very accurate results, since refraction is already incorporated via the choice of coordinate system, and the solution on a single Riemann sheet does not involve interference of quasi-plane waves. The elliptic equation itself, on a single Riemann sheet, can be preconditioned by its parabolic approximation, with subsequent iterative solution of the preconditioned equation. According to our estimates, two-dimensional problems should be readily dealt with in this manner, and even modest three-dimensional problems should be tractable.

The notions described above can be generalized to deal with acoustic wave packets. The technique is that of Generalized Fourier Integrals. These are continuous superpositions of quasi-plane waves of the geometric acoustics type, and can be employed to fit arbitrary initial conditions. The propagation of the amplitudes can be found approximately by geometric acoustics, and exactly by solving the counterpart of the equation described in the paragraph above. The theory of this is described in detail in [2,3].

The ideas briefly described above lend themselves to the accurate numerical determination of the acoustic field. The inclusion of realistic bottom conditions seems tractable, and the effect of surface waves is investigable. For the case of harmonic waves the accurate treatment of realistic acoustic radiators is presently feasible using integral equation techniques (see [4]), which yield accurate numerical solutions for the acoustic field at distances from the source less than ℓ. They are particularly efficient since they automatically make the solution outgoing and require the numerical determination of the solution only on the surface where it is required to match to the next region of propagation where the information can then be wed to the methods proposed here.

2. HARMONIC WAVES (HELMHOLTZ EQUATION)

Consider the Helmholtz equation

$$\nabla^2 \rho + \mu(r)^2 \rho = 0 \tag{1}$$

Let us seek solutions of the form

$$\rho = ae^{i\psi} \tag{2}$$

and define

$$k = \nabla\psi \tag{3}$$

Then (1) is carried into

$$(\mu^2 - k^2)a + i(2k \cdot \nabla a + a\nabla \cdot k) + \nabla^2 a = 0 \qquad (4)$$

Note that $\lambda = 2\pi/\mu$ is the local wave length associated with ψ. Let ℓ be the next smallest scale length, typically that characterizing temperature or salinity variation. We wish a theory in which a is a slowly varying object, and where the rapid variation is in the factor $e^{i\psi}$. This suggests that we choose $k^2 = \mu^2$, which is a first order nonlinear partial differential equation for ψ. It can be solved by the method of characteristics. Namely, one defines space curves via the system or ordinary differential equations and conditions

$$dr'/d\sigma' = k'$$
$$r'(0) = r_0 \qquad (5)$$
$$r'(\sigma) = r$$

$$dk'/d\sigma' = -\nabla'(\mu(r')^2)$$
$$k'(0) = k_0(r_0) \qquad (6)$$
$$k'(\sigma) = k$$

where r_0 denotes a point on the surface where the launch information is given and $k_0(r_0)$ is the value of $\nabla\psi$ on that surface. Then ψ is determined via the ordinary differential equation and initial condition

$$d\psi'/d\sigma' = k'^2$$
$$\psi'(0) = \psi_0(r_0) \qquad (7)$$

Equation (2) then reduces to

$$i(2k \cdot \nabla a + a\nabla \cdot k) + \nabla^2 a = 0 \qquad (8)$$

In (2) ostensibly $|\nabla^2 a/(ka)| \sim \lambda/\ell$. This suggests that we neglect $\nabla^2 a$ in (8) in which event the result is equivalent to

$$da'/d\sigma' = -a'\nabla' \cdot k'(r')$$
$$a'(0) = a_0(r_0) \qquad (9)$$

The coefficient $\nabla \cdot k$ too can be generated by integration along trajectories. One defines the dyadic $\mathbf{A} = \nabla k$. Then $\nabla(k) = Tr(\mathbf{A})$, and $\mathbf{A}$ is determined via

$$d\mathbf{A}'/d\sigma' = -\nabla'\nabla'(\mu')^2 + \mathbf{A}'\mathbf{A}'$$
$$\mathbf{A}'(0) = \mathbf{A}_0(r_0) \tag{10}$$
$$\mathbf{A}'(\sigma) = \mathbf{A}$$

Consider a two-dimensional example. Let the equation of a ray be $\chi(x,y) = $ const. Then the orthogonal trajectories of the system of rays are the surfaces of constant phase $\psi(x,y) = $ const. If we use χ and ψ as coordinates in place of x and y, Eq. (8) is carried into

$$i(2P(\partial a/\partial\psi) + a(\partial P/\partial\psi)) + (\partial/\partial\psi)[P(\partial a/\partial\psi)]$$
$$+ (\partial/\partial\chi)[P^{-1}(\partial a/\partial\chi)] = 0 \tag{11}$$

with $P = |\nabla\psi|/|\nabla\chi|$.

Recall that $|\nabla\psi| = \mu$, while $|\nabla\chi|$ depends on the choice of orthogonal trajectory and must be determined numerically. Note that if in (11) we keep only the first order terms, there results the geometric acoustics solution $(\partial/\partial\psi)(a^2 P) = 0$. If we neglect terms involving second order derivatives with respect to ψ, there results the generalized parabolic equation

$$i(2P(\partial a/\partial\psi) + a(\partial P/\partial\psi)) + (\partial/\partial\chi)[P^{-1}(\partial a/\partial\chi)] = 0 \tag{12}$$

The solution of (12) for moderate ranging distances should exhibit little fine structure, although such may gradually develop as the range gets longer and longer. Because the term neglected in (11) in arriving at (12) is small, it should be possible to solve (11) by preconditioning it by (12) and solving the preconditioned equation iteratively.

In practice, the situation is complicated by the feature that because of reflection of waves at boundaries or caustics there may be more than one ray through a given point. In such a case one must solve the counterpart of (12) for each of the associated systems of rays, the solutions being linked by local solutions of the original Helmholtz equation in the neighborhood of the caustics or boundaries.

The acoustic field in the neighborhood of a source of known characteristics can be found readily by adapting exiting solvers for an integral equation equivalent to the Helmholtz equation and outgoing boundary conditions, in a domain around the source of dimension much greater than λ but less than ℓ.

3. NUMERICAL APPARATUS

Numerical implementation of the approach proposed above will involve four steps:

(a) Solving the equation for the phase ψ by the method of characteristics.

(b) Constructing the solution to equation (1) in the vicinity of the radiators.

(c) Solving the parabolic equation (12) numerically.

(d) Solving the elliptic equation (11) numerically.

The following is a brief discussion of the four steps above.

Step (a) requires that a set (7) of Cauchy problems for non-linear ordinary differential equations (ODEs) be solved numerically. The ODEs (7) are not stiff (except in the vicinity of a caustic), and can be solved by means of an explicit scheme, such as a Runge-Kutta method. In the vicinity of a caustic, the equations (7) become stiff, and a certain amount of care should be exercised in their numerical treatment. However, numerical solution of ODEs is an extremely well-developed field, and we do not anticipate serious problems in this area.

Most realistic radiators are considerably smaller than the scale ℓ, which permits us to assume in step (b) that the radiator is imbedded in an infinite medium with a constant index of refraction μ. This results in a boundary value problem for the Helmholtz equation with a constant Helmholtz coefficient in the region exterior to the radiator.

Radiators of practical interest often have complicated shapes, and are several hundred wavelengths in size. Numerical solution of scattering problems of this type has traditionally presented serious difficulties. Indeed, algorithms based on the separation of variables and geometrical perturbation (such as the T-matrix method) cannot be effectively applied to complicated scattering geometries. Finite differences (FD) and finite elements (FE) require that the whole domain of the differential equation be discretized, with at least two nodes of the discretization per wavelength. However, in exterior boundary value problems, the domain is infinite, and an infinite number of nodes would be required to impose the radiation condition at infinity. Hybrid algorithms using FD or FE discretizations in the vicinity of the scatterer, and separation of variables to impose the radiation condition on the far field (such as the unimoment method) are quite satisfactory analytically, and produce good numerical results for small-

scale problems. They tend, however, to result in excessive CPU
time requirements in two dimensions whenever the scatterer is more
than several wavelengths in size, and are almost always prohibitively expensive in three dimensions.

Thus, boundary integral equations (BIEs) become an attractive
alternative for the solution of such problems. In most BIE-based
algorithms, the problem is reduced to a second kind integral equation (SKIE) on the boundary of the scatterer. Discretization of
the resulting SKIE leads to a dense large-scale system of linear
algebraic equations, which in turn is solved by means of some iterative technique. Most of the iterative schemes for the solution
of linear systems of this type require that the matrix of the system be applied to a sequence of recursively generated vectors.
Applying a dense matrix to a vector requires k^2 multiplications
and about as many additions, where k is the dimension of the system, and the dimension of the system is equal to the number of
nodes in the discretization of the boundary of the scatterer. As
a result, the whole process is at least of order k^2, which tends
to be acceptable for small to medium size problems in two dimensions. Unfortunately, for large-scale problems in two dimensions,
and for most problems in three dimensions, this is still prohibitively expensive.

Recently, an algorithm was constructed (see [4]) for the rapid
application of matrices resulting from the discretization of integral equations of scattering theory to arbitrary vectors. The algorithm requires an amount of work proportional to n log(n), where
n is the number of nodes in the discretization of the boundary of
the scatterer. When it is combined with a Generalized Conjugate
Residual type algorithm, the resulting process takes very few iterations to converge, leading to an order n log(n) algorithm for the
solution of the original scattering problem. The actual timings
obtained with this algorithm indicate that for large-scale problems
in two dimensions it leads to trivial CPU time requirements, and
should make medium to large-scale problems in three dimensions
tractable.

Efficient numerical solution of parabolic PDEs normally involves the use of implicit marching schemes, such as the Crank-
Nicolson algorithm. Such schemes eliminate stability problems
associated with the discretizations of parabolic systems, but require that a large-scale system of linear algebraic equations be
solved on each step. In two-dimensional situations, the matrices

of the systems to be solved are usually tri-diagonal (five-diagonal in some algorithms), and the numerical solution of such systems is an extremely efficient process. As a result, solving equation (12) in two dimensions effectively consists of choosing one of several existing numerical schemes (all of which are quite adequate), and implementing it.

In three dimensions, the situation is considerably more complicated. Here, implicit marching schemes involve the solution of linear systems whose matrices are banded, with some bands far removed from the main diagonal. Several approaches have been developed for dealing with such problems, but none of them are as efficient as one would like. Fortunately, the problem associated with equation (12) is not likely to be a very large-scale one, since the fine structure has been removed from the solution (see Section 2 above). Consequently, the system (12) should be treatable by existing methods, in three dimensions as well as in two.

Equations of the form (12) are known to cause serious numerical difficulties, since they tend to develop boundary layers, where the solution varies extrenely rapidly. As a result, most numerical schemes for the solution of equations of the form (12) involve adaptive gridding techniques and fairly sophisticated (and expensive) finite difference or finite element schemes. However, it turns out that by utilizing the technique of volume integral equations (VIEs), extremely efficient algorithms for the solution of the equations of this type can be developed. Below, we describe the two-dimensional version of this approach in some detail.

Integral Equation Formulation. In two dimensions, equation (12) can be rewritten in the form

$$\nabla^2 + 2i(\nabla\psi \cdot \nabla a) + ipa = 0 \tag{13}$$

with

$$p = \nabla^2\psi \tag{14}$$

Observation 1. It is important to note that in (13), $\|\nabla\psi\| = \mu$, while the coefficient p is reasonably small, except in the vicinity of a caustic. This will be important in the qualitative analysis of the equation (13) given below.

As is well known, equation (13) can be converted into a second kind integral equation (SKIE) of the form

$$\sigma(x) + \int K(x,t)\sigma(t)\, dt = f(x) \tag{15}$$

by representing a by an expression

$$a(x) = \int G(x,t)\sigma(t)\ dt \tag{16}$$

with G the Green's function of the equation

$$\nabla^2\phi + 2i\left(\alpha\ \frac{\partial\phi}{\partial x} + \beta\ \frac{\partial\phi}{\partial y}\right) + iq\phi = 0 \tag{17}$$

with some "background" parameters α, β, q, and the kernel K in (15) defined by the formula

$$K((x_0,y_0)(x,y)) = \frac{\partial G}{\partial x}\ ((x_0,y_0),(x,y))\left[\frac{\partial\psi}{\partial x}\ (x,y) - \alpha\right]$$

$$+ \frac{\partial G}{\partial y}\ ((x_0,y_0),(x,y))\left[\frac{\partial\psi}{\partial y}\ (x,y) - \beta\right]$$

$$+ (p(x,y) - q)G((x_0,y_0),(x,y)) \tag{18}$$

Obviously, the closer the actual values of $\partial\psi/\partial x$, $\partial\psi/\partial y$, p to the background parameters α, β, p, the closer the integral equation (13) to its "diagonal" form

$$\sigma(x) = f(x) \tag{19}$$

Green's Function for the Model Equation

We will seek the Green's function of equation (17) in the form

$$G((0,0),(x,y)) = u(x,y)e^{\gamma x+\delta y} \tag{20}$$

with the constant coefficients γ, δ, and the function u to be determined.

Substituting (20) into (17), we obtain

$$\nabla^2 u + 2(\gamma + i\alpha)\ \frac{\partial u}{\partial x} + 2(\delta + i\beta)\ \frac{\partial u}{\partial y}$$

$$+ ((\gamma^2 + \delta^2) + 2i(\alpha\gamma + \beta\delta + q)u = 0 \tag{21}$$

Now, setting

$$\gamma = -i\alpha \qquad \delta = -i\beta \tag{22}$$

we obtain an equation

$$\nabla^2 u + (\alpha^2 + \beta^2 + iq)u = 0 \tag{23}$$

However, (23) is a Helmholtz equation with the Helmholtz coefficient $k_m = \sqrt{(\alpha^2 + \beta^2 + iq)}$, and its Green's function in R^2 is known to be

$$G_m((x_0,y_0),(x,y)) = H_0(k_m\sqrt{((x - x_0)^2 + y - y_0)^2))} \tag{24}$$

where H_0 is the Hankel function of order zero. Combining (24)
with (20), we obtain the following expression for the Green's function of (13) in R^2:

$$G((x_0,y_0),(x,y)) = H_0(k_m\sqrt{((x - x_0)^2 + (y - y_0)^2)})$$
$$\times\, e^{-i(\alpha(x-x_0)+\beta(y-y_0))} \tag{25}$$

Observation 2. As is well known, asymptotically,

$$H_0(z) \sim \sqrt{(2/\pi z)}\, e^{i(z-\pi/r)} \tag{26}$$

and outside the immediate vicinity of (x_0,y_0),

$$G((x_0,y_0),(x,y)) \sim \sqrt{(2/\pi(k_m\sqrt{((x - x_0)^2 + (y - y_p)^2)})}$$
$$\times\, e^{i(k_m\sqrt{((x-x_0)^2+(y-y_0)^2)}-\pi/4)}$$
$$\times\, e^{-i(\alpha(x-x_0)+\beta(y-y_0))} \tag{27}$$

It is easy to make several interesting physical observations
from (27). Namely, upstream from x_0, the function $G_0((x_0,y_0),$
$(x,y))$ oscillates in a "nonphysical" manner, i.e., faster than any
propagating wave in the medium. Downstream from x_0, the function
$G_0((x_0,y_0),(x,y))$ oscillates with a frequency lower than that of a
propagating wave in the medium, and along the ray $x = x_0 + \alpha \cdot t, y =$
$y_0 + \beta \cdot t, t \in (0,\infty)$, the function $G_0((x_0,y_0)(x,y))$ does not oscil-
late at all. On this line, it has phase 1, and decays like
$((x - x_0)^2 + (y - y_0)^2)^{-1/4}$. Since the solution of (25) is known
to be smooth on the scale of a wavelength λ, this latter observation means that equation (25) will display a "Volterra-like" behavior with respect to the variable t. Namely, the solution σ at
the point (x,y) will depend on the values of f upwind from the
point (x,y) but not on the values of f downstream from (x,y).

Numerical Solution of Equation (25)

A standard procedure for the solution of large scale SKIEs
calls for discretizing them via some numerical scheme, such as the
Nystrom algorithm based on an appropriate quadrature formula (see,
for example, [5]), with the subsequent solution of the obtained
linear system by means of a Generalized Conjugate Residual-type
algorithm. As is well known, linear systems resulting from SKIEs
have asymptotically limited condition numbers, and the number of
iterations required by a Generalized Conjugate Residual-type

algorithm to achieve a given accuracy ε for such systems is virtually independent of the dimension n of the system, and is of the order $-\log(\varepsilon) \cdot \log(K)$, where K is the condition number of the system (see [1,5,6]). Furthermore, matrices resulting from discretization of SKIEs of the form (25) can be applied to arbitrary vectors for a cost proportional to $n \cdot \log(n)$, where n is the number of nodes in the discretization (see [4,5]). Thus, the total CPU time estimate for the solution of equation (23) is of the order

$$-n \cdot \log(n) \cdot \log(\varepsilon) \cdot \log(K) \tag{28}$$

Unfortunately, wave propagation problems can lead to SKIEs whose discretizations have fairly high condition numbers, and the estimate (28), while almost optimal in the number of nodes n, can still be prohibitive for large-scale problems. In such cases, observation 2 (see above) can be used to construct a preconditioning for equation (25).

In the preconditioning step, the "upstream" part of the integral (26) is ignored, and equation (25) assumes the form

$$\sigma(x) + \int \tilde{K}(x,t)\sigma(t) \, dt = f(x) \tag{29}$$

with the new kernel $\tilde{K}$ defined by the formulas

$$\tilde{K}((x_0,y_0),(x,y)) = K((x_0,y_0),(x,y))$$
$$\text{for } (x - x_0)\alpha + (x - x_0)\alpha < 0$$
$$\tilde{K}((x_0,y_0),(x,y)) = 0 \tag{30}$$
$$\text{for } (x - x_0)\beta - (x - x_0)\alpha > 0$$

Introducing new coordinates ζ, η by the formulas

$$\zeta(x,y) = \alpha x + \beta y$$
$$\eta(x,y) = \alpha y - \beta x \tag{31}$$

it is easy to see that equation (29) is of Volterra type with respect to the variable ζ, and can be solved by a marching scheme along this variable. Now, equation (25) can be preconditioned by equation (29), with the subsequent solution of the preconditioned equation by means of a Generalized Conjugate Residual-type algorithm.

REFERENCES

1. S. C. Eisenstat, H. C. Elman, and M. H. Schultz, Variational iterative methods for nonsymmetric systems of linear equations, SIAM JOURNAL ON NUMERICAL ANALYSIS, 20 (1983).

2. G. Hazak, I. B. Bernstein, and T. M. Smith, Integral representation for geometric optics solutions, PHYSICS OF FLUIDS, 26 (1984).

3. G. Hazak, L. Friedland, and I. B. Bernstein, A uniform integral representation for geometric optics solutions near caustics, PHYSICS OF FLUIDS, 27 (1984).

4. V. Rokhlin, Rapid solution of Integral equations of scattering theory in two dimensions, Technical Report 440, Dept. of Computer Science, Yale University, 1985.

5. V. Rokhlin, Application of volume integrals to the solution of partial differential equations, COMPUTERS AND MATHEMATICS WITH APPLICATIONS, 11, No. 7/8 (1985).

6. V. Rokhlin, Rapid solution of integral equations of classical potential theory, JOURNAL OF COMPUTATIONAL PHYSICS 6, No. 2 (1985).

COMPUTATIONAL ACOUSTICS: Wave Propagation
D. Lee, R.L. Sternberg, M.H. Schultz (Editors)
Elsevier Science Publishers B.V. (North-Holland)

A FAST, FILTERED, FOURIER-TRANSFORM MARCHING ALGORITHM
FOR WIDE-ANGLE, ONE-WAY WAVE PROPAGATION

Louis Fishman

Department of Civil Engineering
The Catholic University of America
Washington, D.C.

and

Stephen C. Wales

Acoustics Division
Naval Research Laboratory
Washington, D.C.

ABSTRACT

Factorization and path integration methods applied to the n-dimensional Helmholtz equation result in a phase space marching algorithm which generalizes the Tappert/Hardin split-step FFT algorithm to the full one-way (factored Helmholtz) wave equation. In the two-dimensional case, the computational algorithm is ideally suited for computers which provide either a vector or a parallel pipe type of operation. In conjunction with phase space filtering, the code is quite fast. Extensive numerical calculations with the filtered one-way phase space marching algorithm on ocean acoustic, seismological, and extreme model environments designed to establish the range of validity and manner of breakdown have been extremely promising. Of particular significance is the fact that the manner of marching the radiation field is independent of the medium and any approximation to the square root Helmholtz operator, resulting in a modular code architecture and highly versatile propagation program. Moreover, the propagation models constructed and computed through the code correspond to singular integro-differential equation as well as partial differential equation approximations to the one-way wave equation. Indeed, this numerical algorithm represents one of the very few attempts to compute directly with pseudo-differential and Fourier integral operators.

INTRODUCTION

The reduced scalar Helmholtz equation plays a significant role in studies of electromagnetic, seismic, and acoustic direct wave propagation. Even for these linear models, the computation of the wave field and related quantities such as transmission loss and energy density is extremely difficult for multidimensional transversely inhomogeneous environments extending over many wavelengths. For the most part, classical, "macroscopic" methods have resulted in direct wave field approximations (perturbation theory, ray-theory asymptotics, modal analysis, hybrid ray-mode methods), derivations of approximate wave equations (scaling analysis, field splitting techniques, formal operator expansions), and discrete numerical approximations (finite differences, finite elements, spectral methods). However, the "microscopic" phase space analysis developed over the past several decades by mathematicians studying linear partial differential equations and the global functional integral techniques pioneered by Wiener (Brownian motion) and Feynman (quantum mechanics) together provide the framework to both explicitly represent and, subsequently, directly compute the n-dimensional, one-way (half-space) Helmholtz propagator [1-15]. The resulting one-way computational algorithm is based on (1) the marching range step (following from the path integral), (2) a sophisticated symbol analysis (reflecting the detailed study of the Weyl composition equation), and (3) Fourier component, or wave number, filtering in phase space (for increased efficiency, decreased computational time, and reduced error). It provides the generalization of the Tappert/Hardin split-step FFT algorithm to the full one-way (factored Helmholtz) wave equation, and, as such, is ideally suited for present-day and future parallel computing architectures. From a more theoretical viewpoint, the phase space factorization analysis provides for a comprehensive framework for the derivation and analysis of the numerous parabolic (one-way) approximate wave theories [9,14,16-18] presented in recent years. All of these approximate theories can be computed by the phase space marching algorithm with the appropriate symbols.

ONE-WAY COMPUTATIONAL ALGORITHM

The formal one-way n-dimensional Helmholtz equation [9,16-18]

$$(i/\bar{k})\,\partial_x\phi^+(x,\underset{\sim}{x}_\perp) + (K^2(\underset{\sim}{x}_\perp) + (1/\bar{k}^2)\nabla_\perp^2)^{1/2}\phi^+(x,\underset{\sim}{x}_\perp) = 0 \qquad (1)$$

where $K(x_\perp)$ is the refractive index field and $\bar{k}$ is a reference
wave number, can be made explicit through a pseudo-differential
operator construction [3,9,11], leading to a systematic development of full-wave, wide-angle approximate one-way wave equations
[9,11,14,15]. Phase space path integral constructions then provide for an explicit representation of the Helmholtz propagator
[10,11] and, subsequently, through direct computation, for a numerical marching algorithm [11-14]. The marching range step takes
the form [11-14]

$$\phi^+(x + \Delta x, \underset{\sim}{x}_\perp) \approx \int_{R^{n-1}} d\underset{\sim}{p}_\perp \, \exp(i\bar{k}\underset{\sim}{p}_\perp \cdot \underset{\sim}{x}_\perp)$$
$$\times \, (\exp(i\bar{k}\Delta x h_{\mathbf{B}}(\underset{\sim}{p}_\perp, \underset{\sim}{x}_\perp))\hat{\phi}^+(x, \underset{\sim}{p}_\perp)) \tag{2}$$

where

$$h_{\mathbf{B}}(\underset{\sim}{p}_\perp, \underset{\sim}{x}_\perp) = (\bar{k}/\pi)^{n-1} \int_{R^{2n-2}} d\underset{\sim}{s}d\underset{\sim}{t} \, \Omega_{\mathbf{B}}(\underset{\sim}{s}, \underset{\sim}{t})$$
$$\times \, \exp(-2i\bar{k}(\underset{\sim}{x}_\perp - \underset{\sim}{t}) \cdot (\underset{\sim}{p}_\perp - \underset{\sim}{s})) \tag{3}$$

and $\hat{\phi}^+$ is the Fourier-transformed wave field. In Eq. (3), the
operator symbols $\Omega_{\mathbf{B}}(\underset{\sim}{p}, \underset{\sim}{q})$ and $\Omega_{\mathbf{B}}2(\underset{\sim}{p}, \underset{\sim}{q})$ corresponding, respectively,
to the square root Helmholtz operator $\mathbf{B} = (K^2(\underset{\sim}{q}) + (1/\bar{k}^2)\nabla_{\underset{\sim}{q}}^2)^{1/2}$
and its square, $\mathbf{B}^2 = (K^2(\underset{\sim}{q}) + (1/\bar{k}^2)\nabla_{\underset{\sim}{q}}^2)$, are related through the
Weyl composition equation [2,9]

$$\Omega_{\mathbf{B}}2(\underset{\sim}{p}, \underset{\sim}{q}) = K^2(\underset{\sim}{q}) - \underset{\sim}{p}^2 = (\bar{k}/\pi)^{2n-2} \int_{R^{4n-4}} d\underset{\sim}{t}d\underset{\sim}{x}d\underset{\sim}{y}d\underset{\sim}{z}$$
$$\times \, \Omega_{\mathbf{B}}(\underset{\sim}{t} + \underset{\sim}{p}, \underset{\sim}{x} + \underset{\sim}{q})\Omega_{\mathbf{B}}(\underset{\sim}{y} + \underset{\sim}{p}, \underset{\sim}{z} + \underset{\sim}{q})$$
$$\times \, \exp(2i\bar{k}(\underset{\sim}{x} \cdot \underset{\sim}{y} - \underset{\sim}{t} \cdot \underset{\sim}{z})) \tag{4}$$

Equation (2) provides an increasingly better approximation in
the limit of infinitesimal range step size Δx. Further, in the
parabolic limit, the equation reduces to the Tappert/Hardin split-
step FFT algorithm [17,18]. This is to be expected since the
split-step FFT algorithm follows from a direct integration of the
phase space Feynman path integral for the parabolic propagator.
Equations (2)-(4) thus provide the generalizations of (1) the Tap-
pert/Hardin algorithm and (2) a homogeneous half-space algorithm
[11,14] to the full one-way (factored Helmholtz) wave equation.

For a two-dimensional model ocean/bottom propagation environment with a perfectly reflecting ocean surface, the construction
of the computational algorithm is straightforward. Two perfectly
reflecting surfaces, corresponding to the ocean surface and the

bottom of the Fourier transform regime, are introduced. A "false
bottom" with attenuation is provided to absorb the wave field as
it approaches the bottom boundary. With this approximation to the
radiation condition, the Fourier transform of the wave field in
Eq. (2) is replaced by a discrete fast sine transform. The careful construction of the attenuating "false bottom" is essential to
minimize spurious reflection, or wrap-around, effects. The bottom
layer must be sufficiently deep with an attenuation profile which
is sufficiently gradual to avoid "ringing" effects and, ultimately,
sufficiently strong to completely damp out the wave field. Employing a rectangular rule integration for the inverse transform
in Eq. (2) enables the propagated wave field to be expressed in
the matrix form

$$\phi^+(x + \Delta x,\ z_n) = \sum_m A_{nm}\hat{\phi}^+(x,p_m) \tag{5}$$

for each depth point z_n. In Eq. (5), ϕ^+ and $\hat{\phi}^+$ are column vectors
and the matrix $\underset{\approx}{A}$ is defined by its matrix elements

$$A_{nm} = \eta\ \sin(\bar{k}p_m z_n + \bar{k}\Delta x h_{\mathbf{B}}^o(p_m,z_n))\ \exp(i\bar{k}\Delta x h_{\mathbf{B}}^e(p_m,z_n)) \tag{6}$$

where $h_{\mathbf{B}}^e$ and $h_{\mathbf{B}}^o$ are the even and odd parts with respect to p of
$h_{\mathbf{B}}(p,z)$ in Eq. (3) and η is an appropriate transform normalization
constant. The initial start-up field is provided by a normal-mode
start-up program or a Gaussian field when applicable. A detailed
discussion of these and other numerical considerations can be
found elsewhere [18].

   The algorithm is not as computationally efficient as the split-
step FFT algorithm which takes explicit advantage of the separable
form of $h_{\mathbf{B}}(p,z)$ in the parabolic case [13]. Compensating for this,
however, is the far greater range of applicability associated with
this new algorithm. The speed of the algorithm is essentially governed by the time required to (1) compute the matrix $\underset{\approx}{A}$ and (2) advance the wave field one range step. Clearly, the size and speed
of the computer employed effect this performance. The matrix calculation is a function of the number of matrix elements (the square
of the number of depth points) and the computational complexity of
the chosen symbol approximation. The range-incrementing procedure
is just a sequence of matrix multiplications and, thus, easily converted to a sequence of parallel vector operations, resulting in a
substantial increase in speed on any machine which provides either
a vector or parallel pipe type of operation. On any of the current

supercomputers or array processors sufficiently large enough to accommodate the matrix $\underset{\approx}{A}$, advancing the wave field several thousand range steps in a 512 × 512 problem could be expected to take on the order of a few minutes.

Equations (5) and (6) illustrate that the manner of marching the radiation field is independent of the specific properties of the medium and the approximation to the square root Helmholtz operator. This data is contained in the matrix $\underset{\approx}{A}$ which is computed once and forms part of the input to the marching algorithm. For each extended parabolic wave theory (approximate symbol) constructed, there is a subroutine to calculate the corresponding matrix $\underset{\approx}{A}$. The resulting modular code architecture allows for a highly versatile computer program in which different subroutines can be called as different environmental conditions are encountered. Moreover, the propagation models constructed and computed correspond to singular integro-differential equation as well as partial differential equation approximations to the one-way wave equation. Indeed, this numerical algorithm represents one of the very few attempts to compute directly with pseudo-differential and Fourier integral operators [19].

ERROR, ENERGY FLUX CONSERVATION, AND STABILITY

In computing the numerical solution of an approximate wave equation, the phase space marching algorithm introduces two distinct sources of error. The approximation of the square root Helmholtz operator symbol introduces an inherent error at the level of the wave equation, while both the continuous and discretized range steps [Eqs. (2) and (5), respectively] associated with the finite-N realization of the path integral introduce further computational errors. For the exact symbol, the continuous range step (infinitesimal propagator), while exact for the homogeneous medium limit, is, in general, first-order accurate in Δx. This follows immediately upon expanding both sides of Eq. (2) in a Taylor series in the limit of $\Delta x \to 0$ and, subsequently, applying the one-way pseudo-differential wave equation in its standard-ordering form [9,14]. The $0((\Delta x)^2)$ local error is related to a standard pseudo-differential operator with the symbol $h_{\mathbf{B}}2(\underset{\sim}{p},\underset{\sim}{q}) - h_{\mathbf{B}}^2(\underset{\sim}{p},\underset{\sim}{q})$, consistent with the formal construction of the phase space path integral representation [7,14]. For an approximate symbol, the local error associated with Eq. (2) is, in general, $0(\Delta x)$.

Energy conservation for the one-way Helmholtz equation and the marching algorithm is based on the conservation law

$$\frac{d}{dx} \int_{R^{n-1}} dx_{\perp} J_x(x, x_{\perp}) = 0 \tag{7}$$

where

$$J_x(x, x_{\perp}) = \phi^{+}(x, x_{\perp}) \, (i/\bar{k}) \, \partial_x \phi^{+*}(x, x_{\perp})$$
$$- \phi^{+*}(x, x_{\perp}) \, (i/\bar{k}) \, \partial_x \phi^{+}(x, x_{\perp}) \tag{8}$$

is proportional to the energy flux. Equations (7) and (8) follow directly from the continuity equation relating the energy density and flux for the corresponding time-domain wave equation and Gauss' divergence theorem [20]. In combination with the Weyl pseudo-differential wave equation [9,14], the conservation law can be rewritten as

$$\frac{d}{dx} \int_{R^{3n-3}} dx_{\perp} dx_{\perp}' dp_{\perp} \, (\text{Re } \Omega_{\mathbf{B}}(p_{\perp}, (x_{\perp} + x_{\perp}')/2))$$
$$\times \exp(i\bar{k}p_{\perp} \cdot (x_{\perp} - x_{\perp}')) \rho(x, x_{\perp}', x_{\perp}) = 0 \tag{9}$$

where

$$\rho(x, x_{\perp}', x_{\perp}) = \phi^{+}(x, x_{\perp}') \phi^{+*}(x, x_{\perp}) \tag{10}$$

is the coherence function analogous to the quantum mechanical density matrix [21]. The Weyl pseudo-differential wave equation and Eq. (9) further imply that energy-conserving extended parabolic wave theories require the operator $\mathbf{B}^2$ to be selfadjoint. In the Weyl calculus, selfadjoint operators correspond to real symbols (one of its virtues) [1,2], thus requiring

$$\Omega_{\mathbf{B}}^2(p, q) = (\bar{k}/\pi)^{2n-2} \int_{R^{4n-4}} dt dx dy dz \, \Omega_{\mathbf{B}}(t + p, \, x + q)$$
$$\times \Omega_{\mathbf{B}}(y + p, \, z + q) \, \exp(2i\bar{k}(x \cdot y - t \cdot z))$$

to be real. This condition is obviously satisfied by the exact square root Helmholtz operator symbol through the Eq. (4) construction and, for example, by any real approximate symbol $\Omega_{\mathbf{B}}(p, q)$. Extended parabolic wave theories constructed through the factorization analysis, thus, are not a priori energy conserving, and can, in fact, create energy, leading to solutions which grow with increasing range.

The continuous range step in the marching algorithm is not strictly energy conserving. Expressing the conservation law in the form

$$\int_{R^{n-1}} d\underset{\sim}{x}_\perp J_x(x,\underset{\sim}{x}_\perp) = \int_{R^{n-1}} d\underset{\sim}{x}_\perp J_x(x + \Delta x, \underset{\sim}{x}_\perp) \tag{11}$$

in conjunction with Eqs. (9), (10), (2), and (7) results in an $0((\Delta x)^2)$ local error, in general, in the total energy flux conservation for energy-conserving extended parabolic wave theories. For selfadjoint approximations to the square root Helmholtz operator **B**, the continuous range step generally results in a second-order local error in the conservation of the L^2 norm, although, for separable approximations, such as the ordinary parabolic, the L^2 norm is preserved.

The solutions of the one-way Helmholtz initial-value problem are nonincreasing in range [22]. It is, thus, appropriate to require the same boundedness of the corresponding solutions of the fully discretized numerical scheme. This practical stability criterion [23], especially relevant for fixed mesh-size analysis, effectively eliminates the catastrophic growth of errors in the numerical solution. Demanding stability to infinite range, however, is unnecessarily restrictive. Practical computation only requires that the marching algorithm be effectively stable out to a sufficiently large range. The relatively slow onset of uncontrolled growth resulting from the non-energy-conserving properties of the symbol approximation and/or the range and transverse coordinate discretization establishes an effective range of computation rather than marking the complete inapplicability of the algorithm.

The two-dimensional model ocean/bottom fully discretized numerical scheme can be expressed in the matrix form

$$\underset{\sim}{\phi}^+_{j+1} = \underset{\approx}{A} \cdot \underset{\approx}{S} \cdot \underset{\sim}{\phi}^+_j = \underset{\approx}{C} \cdot \underset{\sim}{\phi}^+_j \tag{12}$$

or equivalently,

$$\underset{\sim}{\hat{\phi}}^+_{j+1} = \underset{\approx}{S} \cdot \underset{\approx}{A} \cdot \underset{\sim}{\hat{\phi}}^+_j = \underset{\approx}{S} \cdot \underset{\approx}{C} \cdot \underset{\approx}{S}^{-1} \cdot \underset{\sim}{\hat{\phi}}^+_j \tag{13}$$

where $\underset{\approx}{S}$ is the discrete sine transform matrix, the subscript j denotes the range discretization, and the configuration and Fourier space representations are related by a similarity transformation. For this simple form, the practical stability criterion translates to the condition that the eigenvalues of the matrix $\underset{\approx}{C}$ all lie on or within the unit circle in the complex plane. An analytic characterization of this condition is difficult to state concisely [24].

The Gershgorin circle theorem [25] does not provide a sharp enough bound on the eigenvalues of a matrix to be applicable. Neither the complex extension of Hurwitz's criteria [26] nor an application of Schur's theory [27] appears to be practical in light of the very large dimensional matrices involved. For a specific calculation, however, a numerical determination of the eigenvalues and eigenvectors of the matrix $\underset{\approx}{C}$ would provide the means to estimate the effective range of computation.

SYMBOL ANALYSIS AND PHASE SPACE FILTERING

The principal idea underlying the practical implementation of the phase space marching algorithm is the construction of a small number of approximate operator symbols which, when taken together, allow for wave field computations over a very wide range of model environments and propagation parameters. The symbol constructions can be essentially analytical or numerical since the manner of marching the wave field is independent of the detailed construction of the matrix $\underset{\approx}{A}$. The phase space associated with the model numerical computations is finite, effectively limited in configuration space by an approximate radiation or an exact boundary condition and in Fourier space by finite spatial resolution. Moreover, even for wide-angle propagation problems, only a limited region of this finite phase space contributes significantly to the wave field calculation. This enables the effective use of Fourier component, or wave number, filtering, as is frequently employed in the split-step FFT algorithm [18]. Thus, for fixed mesh-size Δx, a sufficiently accurate symbol approximation over the relevant region of phase space can greatly reduce the magnitude of the first-order local error, resulting in accurate numerical calculations out to realistic ranges.

The standard nonuniform high-frequency asymptotic analysis of the Weyl composition equation [9,11] provides the starting point for the approximate symbol development. Retaining the leading term in the $\bar{k} \to \infty$ limit results in the high-frequency approximation of the operator symbol [9]. This algorithm is equivalent to setting

$$h_{\underset{\sim}{\mathbf{B}}}(\underset{\sim}{p}_{\perp},\underset{\sim}{x}_{\perp}) = (K^2(\underset{\sim}{x}_{\perp}) - \underset{\sim}{p}_{\perp}^2)^{1/2} \tag{14}$$

in Eq. (3), corresponding to a standard pseudo-differential operator with the symbol $(K^2(\underset{\sim}{q}) - \underset{\sim}{p}^2)^{1/2}$. It represents the most naive

generalization of the homogeneous medium result [9,11], corresponding to the locally homogeneous medium approximation of the infinitesimal propagator [10], and is quite distinct from geometric, or
semiclassical, approximations on the wave field. Unlike the ordinary parabolic wave theory [17,18], the high-frequency wave
theory, just like the exact one-way Helmholtz equation, is inde-
pendent of the reference sound speed c_0 ($K(\underset{\sim}{x}_\perp) = c_0/c(\underset{\sim}{x}_\perp)$, $\bar{k} =$
ω/c_0). Further, it follows immediately from Eq. (14) that the
second-order local error associated with the continuous range step
(infinitesimal propagator) is identically zero for the high-frequency approximation. Since the composite Weyl high-frequency
symbol generally has a nonzero imaginary part, the high-frequency
wave theory does not, in general, strictly conserve energy. As
such, the high-frequency propagation algorithm can generally be
expected to break down at sufficiently large ranges, often resulting in numerical solutions which grow with increasing range.

Energy conservation (at the level of the wave equation) can
be restored through the slightly modified algorithm corresponding
to a Weyl pseudo-differential operator with the symbol

$$\Omega_{\mathbf{B}}(\underset{\sim}{p},\underset{\sim}{q}) = \operatorname{Re}(K^2(\underset{\sim}{q}) - \underset{\sim}{p}^2)^{1/2} \tag{15}$$

For this real Weyl high-frequency approximation, in the two-dimensional case, for example, $h_{\mathbf{B}}(p,z)$ in Eq. (3) takes the specific
form

$$h_{\mathbf{B}}(p,z) = \frac{1}{2} \int_{R^1} dt\ t^{-1}\ \exp(2i\bar{k}pt)K(t + z)J_1(2\bar{k}tK(t + z)) \tag{16}$$

where $J_1(y)$ is the appropriate Bessel function. Equation (16) is
efficiently computed by an FFT algorithm and corresponds, again,
to a c_0-independent wave theory. In this approximation, the smoothing property of the one-way Helmholtz propagator [28] is largely
lost due to the reduced damping of the high-frequency Fourier components in the wave field. While the computed wave fields should
generally remain bounded, the increased energy in the high-frequency portion of the spectrum can be expected to result in the
superposition of a fine-scale structure on the smoothed large-
scale wave field envelope.

The square root function represents the leading algebraic
term in an outer-scale asymptotic expansion of the operator symbol
[15]. It provides an inadequate representation of the symbol in
the high-propagating-angle, or near-evanescent, regime, $\Omega_{\mathbf{B}}(\underset{\sim}{p},\underset{\sim}{q}) \approx 0$,

where the oscillatory character of the exact symbol must be properly represented. With increasing sound speed gradient and decreasing frequency, the contribution to the wave field calculation
from this near-evanescent phase space region increases. This results in the earlier onset in range of large-scale phase and amplitude errors in the two high-frequency algorithms. Thus, while an
energy-conserving symbol approximation is desirable, it is not sufficient; accurate calculations require the proper damping of the
high-frequency Fourier components and a detailed treatment of the
near-evanescent regime [14]. The outer-scale high-frequency, or
"classical" limit [9,10] approximation, however, can be supplemented with an appropriate inner-scale solution, resulting in a uniform high-frequency asymptotic symbol approximation [15] which
should lead to a more widely applicable computational algorithm.

For extremely rapidly changing environments and/or very-low
propagation frequencies, the oscillatory character of the symbol
extends significantly throughout the relevant phase space region,
dominating the accurate calculation of the wave field [15]. Under
these conditions, the uniform high-frequency approximation may be
inadequate. While the general inversion of the Weyl composition
equation is of extreme interest and, ultimately, the key to a complete mathematical theory and algorithm, a low-frequency ($\bar{k} \to 0$)
asymptotic development seems appropriate. This suggests an essen-
tially numerical approach based on a $\underset{\sim}{p}$- and $\underset{\sim}{q}$-space complete set
expansion of $\Omega_B(\underset{\sim}{p},\underset{\sim}{q})$, with the expansion coefficients determined
by substitution into the Weyl composition equation and subsequent
numerical solution of the resulting coupled system of equations.
The finite extent of the relevant phase space and the $\bar{k} \to 0$ limit
should allow for reasonable truncation procedures, resulting in a
practical algorithm which is particularly appropriate in regimes
primarily complementary to those addressed by the high-frequency-
based perturbation analysis.

Phase space marching algorithms provide a most natural and
efficient means for introducing filtering techniques. Fourier
component, or wave number, filtering is applicable to all approximations in the two-dimensional algorithm of Eq. (5), and is most
efficiently incorporated by setting

$$A_{nm} = 0 \qquad \text{for } m > m_0 \tag{17}$$

In Eq. (17), m_0 can be chosen to correspond to a propagation
angle through the dispersion relation for a homogeneous medium

characterized by the chosen reference sound speed c_0 [14]. More
sophisticated filters are constructed by appropriate modifications
of the matrix $\underset{\sim}{A}$. Equation (17) reduces both the size of the matrix multiplication and the number of matrix elements initially
computed, in particular, reducing the total range-incrementing computational time by almost an order of magnitude for typical ocean/
bottom calculations. For the high-frequency algorithm, for example, the complex wave field can be computed at 512 (depth) × 1200
(range) points in a total algorithm time of approximately 2 minutes (total set-up time) and 30 seconds (total range-incrementing
time) on a VAX 11/750 computer and FPS-164 array processor system.
With an FPS-264 array processor, the total range-incrementing time
would be reduced to approximately 9 seconds.

NUMERICAL RESULTS

     Numerical results of transmission loss (dB re 1 m) as a func-
tion of range (km) for a number of model ocean/bottom propagation
experiments demonstrate the computational viability of the factor-
ization-/path integration-based phase space marching algorithm [12-
14]. Several propagation experiments are summarized in Figs. 1-3,
with the corresponding transmission loss curves compared with a
reference Fast Field Program (FFP) algorithm [29,30] in Figs. 4-6.

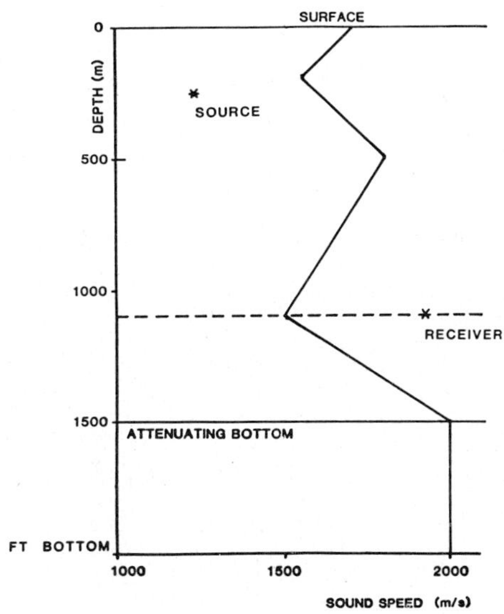

Fig. 1. Model 1 ocean/bottom environment and transmission experiment.

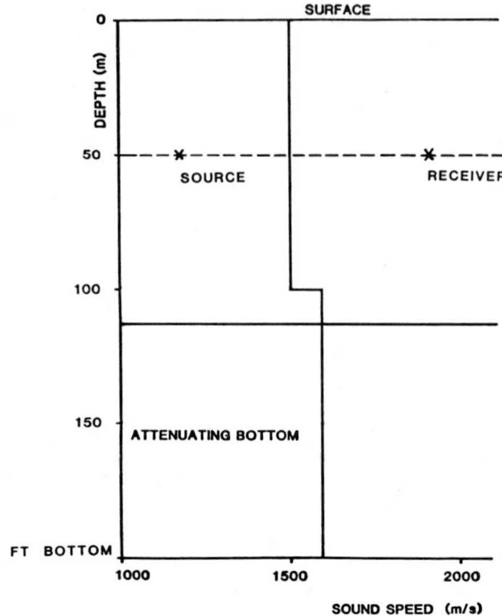

Fig. 2.   Model 2 ocean/bottom environment and transmission experiment.

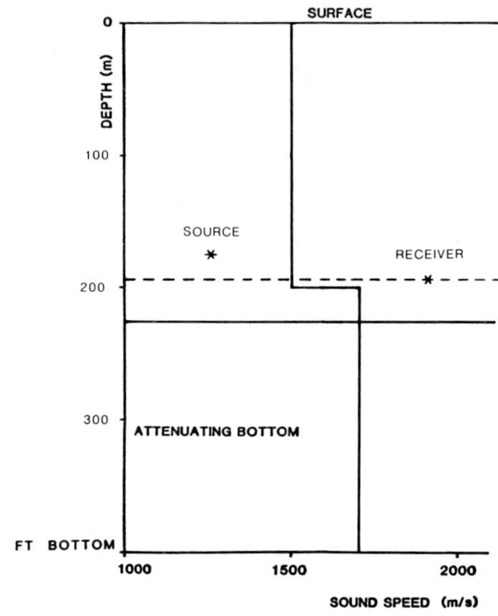

Fig. 3.   Model 3 ocean/bottom environment and transmission experiment.

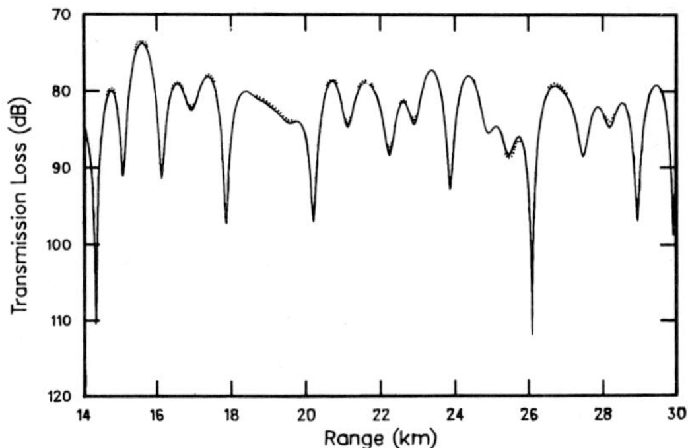

Fig. 4. Transmission loss (dB re 1 m) versus range (km) for model
environment 1 at 25 Hz. (····) FFP algorithm; (———) high-frequency algorithm.

For 25 Hz propagation in the exaggerated double-well model of
Fig. 1, a wide-angle capability well beyond the ordinary parabolic
approximation is required. Figure 4 illustrates the excellent
agreement between the high-frequency and FFP algorithms over ranges
on the order of 500 wavelengths. Figure 5 illustrates the cumulative growth of a phase shift error at long range which characterizes the breakdown of the high-frequency algorithm in the 250 Hz
propagation in the rapidly changing shallow-water model of Fig. 2.
Combining Fourier component, or wave number, filtering with the
high-frequency algorithm leads not only to a more efficient and,
thus, faster algorithm, but also to a more widely applicable numerical scheme. The filtering, in addition to removing Fourier

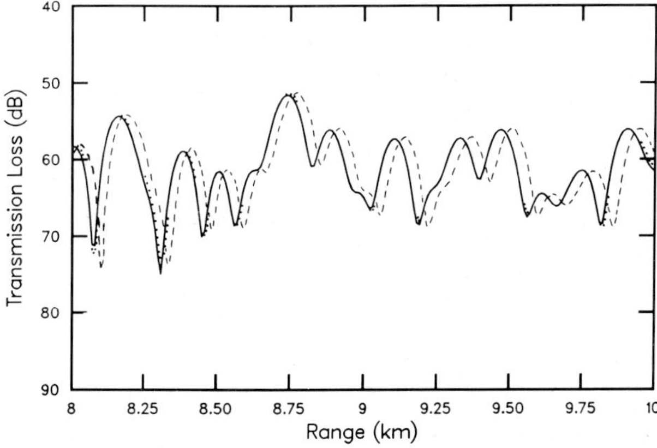

Fig. 5. Transmission loss (dB re 1 m) versus range (km) for model
environment 2 at 250 Hz. (····) FFP algorithm; (----) high-frequency algorithm; (———) high-frequency algorithm (60 deg filter).

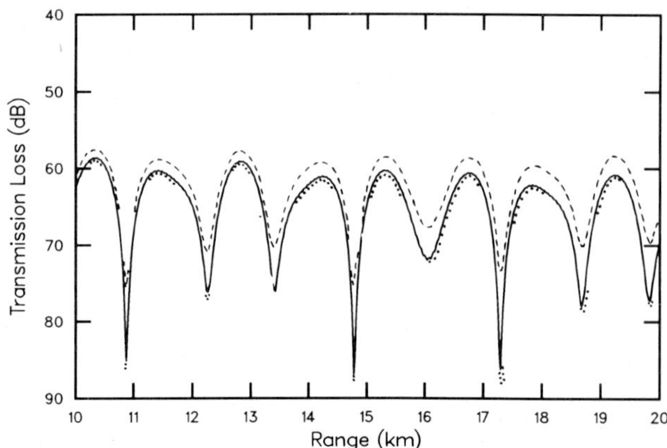

Fig. 6. Transmission loss (dB re 1 m) versus range (km) for model
environment 3 at 25 Hz. (····) FFP algorithm; (----) high-frequency algorithm; (———) real Weyl high-frequency algorithm.

components which, in principle, make no significant contribution
to the computed wave field, eliminates those unnecessary regions
of phase space where the small error in the high-frequency symbol
approximation can lead, in a cumulative manner, to serious discrepancies at sufficiently long ranges. This is particularly well
illustrated in the 60 degree filtered calculation on model environment 2 at 250 Hz which results in the complete elimination of
the cumulative phase shift error (Fig. 5), greatly extending the
effective computational range. Sufficiently decreasing the propagation frequency and increasing the jump discontinuity in the
sound speed, as illustrated in the 25 Hz propagation in the shallow-water model of Fig. 3, demonstrate the violation of energy conservation inherent in the high-frequency wave theory and the now-
rapid decay with increasing range of the corresponding numerical
algorithm. This growth in the wave field, illustrated in Fig. 6,
is eliminated by the real Weyl high-frequency algorithm of Eq. (16),
which effectively restores energy conservation, as is also illustrated in Fig. 6. A more detailed discussion of these and other
points is presented elsewhere [14].

ACKNOWLEDGMENTS

    This work was supported under grants from the Office of Naval
Research (N00014-85-K-0307) and the U.S. Army Research Office
(DAAG 29-85-K-0002).

REFERENCES AND NOTES

1. Taylor, M. E., PSEUDODIFFERENTIAL OPERATORS, Princeton University Press, Princeton, 1981.

2. Hörmander, L., The Weyl calculus of pseudo-differential operators, COMMUNICATIONS ON PURE AND APPLIED MATHEMATICS, 32 (1979), 359.

3. Engquist, B., and A. Majda, Radiation boundary conditions for acoustic and elastic wave calculations, COMMUNICATIONS ON PURE AND APPLIED MATHEMATICS, 32 (1979), 313.

4. Treves, F., INTRODUCTION TO PSEUDODIFFERENTIAL AND FOURIER INTEGRAL OPERATORS. VOLUME 2. FOURIER INTEGRAL OPERATORS, Plenum Press, New York, 1980.

5. Duistermaat, J., FOURIER INTEGRAL OPERATORS, Courant Institute Lecture Notes, New York, 1974.

6. Schulman, L. S., TECHNIQUES AND APPLICATIONS OF PATH INTEGRATION, Wiley, New York, 1981.

7. Langouche, F., D. Roekaerts, and E. Tirapegui, FUNCTIONAL INTEGRATION AND SEMICLASSICAL EXPANSIONS, D. Reidel Publishing Company, Boston, 1982.

8. Besieris, I. M., Wave-kinetic method, phase-space path integrals, and stochastic wave propagation, JOURNAL OF THE OPTICAL SOCIETY OF AMERICA, A2(12) (1985), 2092.

9. Fishman, L., and J. J. McCoy, Derivation and application of extended parabolic wave theories. Part I. The factorized Helmholtz equation, JOURNAL OF MATHEMATICAL PHYSICS, 25(2) (1984), 285.

10. Fishman, L., and J. J. McCoy, Derivation and application of extended parabolic wave theories. Part II. Path integral representations, JOURNAL OF MATHEMATICAL PHYSICS, 25(2) (1984), 297.

11. Fishman, L., and J. J. McCoy, Factorization, path integral representations and the construction of direct and inverse wave propagation theories, IEEE TRANSACTIONS ON GEOSCIENCE AND REMOTE SENSING, GE-22(6) (1984), 682.

12. Fishman, L., and J. J. McCoy, Wave propagation in inhomogeneous media: The factorized Helmholtz equation, in WAVE PROPAGATION IN INHOMOGENEOUS MEDIA AND ULTRASONIC NONDESTRUCTIVE

EVALUATION, G. C. Johnson, ed. (The American Society of Mechanical Engineers, New York, 1984), p. 15.

13. Fishman, L., and J. J. McCoy, A new class of propagation models based on a factorization of the Helmholtz equation, GEOPHYSICAL JOURNAL OF THE ROYAL ASTRONOMICAL SOCIETY, 80 (1985), 439.

14. Fishman, L., J. J. McCoy, and S. C. Wales, Factorization and path integration of the Helmholtz equation: Numerical algorithms, JOURNAL OF THE ACOUSTICAL SOCIETY OF AMERICA, submitted for publication 1986.

15. Fishman, L., and A. Whitman, Exact and uniform perturbation solutions of the Helmholtz composition equation, JOURNAL OF MATHEMATICAL PHYSICS, submitted for publication 1986.

16. Lee, D., The state-of-the-art parabolic equation approximation as applied to underwater acoustic propagation with discussions on intensive computations, Naval Underwater Systems Center Technical Report TD 7247, 1984.

17. Tappert, F. D., The parabolic approximation method, in WAVE PROPAGATION IN UNDERWATER ACOUSTICS, J. B. Keller and J. S. Papadakis, eds., Springer-Verlag, New York, 1977, p. 224.

18. Davis, J. A., D. White, and R. C. Cavanagh, NORDA parabolic equation workshop, Naval Ocean Research and Development Activity Technical Note 143, 1982.

19. Beylkin, G., Imaging of discontinuities in the inverse scattering problem by inversion of a causal generalized Radon transform, JOURNAL OF MATHEMATICAL PHYSICS, 26(1) (1985), 99.

20. Morse, P. M., and K. U. Ingard, THEORETICAL ACOUSTICS, McGraw-Hill Book Company, New York, 1968.

21. DeGroot, S. R., and L. G. Suttorp, FOUNDATIONS OF ELECTRODYNAMICS, North-Holland, Amsterdam, 1972.

22. Polyanskii, E. A., Relationship between the solutions of the Helmholtz and Schrödinger equations, SOVIET PHYSICS ACOUSTICS, 20(1) (1974), 90.

23. Richtmyer, R. D., and K. W. Morton, DIFFERENCE METHODS FOR INITIAL-VALUE PROBLEMS, Wiley Interscience, New York, 1967.

24. This difficulty can be appreciated by observing that for $\Delta x = 0$, $\underset{\sim}{C}$ reduces to the identity matrix with the eigenvalue

1 on the unit circle. For finite Δx, the eigenvalues spread out from this degenerate value and their detailed motion is crucial in light of the stability requirement. An analytic characterization would be easier if, for example, the eigenvalues had started out about zero in the complex plane rather than right on the unit circle.

25. Conte, S. D., and C. deBoor, ELEMENTARY NUMERICAL ANALYSIS: AN ALGORITHMIC APPROACH, McGraw-Hill Book Company, New York, 1972.

26. Bellman, R. E., INTRODUCTION TO MATRIX ANALYSIS, McGraw-Hill Book Company, New York, 1960.

27. Chan, T. F., and L. Shen, Difference schemes for equations of Schrödinger type, Research Report, Yale University, Dept. of Computer Science, RR-320, 1984.

28. Shewell, J. R., and E. Wolf, Inverse diffraction and a new reciprocity theorem, JOURNAL OF THE OPTICAL SOCIETY OF AMERICA, 58(12) (1968), 1596.

29. DiNapoli, F. R., and R. L. Deavenport, Theoretical and numerical Green's function solution in a plane multilayered medium, JOURNAL OF THE ACOUSTICAL SOCIETY OF AMERICA, 67(1) (1980), 92.

30. Kutschale, H. W., Rapid computation by wave theory of propagation loss in the Arctic Ocean, Lamont-Doherty Geological Observatory of Columbia University, Technical Report No. 8 (CU-8-73), 1973.

COMPUTATIONAL ACOUSTICS: Wave Propagation
D. Lee, R.L. Sternberg, M.H. Schultz (Editors)
Elsevier Science Publishers B.V. (North-Holland)
© IMACS, 1988

MATHEMATICAL MODELING OF DIFFERENTIAL
EQUATIONS BY DIFFERENCE EQUATIONS*

Ronald E. Mickens

Department of Physics
Atlanta University
Atlanta, Georgia

ABSTRACT

We discuss general problems related to the modeling of differential equations by difference equations. We prove a theorem which states that corresponding to every ordinary differential equation there is an "exact" finite-difference scheme, i.e., the local truncation error is zero. Next, we show how to construct a linear difference equation that has a given set of solutions. We then relate these results to the modeling of differential equations by difference equations. Several general modeling rules are presented. A number of explicit examples to illustrate the application of these rules will be given.

I. INTRODUCTION

A major problem in acoustics [1,2], in particular, and dynamic systems [3,4], in general, is the numerical integration of the linear and nonlinear differential equations which arise [5,6,7]. Finite-difference schemes have proven to be of great utility [8]. However, an issue of great importance is the relationship between the obtained numerical solutions and the actual solutions to the problem of interest [6,7,8]. Questions arise concerning the local truncation errors, numerical instabilities, etc. [8,9]. The purpose of this paper is to discuss general problems related to the modeling of differential equations by finite-difference schemes.

In the next section, we show that corresponding to every system of ordinary differential equations that has a set of solutions, there exists an exact finite-difference scheme, i.e., the local truncation errors are zero. We illustrate this principle by several examples.

We demonstrate, in the third section, that certain limited results can be obtained for partial differential equations.

---

*Research supported in part by grants from DOE and NASA.

However, we do not, in general, expect exact finite-difference schemes to exist in this case.

Next, we present a procedure for constructing a linear difference equation having a given set of solutions.

Finally, in the last section we discuss several general rules for modeling differential equations by difference equations.

II.   ORDINARY DIFFERENTIAL EQUATIONS

Assume that the first-order ordinary differential equation (ODE)

$$\frac{dx}{dt} = f(t,x) \qquad x(t_0) = x_0 \tag{1}$$

has the solution

$$x(t) = \phi(t, t_0, x_0) \tag{2}$$

We now show there exists, corresponding to Eq. (1), an exact difference scheme that has the <u>same general solution</u>. If this difference scheme is

$$x_{k+1} = g(k, x_k) \qquad k = \text{integer} \tag{3}$$

then Eqs. (2) and (3) have the same general solution if, for arbitrary constant step-size h, the values x_k satisfying Eq. (3) are related to the solution x(t) of Eq. (1) by the relation

$$x_k = x(hk) \tag{4}$$

Geometrically, we can interpret Eq. (4) as stating that the x_k satisfying Eq. (3) will be points on the solution curve x(t) of Eq. (1) at successive values $t_k = hk$.

We now show that a finite-difference scheme having this property does exist. Such a scheme is an exact finite-difference scheme in the sense that the local truncation error is zero [7,8,9].

Let h be an arbitrary, positive constant. The "group property" of the solutions to Eq. (1) gives the result [10]

$$x(t + h) = \phi[t + h, t, x(t)] \tag{5}$$

Let $t = t_k = hk$ and define $x_k = x(hk)$; then, from Eq. (5), we conclude

$$x_{k+1} = \phi[h(k + 1), hk, x_k] \tag{6}$$

Thus, by construction, this is the explicit form of the finite-difference scheme that has the same general solution as Eq. (1).

This procedure can be easily generalized to cover the case of systems of first-order differential equations.

To illustrate the above, consider

$$\gamma \frac{dx}{dt} = \frac{d^2x}{dt^2} \tag{7}$$

Under a redefinition of variables, an equation of this form arises in the problem of steady-state one-dimensional convection and diffusion [11] and the steady-state transfer of heat in a one-dimensional moving medium without internal heat sources [12]. The exact solution to Eq. (7) is

$$x(t) = c_1 + c_2 e^{\gamma x} \tag{8}$$

where c_1 and c_2 are arbitrary constants. An easy calculation, using the above procedure, shows that the exact difference scheme corresponding to Eq. (7) is

$$\gamma \left[\frac{x_k - x_{k-1}}{h} \right] = \frac{x_{k+1} - 2x_k + x_{k-1}}{\left(\frac{e^h - 1}{\gamma} \right)} \qquad t = h \tag{9}$$

Since Eq. (9) is a second-order, linear difference equation with constant coefficients, it can be easily solved to give the solution

$$x_k = c_1 + c_2 e^{\gamma hk} \tag{10}$$

Comparison of Eqs. (8) and (10) shows that the relationship of Eq. (4) is satisfied.

As a second example, the equation

$$\frac{dx}{dt} = -x^2 \qquad x(t_0) = x_0 \tag{11}$$

has the solution

$$x(t) = \frac{x_0}{1 + x_0(t - t_0)} \tag{12}$$

The corresponding exact finite-difference is

$$\frac{x_{k+1} - x_k}{h} = -x_{k+1} x_k \tag{13}$$

From Eq. (6), it is seen that the above procedure leads to an explicit finite difference model for the corresponding differential equation. Unfortunately, we only have an "existence theorem;"

consequently, while we can conclude that such schemes always exist, in general, we have no a priori means to construct them.

III. PARTIAL DIFFERENTIAL EQUATIONS

We should not expect exact finite-difference schemes to exist, in general, for partial differential equations (PDE). The major reason being the lack of an unambiguous definition of the general solution to a PDE [9]. This is in part reflected in the fact that formally one can derive finite-difference schemes which have zero local truncation errors but are of infinite order [5]. If such schemes do exist, then one or more (depending on the dimensionality of the equation) relations should hold between the space and time step sizes.

Exact difference schemes have been obtained for the following two equations:

Unidirectional (linear) wave equation [9]:

$$u_t + u_x = 0 \tag{14a}$$

$$\frac{u_m^{n+1} - u_m^n}{\Delta t} + \frac{u_m^n - u_{m-1}^n}{\Delta x} = 0 \qquad \Delta t = \Delta x \tag{14b}$$

Unidirectional (nonlinear) wave equation [13]:

$$u_t + u_x = \lambda u^2 \tag{15a}$$

$$\frac{u_m^{n+1} - u_m^n}{\Delta t} + \frac{u_m^n - u_{m-1}^n}{\Delta x} = \lambda u_m^{n+1} u_{m-1}^n \qquad \Delta t = \Delta x \tag{15b}$$

In the above, we have $u_m^n = u[(\Delta x)m, (\Delta t)n]$. Note, in both cases, we have one relationship between the step-sizes, i.e., $\Delta t = \Delta x$.

IV. LINEAR DIFFERENCE EQUATIONS WITH GIVEN SOLUTIONS

In Section II, we demonstrated that to each ODE there corresponds an "exact" finite-difference scheme. We now show how to construct finite-difference schemes using an alternative procedure.

Assume that n linearly independent functions are given, i.e., $\{y^{(i)}(t)\}$ where $i = 1, 2, \ldots, n$. Define, for fixed, positive constant h

$$y_k^{(i)} = y^{(i)}(hk) \tag{16}$$

An nth order, linear difference equation having these functions as solutions is given by the following determinant equation

$$
\begin{vmatrix}
y_k & y_k^{(1)} & \cdots & y_k^{(n)} \\
y_{k+1} & y_{k+1}^{(1)} & \cdots & y_{k+1}^{(n)} \\
\vdots & & & \vdots \\
y_{k+n} & y_{k+n}^{(1)} & \cdots & y_{k+n}^{(n)}
\end{vmatrix} = 0
\tag{17}
$$

The proof of this follows directly from the properties of the determinant.

As an illustration, consider the two functions: $\{1, \exp(\gamma t)\}$. Therefore, $y_k^{(1)} = 1$, $y_k^{(2)} = \exp(\gamma hk)$, and

$$
\begin{vmatrix}
y_k & 1 & e^{\gamma hk} \\
y_{k+1} & 1 & e^{\gamma h(k+1)} \\
y_{k+2} & 1 & e^{\gamma h(k+2)}
\end{vmatrix} = 0
\tag{18}
$$

which on expansion gives Eq. (9) [see Eq. (10)].

V. DISCUSSION

With the above details out of the way, we are now in a position to discuss the mathematical modeling of differential equations by finite-difference schemes. Unfortunately, space restrictions will limit the discussion to a broad overview without the supporting details. The rules to follow are based on analytical and numerical studies of a rather large set of linear and nonlinear differential equations.

Rules of Modeling

(a) The order of the finite-difference scheme should be equal to the order of the differential equation.

If, in fact, the order of the difference scheme is larger than the order of the differential equation, then "ghost" solutions can appear [9,14]. In certain instances, the "ghost" solutions (numerical instabilities) can lead to chaotic behavior of the numerical solution [14,15].

(b) Our analysis clearly indicates that nonlinear terms should be modeled nonlocally on the lattice.

Examples of this rule are shown in the comparison of Eqs. (11) and (13), and Eqs. (15a) and (15b). We now give two additional examples where this rule has been applied; they are

$$
u_t + uu_x = 0
\tag{19a}
$$

$$\frac{u_m^{n+1} - u_m^n}{\Delta t} + \frac{u_m^{n+1}(u_{m+1}^n - u_m^n)}{\Delta x} = 0 \tag{19b}$$

and

$$u_t = uu_{xx} \tag{20a}$$

$$\frac{u_m^{n+1} - u_m^n}{\Delta t} = u_m^{n+1}\left[\frac{u_{m+1}^n - 2u_m^n + u_{m-1}^n}{(\Delta x)^2}\right] \tag{20b}$$

Note that these two finite-difference schemes are not exact, i.e.,
they have respective truncation errors of $O(\Delta t) + O(\Delta x)$ and
$O(\Delta t) + O((\Delta x)^2)$. However, one can show directly that these pairs
of equations [(19a) and (19b), (20a) and (20b)] have solutions
that can be determined by the method of separation of variables
and that the corresponding solutions are _exactly_ equal to each
other for arbitrary positive values of Δx and Δt. Numerical experiments show that the differential and corresponding difference
equations have solutions with the same quantative behaviors.

   (c)   More general expressions involving the time-space steps
may be needed in constructing finite-difference models of differential equations. For example, see Eqs. (7) and (9), and the
"exact" scheme for the linear, oscillator equation

$$\frac{d^2x}{dt^2} + \omega^2 x = 0 \qquad \frac{x_{k+1} - 2x_k + x_{k-1}}{4\sin^2(\frac{\omega h}{2})} + x_k = 0 \tag{21}$$

   (d)   Suppose a nonlinear differential equation has a linear
part that can be modeled "exactly" by a finite-difference scheme.
Then the complete equation can be modeled by adding to the model
of the linear part a nonlocal finite-difference model of the nonlinear terms.

   In summary, we feel that application of the above procedures
and rules will allow the construction of finite-difference schemes
such that issues concerning consistency, stability, and convergence will either not arise or at least their (bad) effects will
be minimized.

REFERENCES

1.   R. T. Beyer, NONLINEAR ACOUSTICS, Naval Ship Systems Command,
     United States Department of the Navy, Washington, D.C., 1974.

2.   K. W. Morton and M. J. Baines, editors, NUMERICAL METHODS FOR
     FLUID DYNAMICS, Academic, London, 1982.

3. D. Potter, COMPUTATIONAL PHYSICS, Wiley, New York, 1973.

4. G. B. Whitham, LINEAR AND NONLINEAR WAVES, Wiley-Interscience, New York, 1974.

5. A. R. Mitchell, COMPUTATIONAL METHODS IN PARTIAL DIFFERENTIAL EQUATIONS, Wiley, New York, 1969.

6. J. M. Hyman, Numerical Methods for Nonlinear Differential Equations, Los Alamos National Laboratory, Report LA-8927-MS, January 1982.

7. M. K. Jain, NUMERICAL SOLUTION OF DIFFERENTIAL EQUATIONS, Halsted Press, New York, 1984.

8. G. D. Smith, NUMERICAL SOLUTION OF PARTIAL DIFFERENTIAL EQUATIONS: FINITE DIFFERENCE METHODS, Oxford University Press, Oxford, 1978.

9. F. B. Hildebrand, FINITE-DIFFERENCE EQUATIONS AND SIMULATIONS, Prentice-Hall, Englewood Cliffs, N.J., 1968.

10. V. V. Nemytski and V. V. Stepanov, QUALITATIVE THEORY OF DIFFERENTIAL EQUATIONS, Princeton University Press, Princeton, N.J., 1969.

11. G. D. Stubley, G. D. Raithby, and A. B. Strong, Proposal for a new discrete method based on an assessment of discretization errors, NUMERICAL HEAT TRANSFOR, 3 (1980), 411-428.

12. D. B. Spalding, A novel finite difference formulation for differential expressions involving both first and second derivatives, INTERNATIONAL JOURNAL OF MATHEMATICS AND ENGINEERING, 4 (1972), 551-559.

13. R. E. Mickens, Exact finite difference schemes for the nonlinear unidirectional wave equation, JOURNAL OF SOUND AND VIBRATION, 100 (1985), 452-455.

14. M. Yamaguti and S. Ushiki, Chaos in numerical analysis of ordinary differential equations, PHYSICA, 3D (1981), 618-626.

15. A. R. Mitchell and J. C. Bruch, Jr., A numerical study of chaos in a reaction-diffusion equation, NUMERICAL METHODS FOR PARTIAL DIFFERENTIAL EQUATIONS, 1 (1985), 13-23.

COMPUTATIONAL ACOUSTICS: Wave Propagation
D. Lee, R.L. Sternberg, M.H. Schultz (Editors)
Elsevier Science Publishers B.V. (North-Holland)
© IMACS, 1988

BENCHMARKS—PURPOSE AND PERSPECTIVES

L. B. Felsen

Department of Electrical Engineering
and Computer Science
Weber Research Institute
Polytechnic University, Farmingdale, New York

ABSTRACT

This brief summary describes the motivation for, and preliminary consequences of, the Special Session on Benchmarks.

In underwater acoustic propagation, successively more emphasis is being placed on modeling the ocean-bottom environment, with lateral range dependence, over long distances. This puts a severe strain on present analytical and numerical capabilities. Analytical techniques are strained because the problem is generally non-separable, thereby eliminating direct applicability of the traditional and explicit workhorse method of separation of variables. Numerical techniques are strained because the problem is computationally large. The burden on the analytical side is not merely to implement numerically an analytically exact explicit solution but to render tractable a generally implicit formal result. Thus, the accuracy of the reduced analytic form itself is in question, apart from its numerical implementation. On the purely numerical side, truncation and roundoff errors for such large problems are usually not well in hand. Even if analytical error bounds could be found, these are likely to be conservative, and an approximate analytic solution or a numerical algorithm may actually work well beyond these stated criteria, if they are available. Evidently, because of the complexity and size of the problem, there is a tendency to push each method as far as possible.

The above scenario highlights the need for reference data with <u>predictable error</u>, usually referred to as <u>benchmarks</u>, with which other solutions (generated analytically and implemented numerically, or purely numerical) can be compared. The selection of benchmark problems poses a new dilemma: if the problem is sufficiently broad, it is unlikely, as stated above, that tight error criteria can be stated; if it is restricted so as to render error prediction feasible, the problem scope may be too limited to test a solution procedure in the large. Yet the need for reliability tests increases with the number of new analytical and numerical schemes, or extensions of older schemes into previously untrod territory, which are making their appearance. If they are to be successful, analytic forward propagation models, and models for data inversion, need to be parametrized around physical observables, which must be embedded in an analytical framework by approximate procedures whose validity can really be tested only by numerical comparison. If the comparison data have no error bounds, discrepancies with respect to the model data cannot be ascribed with certainty to deficiencies of the model, thereby inhibiting systematic model improvement. Particularly disturbing with respect to numerical solutions is the appearance of spurious periodicities and artifacts, which are not always excised from the presented data and may leave the analytic modeler with the feeling that an important wave phenomenon has been overlooked.

These observations are certainly not new but they continue to be relevant. By bringing them again to the attention of the underwater acoustics community, it is hoped that new efforts will be made to create a reliable data base, with which results generated by various techniques can be compared. How to assemble such a base, and what "standard" configurations to select as benchmarks, will require much deliberation. In his presentation at the symposium, the author focused on the overall need, made suggestions, and cited examples of how one might proceed. His remarks were intended to set the tone for the discussion period, which followed. There is no need to record these statements for posterity since their substance was tentative and exploratory but hopefully sufficiently motivating to pursue this matter further.

Indeed, there emerged a consensus that "something should be done." Something _is_ being done. At the 112th meeting of the Acoustical Society of America, Anaheim, California, December 8-12, 1986, a two-part special session entitled "Quality Assessment of Numerical Codes; Part I: Codes; Part 2: Benchmarks" has been organized by the writer. Contributions have been solicited on topics brought forth at the present IMACS Symposium, and a panel discussion at the end of the special session will explore "What Benchmarks Are Relevant?" A follow-up special session, hopefully concentrating on benchmark strategy implementation, is scheduled for the 113th meeting of the Acoustical Society of America, to be held in Indianapolis, Indiana, May 11-15, 1987. Thus, there may be progress to report at the next IMACS Symposium.

COMPUTATIONAL ACOUSTICS: Wave Propagation
D. Lee, R.L. Sternberg, M.H. Schultz (Editors)
Elsevier Science Publishers B.V. (North-Holland)
© IMACS, 1988

MARCHING METHODS FOR ELLIPTIC MODELS
OF UNDERWATER SOUND PROPAGATION

George H. Knightly and Donald F. St. Mary

Center for Applied Mathematics and
Mathematical Computation
Department of Mathematics and Statistics
University of Massachusetts
Amherst, Massachusetts

1. INTRODUCTION

In this work we begin an investigation of marching methods
for the far field elliptic problems of underwater sound propagation. Although initial-value problems for elliptic equations
have long been known to be unstable (e.g., see Hadamard [2]), the
work of Baumeister (see [1] and the references cited therein) and
that of Zavadskii and Kryukov [3] indicate that such methods may
be appropriate under certain restrictions on the parameters. Our
objective here is to demonstrate two marching schemes for computing approximate solutions, to show that these methods perform
well in some situations, and to initiate a program to determine
parameter domains within which the schemes yield good results.

We consider the problem of determining the time-harmonic far
field pressure, $p = p(r,z)$, due to a point source in an ocean having a horizontal, hard bottom, pressure-release surface, and wave
speed depending only on (r,z), where r is the horizontal distance
from the vertical through the source and z is the depth coordinate.
We assume that $p(r,z)$ is known near the source to a distance $r =
r_0$. Thus, we are concerned with the problem

$$\nabla^2 p + k^2(r,z)p = 0 \qquad 0 < z < H \qquad r_0 < r \qquad (1a)$$

$$p(r,0) = \frac{\partial p}{\partial z}(r,H) = 0 \qquad r_0 < r \qquad (1b)$$

$$p \text{ outgoing as } r \to \infty \qquad (1c)$$

$$p \text{ given for } r \leq r_0 \qquad 0 \leq z \leq H \qquad (1d)$$

If we make the substitution $p(r,z) = w(r,z)H_0^{(1)}(k_0 r)$ and use the
far field properties of the Hankel function $H_0^{(1)}$, then (1) leads
to a problem for w:

$$w_{rr} + 2ik_0 w_r + w_{zz} + k_0^2 \phi w = 0 \qquad r_0 < r, \ 0 < z < H \qquad (2a)$$

$$w(r,0) = \frac{\partial w}{\partial z}(r,H) = 0 \qquad r_0 < r \tag{2b}$$

$$wH_0^{(1)} \text{ outgoing as } r \to \infty \tag{2c}$$

$$w(r,z) \text{ given for } r \le r_0 \qquad 0 \le z \le H \tag{2d}$$

In (2), ϕ is given by

$$\phi(r,z) = [k(r,z)/k_0]^2 - 1 \tag{3}$$

and is a measure of the amount by which $k(r,z)$ differs from k_0.

When $k(r,z) \equiv k_0$ is constant one easily obtains as solutions of (2)

$$w_m = \frac{\varepsilon}{a_m} e^{a_m(r-r_0)} \sin \lambda_m z \tag{4}$$

where $\lambda_m = (m + \frac{1}{2})\pi/H$ and $a_m = -ik_0 + (\lambda_m^2 - k_0^2)^{1/2}$. Note that $|w_m(r_0,z)|$ and $|\partial w_m/\partial r(r_0,z)|$ are dominated by $\varepsilon > 0$ for $0 \le z \le H$ and large m, and that, e.g., $w_m(r,\pi/2\lambda_m)$ grows exponentially as m increases. Thus, we see that the "initial value" problem [obtained from (2) by setting $\phi = 0$ and dropping (2c)]

$$w_{rr} + 2ik_0 w_r + w_{zz} = 0 \qquad r_0 < r \qquad 0 < z < H \tag{5a}$$

$$w(r,0) = \frac{\partial w}{\partial z}(r,H) = 0 \qquad r_0 < r \tag{5b}$$

$$w(r_0,z), \; w_r(r_0,z) \text{ given} \tag{5c}$$

is not well posed since there are solutions with arbitrarily small initial values (i.e., at $r = r_0$) but having arbitrarily large values on any given strip $r_0 \le r \le r_1$, $0 \le z \le H$.

Despite this ill-posedness associated with (5), the work in [1,3] suggests, at least in some cases, that marching methods based on the differential equation (5a) can yield useful computational results. Nevertheless, it is evident that a stable scheme for marching the solution of (2) in range must provide a way to implement condition (2c), i.e., to avoid the effects of incoming modes [such as (4)]. Here we describe two marching procedures for problem (2) and we present related computations showing success for these methods in certain circumstances.

2. THE MARCHING SCHEMES

The marching methods for (2) considered here rely upon the following finite difference scheme for (2). We introduce mesh

widths h in r and s in z, s = H/M, where M is the number of vertical steps, and consider mesh points

$$(r_0 + nh,\ ms) \qquad m = 0,\ 1,\ \ldots,\ M;\ n = -1,\ 0,\ 1,\ \ldots \qquad (6)$$

If w = u + iv is a complex valued function of (r,z), we denote its
values at mesh points by

$$w(r_0 + nh,\ ms) = w_n^m = u_n^m + iv_n^m$$

The vectors $u_n = (u_n^m)_{m=1}^M$ and $v_n = (v_n^m)_{m=1}^M$ lie in $\mathbb{R}^M$ so that the
vector $w_n = (u_n, v_n)$ lies in $\mathbb{R}^{2M}$. We introduce the usual central
difference approximations

$$\left(\frac{\partial w}{\partial r}\right)_n^m = \frac{1}{2h}\ (w_{n+1}^m - w_{n-1}^m) \qquad (7a)$$

$$\left(\frac{\partial^2 w}{\partial r^2}\right)_n^m = \frac{1}{h^2}\ (w_{n+1}^m - 2w_n^m + w_{n-1}^m) \qquad (7b)$$

$$\left(\frac{\partial^2 w}{\partial z^2}\right)_n^m = \frac{1}{s^2}\ (w_n^{m+1} - 2w_n^m + w_n^{m-1}) \qquad (7c)$$

The finite difference approximation to (2a) at the mesh point (6)
for m = 1, 2, ..., M and n = 0, 1, 2, ... is

$$0 = \frac{1}{h^2}\ (w_{n+1}^m - 2w_n^m + w_{n-1}^m) + 2ik_0\ \frac{w_{n+1}^m - w_{n-1}^m}{2h}$$

$$+ \frac{1}{s^2}\ (w_n^{m+1} - 2w_n^m + w_n^{m-1}) + k_0^2 \phi_n^m w_n^m \qquad (8)$$

In order that (8) make sense at m = 1 and m = M we apply the
boundary conditions at z = 0, H in the form

$$w_n^0 = 0 \qquad w_n^{M+1} = w_n^{M-1} \qquad n = -1,\ 0,\ 1,\ \ldots \qquad (9)$$

We may separate the real and imaginary parts in (8) and solve
for u_{n+1} and v_{n+1} to get

$$v_{n+1} = V_n [w_n, w_{n-1}] \qquad (10a)$$

$$u_{n+1} = U_n [w_n, w_{n-1}, v_{n+1}] \qquad (10b)$$

Here the mth components of V_n and U_n are given by

$$V_n^m[w_n, w_{n-1}] = (1 + k_0^2 h^2)^{-1} [-k_0 h T_n^m u_n^m + h^3 s^{-2} k_0 (u_n^{m+1} + u_n^{m-1})$$

$$+ T_n^m v_n^m - h^2 s^{-2} (v_n^{m+1} + v_n^{m-1})$$

$$- (1 - h^2 k_0^2) v_{n-1}^m + 2k_0 h u_{n-1}^m] \tag{11a}$$

$$U_n^m[w_n, w_{n-1}, v_{n+1}] = T_n^m u_n^m - h^2 s^{-2} (u_n^{m+1} + u_n^{m-1})$$

$$+ k_0 h (v_{n+1}^m - v_{n-1}^m) - u_{n-1}^m \tag{11b}$$

and

$$T_n^m = 2(1 + h^2 s^{-2}) - h^2 k_0^2 \phi_n^m \tag{12}$$

Equations (10) determine w_{n+1} in terms of w_n and w_{n-1}; v_{n+1} is
first obtained from (10a), then used with w_n and w_{n-1} in (10b) to
generate u_{n+1}.

The marching scheme just described and based directly on (10)
will be called Method I. It is the counterpart for the hard bottom problem of the scheme used in [3] for the soft bottom case.
A second scheme, Method II, is obtained by making the substitution (10a) for v_{n+1} on the right side of (10b) and introducing
vectors

$$W_n = (w_n, w_{n-1}) \in \mathbb{R}^{4M} \tag{13}$$

This leads to the process

$$W_{n+1} = M_n W_n \qquad W_0 \text{ given} \tag{14}$$

where the 4M × 4M marching matrix M_n has the form

$$M_n = \begin{bmatrix} A_n & B \\ C & O_{2M} \end{bmatrix}$$

with 2M × 2M blocks

$$A_n = (1 + \varepsilon^2)^{-1} \begin{bmatrix} T_n & \varepsilon T_n \\ -\varepsilon T_n & T_n \end{bmatrix}$$

$$B = (1 + \varepsilon^2)^{-1} \begin{bmatrix} (\varepsilon^2 - 1)I & -2\varepsilon I \\ 2\varepsilon I & (\varepsilon^2 - 1)I \end{bmatrix}$$

$$C = \begin{bmatrix} I & O_M \\ O_M & I \end{bmatrix}$$

Here $\varepsilon = k_0 h$, I is the M × M identity matrix, O_j is the j × j zero matrix, and T_n is an M × M tridiagonal matrix. T_n has diagonal elements

$$T_n^m = 2 + 2(h/s)^2 - k_0^2 \phi_n^m h^2 \qquad m = 1, 2, \ldots, M$$

and all nonzero off-diagonal elements equal to $-h^2 s^{-2}$ except that the last element in the subdiagonal is $-2h^2 s^{-2}$.

Some computations based on Methods I and II are presented in Section 4. In Section 3 we give a brief analysis of the methods.

3. ANALYSIS OF METHODS I AND II

Just as the continuous problem (5) has states such as those in (4) that grow with r so also does the finite difference problem (8), (9) have solutions that grow with n. In fact, if the von Neumann method is investigated, in which one "freezes" the coefficient ϕ at some constant value and seeks solutions of (8) of the form

$$w_n^m = [\zeta(j)]^n \sin \lambda_j ms \qquad (15)$$

with $\lambda_j = (j + \frac{1}{2})\pi/H$, then w_n^m satisfies the boundary conditions (9) and equation (8) reduces to

$$0 = \alpha \zeta^2 + \beta \zeta + \bar{\alpha} \qquad (16)$$

with

$$\alpha = 1 + ik_0 h \qquad (17)$$

$$\beta = -2 + h^2 k_0^2 \phi = 4 \frac{h^2}{s^2} \sin^2 \frac{\lambda_j s}{2} \qquad (18)$$

Since ζ^{-1} satisfies (16) whenever ζ satisfies (16), it follows that the difference equation (8) always has solutions of the form (15) with $|\zeta| \geq 1$. Thus, one sees that Methods I and II cannot generally be expected to be stable without some restrictions on the parameters and data. A stability analysis in [3] for the soft bottom problem for (8) could be repeated here, however, the arguments are insufficient for useful bounds on h and s or other parameters. Here we attempt to ensure stable calculations in Method I by selecting accurate starting data corresponding to modes known to propagate outward and by experimenting with the step sizes s and h.

For Method II one determines the eigenspace of the marching matrix M_n and works in the subspace spanned by eigenvectors corresponding to eigenvalues associated with outgoing modes. One can show that the matrix T_n has M eigenvalues x_j and eigenvectors X_j. Each pair (x_j, X_j) generates four eigenpairs (λ_{jq}, Y_{jq}), $q = 1, \ldots, 4$, of M_n; two of these, say (λ_{j1}, Y_{j1}), (λ_{j2}, Y_{j2}), are associated with outgoing modes. (Here and in the sequel, the dependence of these eigenpairs on n is suppressed.) Eigenvalues λ_{jp} of M_n corresponding to outgoing propagating modes satisfy $|\lambda_{jp}| = 1$; those for decaying modes satisfy $|\lambda_{jp}| < 1$. For each n, we then modify the process (14), if necessary, to require that W_{n+1} have the form

$$W_{n+1} = \sum_{j=1}^{M} (a_j Y_{j1} + b_j Y_{j2}) \qquad (19)$$

with scalars a_j, b_j, $j = 1, \ldots, M$; that is, at each step we require that the iterates lie in the "outgoing" subspace. Given the nth approximate solution W_n, we project W_n into the outgoing eigensubspace of M_n as

$$\tilde{W}_n = \sum_{j=1}^{M} \tilde{a}_j Y_{j1} + \tilde{b}_j Y_{j2}$$

Then we compute W_{n+1} by

$$W_{n+1} = M_n \tilde{W}_n = \sum_{j=1}^{M} (\tilde{a}_j \lambda_{j1} Y_{j1} + \tilde{b}_j \lambda_{j2} Y_{j2})$$

In particular, when the wave speed is independent of range, the M_n are independent of n. In this case the (λ_{jp}, Y_{jp}) are independent of n, so that if

$$W_0 = \sum_{j=1}^{M} a_j Y_{j1} + b_j Y_{j2}$$

then

$$W_{n+1} = \sum_{j=1}^{M} a_j \lambda_{j1}^n Y_{j1} + b_j \lambda_{j2}^n Y_{j2}$$

and the iteration is obviously stable, since $|\lambda_{jp}| \leq 1$, $p = 1, 2$, $j = 1, 2, \ldots, M$.

4. CALCULATIONS

Methods I and II were tested on three sample problems, Test Problems A, B, and C. In Test Problems A and B the wave speed $k(r,z) \equiv k_0$ is constant so that exact solutions are available for comparison with the computed solution. In Test Problem C, the wave speed $k = k(z)$ is independent of range and piecewise linear in depth; exact solutions are not readily available in this case.

<u>Test Problem A</u>. In this problem the ocean depth is $H = 400$ m, the source depth is 300 m, and the receiver depth is 100 m. The wave number $k \equiv k_0 = \pi$ is assumed constant, corresponding to frequency $f = 750$ Hz at the sound speed $c_0 = 1500$ m/sec. Then problem (1) has the well-known outgoing modes

$$P_m(r,z) = H_0^{(1)}(\sqrt{k_0^2 - \lambda_m^2}\, r)\, \sin \lambda_m z$$

$$\lambda_m = (m - \tfrac{1}{2})\frac{\pi}{H} \qquad m = 1,\ 2,\ \ldots$$

The Green's function for a source at $(r,z) = (0,z_s)$ is

$$G(r,z;z_s) = \frac{i}{2H} \sum_{m=1}^{\infty} H_0^{(1)}(\sqrt{k^2 - \lambda_m^2}\, r)\, \sin \lambda_m z\, \sin \lambda_m z_s \qquad (20)$$

For the test problem, we use as exact solution a finite partial sum of the series in (20)

$$G_J(r,z;z_s) = \frac{i}{2H} \sum_{m=1}^{J} H_0^{(1)}(\sqrt{k^2 - \lambda_m^2}\, r)\, \sin \lambda_m z\, \sin \lambda_m z_s \qquad (21)$$

We also use (21) to generate the starting data for the computed solution; that is, to compute the approximate solution by Methods I and II starting at $r = r_0$, we choose $h = \Delta r$, $s = \Delta z$, and determine u_{-1} and u_0 by

$$u_{-1}^m = G_J(r_0 - h,\ ms, z_s)$$
$$u_0^m = G_J(r_0, ms, zs) \qquad\qquad m = 0,\ 1,\ \ldots,\ M \qquad (22)$$

<u>Test Problem B</u>. The parameters are now $H = 3141$ m, the source depth is 1570 m, the receiver depth is 1130 m, $c_0 = 1500$ m/sec, $f = 238.7$ Hz, and $k_0 = 1$. Again $k \equiv k_0$ is constant, so that the exact solution (21) and starting data (22) are still appropriate.

<u>Test Problem C</u>. Here the parameters H, c_0, f, and k_0 are those of Test Problem B but the sound speed is no longer assumed

constant. Instead, a depth dependent profile $c = c_1(z)$ used in
[3] is assumed; that is, we suppose that $c_1(z)$ is given at the
23 depths specified in Table 1 and obtained by linear interpolation at other values of z.

TABLE 1. Wave Speed Profile

z (meters)	$c_1(z)$ (meters/sec)
0	1507.57
150	1506.27
300	1505.08
450	1503.91
600	1502.94
750	1502.12
900	1501.43
1050	1500.87
1200	1500.45
1350	1500.17
1500	1500.06
1570	1500.00
1650	1500.06
1800	1500.15
1950	1500.42
2100	1500.84
2250	1501.38
2400	1502.06
2550	1502.87
2700	1503.83
2850	1504.94
3000	1506.16
3141	1507.57

The results of some sample calculations are shown in Figures
1, 2, and 3. The calculations were performed on a CYBER 175 computer at the University of Massachusetts. In each case, as a
measure of propagation loss, TL is plotted as a function of range,
where

$$TL = -20 \log_{10}|p|$$

and $p = w(r, z_R) H_0^{(1)}(k_0 r)$ is the computed pressure at receiver
depth z_R. If the nominal wave number is k_0 then the (half) angle

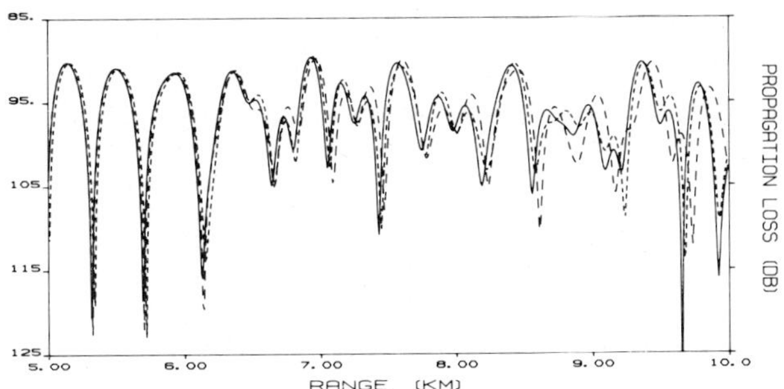

Fig. 1. Test problem A with J = 60 modes. Exact (solid curve),
Method I (large dashes), and Method II (small dashes) solutions
are shown.

θ_m, for a calculation associated with the first m propagating
modes (m $\leq$ ($k_0 H/\pi$) + 1/2) is given by

$$\sin \theta_m = (m - \tfrac{1}{2})\pi/k_0 H$$

Some experimentation with the calculations indicated stability of Method I for step sizes related roughly by h = const s^2;
the precise relationship requires further investigation. Once a
suitable pair (s,h) of step sizes was located, Method I remained
stable to all values of range attempted (5-10 km in Problem A;
1-300 km in Problems B and C). Method II is unconditionally stable, as observed in the previous section.

In Figure 1 the solid curve represents the exact solution,
the curve with large dashes represents the approximate solution
given by Method I, and the curve with small dashes represents the
approximate solution given by Method II for test problem A with
J = 60 modes. In this case 400 modes propagate so that the angle
associated with this solution is θ_m = 8.6°. Both Method I and
Method II use step sizes s = 1 m, h = 1 m. Here r_0 = 5 km, and
the computations extend to r = 10 km. Methods I and II both show
excellent approximation of the exact propagation loss curve near
r = r_0 with Method I perhaps slightly better than Method II. As
the range increases Method II remains in phase with the exact
curve; Method I has gone out of phase by r = 10 km.

In Figure 2 the computations obtained by Method I (dashed
curve) are compared with the exact solution (solid curve) of problem B with J = 100 modes and r_0 = 1 km. In this case 1000 modes
propagate, so that the angle associated with the solution is θ_m =
5.7°. The stability of Method I (s = h = 1 km) indicated in the

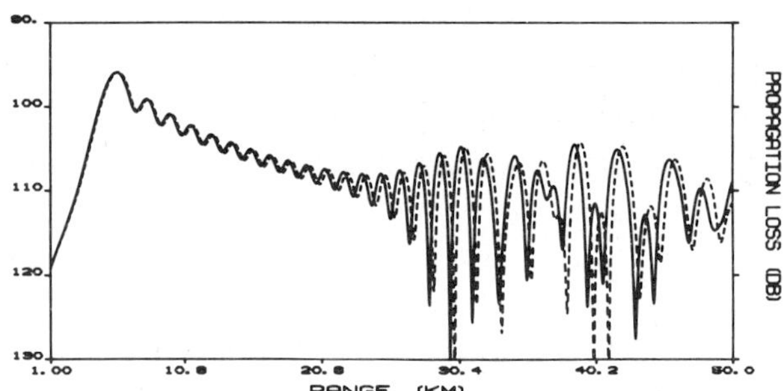

Fig. 2.   Test problem B with J = 100 modes.   Exact (solid curve)
and Method I (dashed curve) solutions are shown.

figure actually persisted to the largest range calculated (301 km).

For the purpose of the remainder of the discussion, we denote
by PE the standard parabolic equation

$$w_r = \frac{ik_0}{2} \, (n^2(r,z) - 1)w + \frac{i}{2k_0} \, w_{zz} \tag{PE}$$

utilizing the Crank-Nicolson implicit finite difference discretization.

Figure 3 shows the results of Method I (solid curve) with
(s,h) = (3.9 m, 5 m) and PE (dashed curve) with (s,h) = (7.8 m,
20 m) for problem C with J = 100 modes and variable sound speed
profile determined by Table 1.   In this case the trapping angle
$\cos^{-1}$[min c(z)/max c(z)] is 5.5°, so that essentially all components of the starting field are trapped.   The pattern obtained is
similar to those found in [3] for the case of a soft bottom.
Small oscillations of high frequency have been smoothed somewhat
in Figure 3 by averaging the data over one kilometer in range.

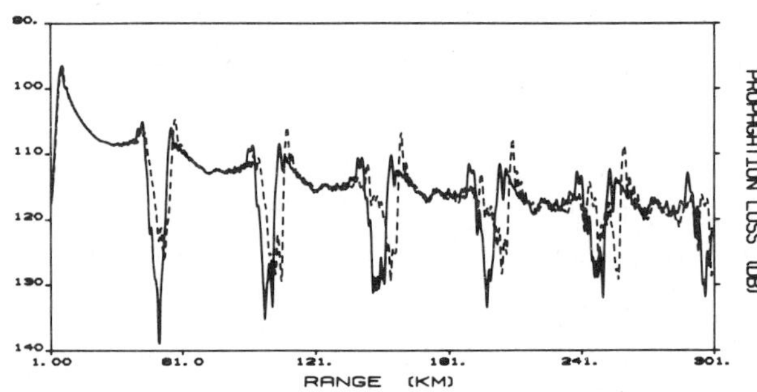

Fig. 3.   Test problem C with J = 100 modes.   Method I (solid
curve) and PE (dashed curve) solutions are shown.

When the step sizes in Method I were increased to (s,h) = (7.8 m,
20 m) the solid curve experienced slightly the usual shift "to
the right," but not all the way to the dashed curve. The latter
observation is consistent with the indication in [3] that the
elliptic marching technique may be more accurate than PE for this
problem.

The accuracy of the calculations by Method I in Figures 1-3
is about the same as that obtained using standard parabolic equation (PE) methods. At stable step sizes, however, Method I appears to be much faster than PE. Furthermore, for problems in
which the second derivative w_{rr} is important, Method I has the
advantage of including these derivatives in its formulation.
Method II is more accurate and is applicable for wide-angle problems, although it is slower than Method I.

ACKNOWLEDGMENTS

This research is supported in part by ONR Contract N-00014-
85-K-0743 and in part by NSF Grant No. RII-8115410-01.

REFERENCES

1. Baumeister, K. J., Numerical spatial marching techniques in
    duct acoustics, JOURNAL OF THE ACOUSTICAL SOCIETY OF AMERICA
    66 (1979), 297-306.

2. Hadamard, J., LECTURES ON CAUCHY'S PROBLEM IN LINEAR PARTIAL
    DIFFERENTIAL EQUATIONS, Dover, New York, 1952.

3. Zavadskii, V. Y., and Y. S. Kryukov, Finite-difference calcu-
    lation of sound fields in an irregular ocean sound channel,
    SOVIET PHYSICS ACOUSTICS 29 (1983), 449-453.

COMPUTATIONAL ACOUSTICS: Wave Propagation
D. Lee, R.L. Sternberg, M.H. Schultz (Editors)
Elsevier Science Publishers B.V. (North-Holland)
© IMACS, 1988

ACCURATE COMPUTATION OF THE WIDE-ANGLE WAVE EQUATION

D. F. St. Mary

Center for Applied Mathematics and
Mathematical Computation
Department of Mathematics
University of Massachusetts
Amherst, Massachusetts

Ding Lee

Naval Underwater Systems Center
New London, Connecticut

ABSTRACT

The pseudo-differential operator or dispersion relation approach to the derivation of parabolic-like approximating partial differential equations to the reduced wave equation is presented. This approach leads in a natural way to the derivation of large classes of parabolic-like approximating equations. Several criteria are presented for purposes of making comparisons among the equations in one of these classes, and examples are displayed that demonstrate the results of comparisons made.

1. INTRODUCTION

The approximation of the wave equation by a one-way wave equation is an important and useful tool. This technique is the primary one used in modern reflection seismology, it is used in the focusing of lasers, in the study of absorbing boundary conditions, in the calculation of transformation and propagation of Stokes waves, and is the main tool in the study of long-range underwater propagation.

Traditionally, the principal parabolic-like equations approximating the reduced wave equation which have been used in underwater wave propagation (and in fact the other areas as well) have been the "small angle" equation

$$u_r = \frac{ik_0}{2} (n^2(r,z) - 1)u + \frac{i}{2k_0} u_{zz} \tag{PE}$$

and the "wide angle" equation

$$\left(1 + \frac{1}{4} L\right)u_r = \frac{ik_0}{2} Lu \qquad L = (n^2(r,z) - 1) + \frac{1}{k_0^2}\frac{\partial^2}{\partial z^2} \qquad \text{(WA)}$$

Recently, a large number of additional approximating equations have been developed. They have arisen largely because of the development of a new approach to the derivation of approximating equations, the so-called pseudo-differential operator or dispersion relation approach [5,6]. A number of these equations are of the same order as (WA) and it is our intention to discuss and begin to compare these equations in this paper. Implicit finite difference techniques and other techniques [4,7,8] for the numerical solution of partial differential equations of the general form of (WA) have existed for a number of years and are well understood, and thus it is possible to make comparisons of equations of that form from a numerical standpoint as well as via other methods. We remark that the new approximating methods have also yielded equations of higher order than (WA) and efforts have begun to develop numerical solutions to these equations [13,17,18].

2. DERIVATIONS OF OUTGOING WAVE EQUATIONS

The propagation of acoustic energy underwater corresponding to a time harmonic point source is modeled by the reduced wave equation in cylindrical coordinates

$$\frac{\partial^2 p}{\partial r^2} + \frac{1}{r}\frac{\partial p}{\partial r} + \frac{\partial^2 p}{\partial z^2} + \left[\frac{\omega}{c(r,z)}\right]^2 p = -\frac{2}{r}\delta(z - z_s)\delta(r) \qquad (1)$$

Here, the azimuthal variation $\partial p/\partial\theta$ has been suppressed, p is the acoustic pressure, c is the sound speed, and ω is the angular frequency of the source. We introduce a constant reference sound speed c_0 and define the index of refraction n as $n(r,z) = c_0/c(r,z)$ and a reference wave number as $k_0 = \omega/c_0$. The source is located at $r = 0$, $z = z_s$, with a pressure release boundary condition, $p = 0$ at the surface $z = 0$, and a soft bottom condition $p = 0$ at $z = B$, or a rigid bottom condition $\partial p/\partial z = 0$ at $z = B$. We seek an outgoing wave solution in the range variable r, and so the Sommerfield radiation condition at infinity is applied,

$$\lim_{r\to\infty} \sqrt{r}\left[\frac{\partial p}{\partial r} - ik_0 p\right] = 0$$

The classical transformational approach to the derivation of one-way wave equations, due to Clairbout [1] (see also [19]), is

instructive and provides motivation for the modern approach. One
of the main assumptions is that, in the application of interest,
the index of refraction is such that waves generally propagate in
a preferred direction. In the case of underwater acoustics, energy propagating in the far field is generally concentrated in a
narrow band about the horizontal. Thus it is natural to consider
the case of constant index of refraction in which one can study
plane waves traveling in the positive r direction. This leads to
$H_0^{(1)}$, the Hankel function of the first kind of order zero. $H_0^{(1)}(k_0 r)$
satisfies

$$\left(\frac{d^2}{dr^2} + \frac{1}{r}\frac{d}{dr} + k_0^2\right) H_0^{(1)}(k_0 r) = -\frac{2i}{\pi r}\,\delta(r) \tag{2}$$

Then one asks, can a more general outgoing wave solution of (1) be
represented as a perturbation of this "horizontal" wave, i.e., one
seeks p of the form

$$p(r,z) = p^+(r,z) H_0^{(1)}(k_0 r) \tag{3}$$

where at some distance from the source it is assumed that p is
approximately planar and hence p^+ is "slowly variable." Now, as
Clairbout [1] points out, presumed inhomogeneities in a more general sound speed would cause reflected waves to arise and thus a
more appropriate solution to (1) would be

$$p(r,z) = p^+(r,z) H_0^{(1)}(k_0 r) + p^-(r,z)$$

where p^- is an incoming wave. Hence there is a need to seek an
equation which governs the outgoing wave only, i.e., an equation
for which p^- is not a solution. Thus, one proceeds to substitute
p from (3) into (1) and uses (2) to obtain

$$p_{rr}^+ + 2ik_0 p_r^+ + p_{zz}^+ + k_0^2(n^2(r,z) - 1)p^+ = 0$$

Clairbout [1] comments "...if the amplitude modulation is slow,
the $[p_{rr}^+]$ term will be much smaller than the $[2ik_0 p_r^+]$ term, so
that we might consider dropping the $[p_{rr}^+]$ term...." This then
yields the very well-known parabolic approximating equation

$$p_r^+ = \frac{ik_0}{2}(n^2(r,z) - 1)p^+ + \frac{i}{2k_0} p_{zz}^+ \tag{PE}$$

The pseudo-differential operator approach to the derivation
of parabolic-like approximating equations is best demonstrated in

the work of Engquist and Majda [5,6] where the problem they consider has x and y space variables and is time dependent with the designated direction of propagation being the negative x direction. The context of the derivation is to obtain absorbing boundary conditions at the "wall" x = 0. Thus equations which annihilate, at x = 0, waves traveling in the negative x direction are sought. The comparable situation in our context is to seek equations which annihilate waves traveling in the positive r direction at r = ∞. In the ensuing we interpret this as the equation obtained in the limit as r → ∞.

The comparable derivation of parabolic-like approximating equations to the reduced wave equation (1) begins with the observation that, in the constant coefficient case, a family of outgoing plane wave solutions to (1) is given by

$$p(r,z) = H_0^{(1)}(k_r r)e^{ik_z z}, \qquad k_r > 0 \tag{4}$$

as long as k_r, k_z satisfy the dispersion relation

$$k_r^2 + k_z^2 = \left(\frac{\omega}{c}\right)^2 \tag{5}$$

i.e.,

$$k_r = +\sqrt{\left(\frac{\omega}{c}\right)^2 - k_z^2} \tag{6}$$

where k_r and k_z are the horizontal and vertical wave numbers, respectively. These spatial wave numbers are used to define the angle of propagation, say α,

$$\alpha \equiv \tan^{-1}\left(\frac{k_z}{k_r}\right)$$

The angle is measured from the horizontal and thus $|\alpha| \leq \pi/2$. It follows that the total angular region of dispersing energy has measure twice α.

Returning to (4), in particular for a fixed k_z, with $(\omega/c)^2 - k_z^2 > 0$,

$$p(r,z) = H_0^{(1)}\left(\sqrt{\left(\frac{\omega}{c}\right)^2 - k_z^2}\, r\right)e^{ik_z z}$$

is an outgoing solution to (1). Now a first order differential equation which annihilates p at r = ∞ is given by

$$\left[\frac{\partial}{\partial r} - i\sqrt{\left(\frac{\omega}{c}\right)^2 - k_z^2}\,\right]p\,\bigg|_{r=\infty} = 0 \tag{7}$$

This follows since for $\zeta > 0$ and r large

$$H_0^{(1)}(\zeta r) \approx \left[\frac{2}{\pi\zeta r}\right]^{1/2} e^{i(\zeta r - (\pi/4))}$$

implies that

$$\frac{\partial}{\partial r} H_0^{(1)}(\zeta r) - i\zeta H_0^{(1)}(\zeta r) = -\frac{1}{2r} H_0^{(1)}(\zeta r) \tag{8}$$

which is of the order of $r^{-3/2}$ for large r and hence (7) is satisfied. Thus (7) is a "perfect" outgoing wave extractor. Now a pseudo-differential operator is defined corresponding to (7) via the recipe, ik_z is replaced by $\partial/\partial z$, $(ik_z)^2$ is replaced by $\partial^2/\partial z^2$, etc., yielding a perfect outgoing wave extractor

$$\left[\frac{\partial}{\partial r} - i\,\frac{\omega}{c}\sqrt{1 + \left(\frac{c}{\omega}\right)^2 \frac{\partial^2}{\partial z^2}}\,\right]p\,\bigg|_{r=\infty} = 0 \tag{9}$$

For this development, the boundary condition in the z variable is $p = 0$ at $z = \pm\infty$ ($B = +\infty$), and p is defined for $z \in (-\infty, 0)$ via odd reflections about $z = 0$. Unfortunately, the condition (9) is non-local in the space variable z. Thus it is necessary to develop local approximations which necessarily would not be perfect outgoing wave extractors but which would be, we hope, "very good" extractors.

The "symbol" for (9) is given by

$$\frac{\partial}{\partial r} - i\,\frac{\omega}{c}\sqrt{1 - \left(\frac{c}{\omega}\right)^2 k_z^2} \tag{10}$$

and it is approximated at $k_z = 0$, which is the case of a wave traveling in the horizontal direction. Rewriting the symbol in terms of k_0, n, the reference wave number and the index of refraction, we obtain

$$\frac{\partial}{\partial r} - i\sqrt{n^2 k_0^2 - k_z^2} = \frac{\partial}{\partial r} - ink_0\sqrt{1 - (k_z/k_0 n)^2}$$

$$= \frac{\partial}{\partial r} - ik_0\sqrt{1 - \left[\frac{1}{k_0^2} k_z^2 - (n^2 - 1)\right]} \tag{11}$$

Several approaches to the approximation of the pseudo-differential operator exist. A systematic approach developed by F. Muir (see [2] and [3]) uses a continued fraction expansion to the square root operator $\sqrt{1 - X}$. Namely,

$$R_0(X) = 1, \qquad R_{k+1}(X) = 1 - \frac{X}{1 + R_k(X)}$$

It is not difficult to prove that the sequence $\{R_k(X)\}$ converges to $\sqrt{1 - X}$ for all $X \leq 1$. Some analysis of this continued fraction expansion appears in [6]. We shall make use of four of these approximations and we list them here for future reference:

$$R_1(X) = 1 - \frac{1}{2} X$$

$$R_2(X) = \frac{1 - \frac{3}{4} X}{1 - \frac{1}{4} X}$$

$$R_3(X) = \frac{1 - X + \frac{1}{8} X^2}{1 - \frac{1}{2} X}$$

$$R_4(X) = \frac{1 - \frac{5}{4} X + \frac{5}{16} X^2}{1 - \frac{3}{4} X + \frac{1}{16} X^2}$$

In approximating (11) we shall choose

$$X = \frac{1}{k_0^2} k_z^2 - (n^2 - 1)$$

Then, using $R_1(X)$ as the approximation to $\sqrt{1 - X}$ and replacing k_z^2 by $-\partial^2/\partial z^2$, we obtain the approximating equation at $r = \infty$,

$$\frac{\partial p}{\partial r} - ik_0 p = \frac{ik_0}{2} (n^2 - 1)p + \frac{i}{2k_0} \frac{\partial^2 p}{\partial z^2} \tag{12}$$

Now, it is customary in underwater acoustics to use the envelope u of the complex acoustic pressure p, where $p(r,z) = u(r,z)H_0^{(1)}(k_0 r)$. Thus, transforming (12) and using (8), and the fact that p, a solution of (12), is easily seen to be bounded and hence $(1/2r)u \to 0$ as $r \to \infty$, we obtain

$$\frac{\partial u}{\partial r} = \frac{ik_0}{2} (n^2 - 1)u + \frac{i}{2k_0} \frac{\partial^2 u}{\partial z^2}$$

which is (PE). A similar analysis as in the above using $R_2(X)$ yields (WA). Further, in the case of $R_3(X)$ and $R_4(X)$ additional higher order approximating equations are obtained

$$\left[1 + \frac{1}{2} L\right] \frac{\partial u}{\partial r} = \frac{ik_0}{2} Lu + \frac{ik_0}{8} L^2 u \tag{H01}$$

and

$$\left[1 + \frac{3}{4} L + \frac{1}{16} L^2\right] \frac{\partial u}{\partial r} = \frac{ik_0}{2} Lu + \frac{ik_0}{4} L^2 u \tag{H02}$$

respectively.

It is not difficult to demonstrate the different capabilities of the various approximating partial differential equations. In the following sequence of figures we compare an exact solution, generated via a finite sum of propagating modes in a homogeneous medium (see, e.g., [18]) with the computed solution using (PE), (WA), and (H01), respectively. The exact solution is constructed so that the maximum angle at which energy is propagated, measured from the horizontal, is a control parameter. We shall be adjusting this parameter in the ensuing development and thus we have taken note of its value in these figures. In addition, ZO is the depth of the flat bottom ocean, and ZS, ZR are the depths of the source and receiver. In these computations the hard bottom condition $\partial p/\partial z = 0$ at $z = ZO = B$ is used and the exact solution is employed as a starter. We remark that we have recently demonstrated the solution of higher order equations such as (H01), (H02) in [13].

One immediately observes in Fig. 1 and Fig. 2 the increasing goodness of fit in moving successively from (PE), to (WA), to (H01).

It is interesting to note that a natural choice for X in (11), namely $X = (k_z/k_0 n)^2$, leads to a slightly different set of equations. For example, if one uses $R_1(X)$, $X = (k_z/k_0 n)^2$, and the same analysis employed above, we obtain

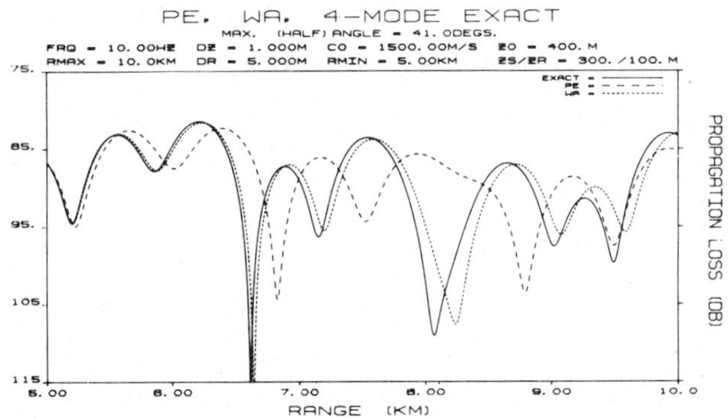

Figure 1

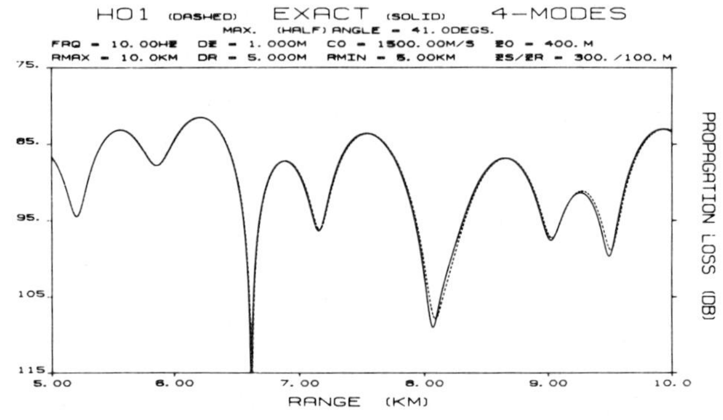

Figure 2

$$\frac{\partial u}{\partial r} = ik_0 (n - 1)u + \frac{i}{2k_0 n} \frac{\partial^2 u}{\partial z^2}$$

instead of (PE), and correspondingly, using $R_2(X)$, we obtain

$$\left[1 + \frac{1}{4k_0^2 n^2} \frac{\partial^2}{\partial z^2}\right] \frac{\partial u}{\partial r} = \frac{i}{2k_0 n^2} \frac{\partial^2 u}{\partial z^2}$$

in place of (WA).

A second approach to the approximation of the pseudo-differential operator (11) is to utilize ad hoc approximations of $\sqrt{1 - X}$. We note that the previous approximations $R_1, \ldots, R_4$ may be considered as having arisen from Taylor or Padé approximation, e.g., R_1 is a two-term Taylor approximation and R_3 is a Padé (2,1) approximation. A three-term Taylor approximation is considered in [18]. To this end, let $f_a(X)$ denote a function which approximates $\sqrt{1 - X}$, then $f_a(X)$ can be used to generate approximating equations as have been the $R_k(X)$. In view of the relationships between the angle of propagation α, the dispersion relation (6), and (10), it is natural to consider "deriving functions," $f_a(X)$, as a function of the angle of propagation α, where $\sin \alpha = k_z/(\omega/c) = k_z/k_0$. Thus, R. Greene [4,10] derives a "40° equation" by considering functions f_a of the form

$$f_a(X) = \frac{a_0 - a_1 X}{1 - b_1 X} \tag{13}$$

and using a Chebyshev approximation method, namely, determine a_0, a_1, b_1 which

$$\min_{} \left(\max_{0 \leq \alpha \leq 40°} \left| \sqrt{1 - \sin^2\alpha} - f_a(\sin^2\alpha) \right| \right)$$

The coefficients obtained by Greene are

$$a_0 = 0.9998707 \qquad a_1 = 0.7962434 \qquad b_1 = 0.301016 \qquad (14)$$

Now these ideas can be extended in a number of directions, e.g., one can choose to use other approximation methods than those of Chebyshev for functions f_a of the form (13), or one might choose other forms for the function f_a. Halpern and Trefethen [11] have explored both of these approaches. In particular, we shall call the class of approximating equations, generated by using functions f_a of the form (13), the "wide angle" equations since (WA) lies in this class. We remark that the class also encompasses (PE). The parabolic-like partial differential equations in this class have the general form

$$(1 + b_1 L)u_r = ik_0 (a_0 - 1 + (a_1 - b_1)L)u \qquad (15)$$

The technique used by Greene to derive a "40°-equation" is certainly effective as is demonstrated by comparing Fig. 2 and Fig. 3. In Fig. 3 G40 refers to the use of coefficients (14) in equation (15). We think of the 40°-equation as an optimized element of the class of wide angle equations. We remark that in the discretization of each of the equations the standard second order accurate centered difference approximations are used in the space variable z.

In Fig. 4, we observe that the optimized 40°-equation displays somewhat more error than does the nonoptimized higher-order equation

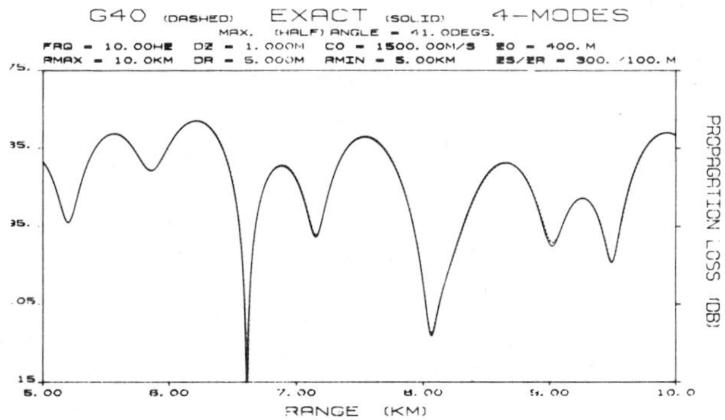

Figure 3

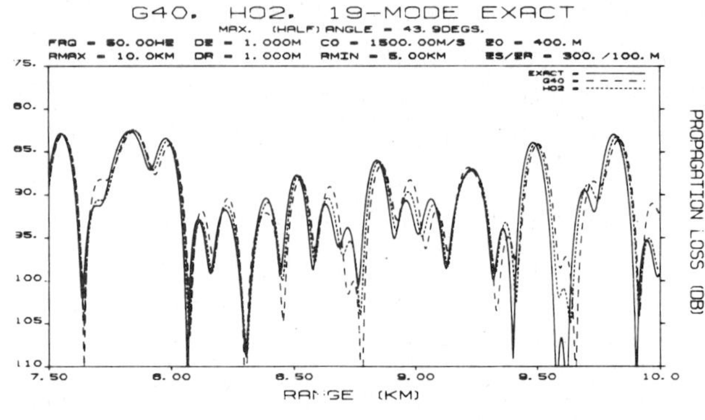

Figure 4

(H02), at the 44° angle level. Of course, the higher-order equations require the solution of a pentadiagonal linear system as opposed to a triadiagonal one for the lower-order equations. We remark that in Fig. 4 the computations were begun at range 5.0 km, but we have plotted only from 7.5-10.0 km. Optimized versions of the higher-order equations appear in [17].

3. COMPARISON AMONG EQUATIONS

One approach to the comparison of the equations in the form (15) might be to study the normal mode solutions of these equations and compare them to the corresponding normal mode solution of the Helmholtz equation. In fact, Fitzgerald [9] and McDaniel [15,16] have pursued this approach for specific equations in the class. In keeping with the earlier development, we wish to classify these equations based upon the approximating function used in their derivation. In addition, we shall consider the angle of propagation α as the main parameter. Halpern and Trefethen [11] derive a large number of equations and present L^2 and L^∞ comparisons among them. Here we shall make additional comparisons but in each case we shall take as benchmark standards the values obtained from applying the various comparison criteria to the Greene 40°-equation.

The first comparison is based on an L^∞ relative error, namely,

$$\max_{0 \leq \alpha \leq \alpha_0} \left| \frac{\sqrt{1 - \sin^2\alpha} - f_a(\sin^2\alpha)}{\sqrt{1 - \sin^2\alpha}} \right| \tag{16}$$

In the case of (14) with $\alpha_0 = 40°$, we obtain the benchmark tolerance 0.00017 for (16). We then seek angles α_0 associated with the other equations, e.g., (PE), (WA), for which the corresponding (16) is less than or equal to 0.00017. In this manner (PE) is associated with an angle between 10° and 11°, and (WA) with an angle between 23° and 24°. This comparison criterion is not a meaningful one for the equations in [11] in our wide-angle class for in essentially every case this L^∞ benchmark is violated at $\alpha = 0°$. We remark that in general the derivations in [11] place no emphasis on keeping the approximations close initially and, in fact, except for one case, L^∞_α, $\alpha = 45°$, rather the emphasis is on balancing the error throughout $0° \leq \alpha \leq 90°$. In fact, in almost all cases the majority of the interpolation points in the derivations used in [11] lie outside the interval $[0°,40°]$. This causes the equations in [11] not to be competitive in the criteria we present since the emphasis is on the interval $[0°,40°]$. Figures 5 and 6 demonstrate the application of (PE) and (WA) to 12° and 25° maximum angles of propagation, respectively. In addition, the L^∞_α, $\alpha = 45°$, curve, denoted by L-A, is plotted in Figs. 5 and 6.

The second comparison method is based on the notion of the approximate dispersion relation

$$\left| [f_a(\sin^2\alpha)]^2 + \sin^2\alpha - 1 \right| \leq \text{tol}. \tag{17}$$

In the case of the dispersion relation (6), $f_a(X)$ would be $\sqrt{1 - X}$, the tolerance, tol = 0, and the equal sign would be attained. The use of the approximating function thus gives rise to (17). As before, we select a benchmark value of tol based upon the maximum value of the left side of (17) associated with (14) for $0° \leq \alpha \leq 40°$. The value thus selected for tol is 0.00026. Now for each of the various approximating functions f_a, we seek the largest angle

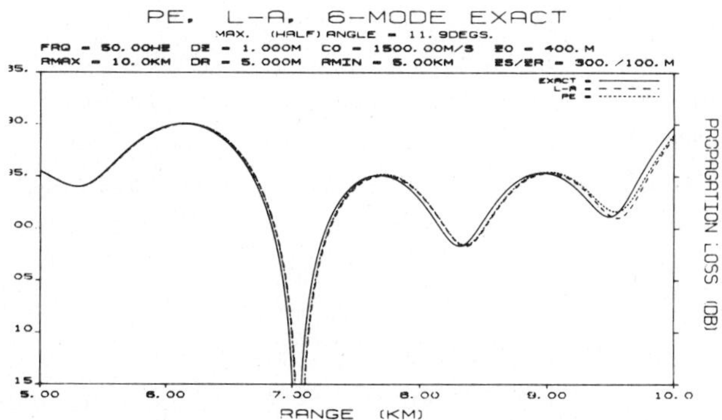

Figure 5

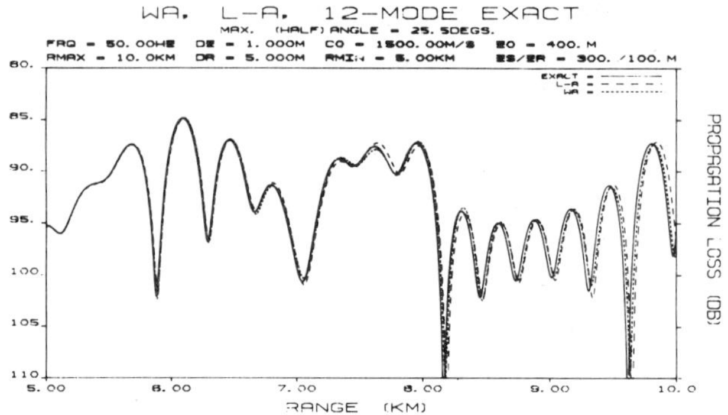

Figure 6

α_0, such that (17) is satisfied for all α, $0 \le \alpha \le \alpha_0$, with tol = 0.00026. Precisely the same conclusions are drawn for (PE), (WA), and the equations in [11] using this criterion as was stated in the previous paragraph for the L^∞ criterion.

Finally, we shall make comparisons based upon an L^2 relative error,

$$\left[\int_0^{\alpha_0} \left(\sqrt{1 - \sin^2\alpha} - f_a(\sin^2\alpha)\right)^2 d\alpha \Big/ \int_0^{\alpha_0} (1 - \sin^2\alpha) d\alpha\right]^{1/2} \qquad (18)$$

As before, we begin by choosing f_a using (14) and $\alpha_0 = 2\pi/9$ to obtain a benchmark tolerance for (18). The value of the tolerance used in (18) is 9.7×10^{-9}. Similar results are obtained in this test as in the above two, except that (PE) tests at between 12° and 13°, and (WA) between 27° and 28°.

REFERENCES

1. Clairbout, J. F., Coarse grid calculations of waves in inhomogeneous media with application to delineation of complicated seismic structure, GEOPHYSICS 35(3) (1970), 407-218.

2. Clairbout, J. F., IMAGING THE EARTH'S INTERIOR, Blackwell Scientific Publications, Oxford, 1985.

3. Clayton, R., and B. Engquist, Absorbing boundary conditions for acoustic and elastic wave equations, BULLETIN OF THE SEISMOLOGICAL SOCIETY OF AMERICA 67 (6) (1977), 1529-1540.

4. Davis, J. A., D. White, and R. C. Cavanagh, NORDA PARABOLIC EQUATION WORKSHOP, Tech. Note No. 143, NORDA, NSTL Station, Miss., 1982.

5. Engquist, B., and A. Majda, Absorbing boundary conditions for the numerical simulation of waves, MATHEMATICS OF COMPUTATION 31(139) (1977), 629-651.

6. Engquist, B., and A. Majda, Radiation boundary conditions for acoustic and elastic wave calculations, COMMUNICATIONS IN PURE AND APPLIED MATHEMATICS 32 (1979), 313-357.

7. Fishman, L., and J. J. McCoy, Derivation and application of extended parabolic wave theories. Part I. The factorized Helmholtz equation, JOURNAL OF MATHEMATICAL PHYSICS 25(2) (1984), 285.

8. Fishman, L., J. J. McCoy, and S. C. Wales, Factorization and path integration of the Helmholtz equations: Numerical algorithms, PROCEEDINGS 11TH IMACS WORLD CONGRESS ON SYSTEM SIMULATION AND SCIENTIFIC COMPUTATION 2 (1985), 139-142.

9. Fitzgerald, R. M., Helmholtz equation as an initial value problem with application to acoustic propagation, JOURNAL OF THE ACOUSTICAL SOCIETY OF AMERICA 57(4) (1975), 839-842.

10. Greene, R. R., The rational approximation to the acoustic wave equation with bottom interaction, JOURNAL OF THE ACOUSTICAL SOCIETY OF AMERICA 76 (1984), 1764-1773.

11. Halpern, L., and L. N. Trefethen, Wide-angle one-way wave equations (submitted).

12. Lee, D., G. Botseas, and J. S. Papadakis, Finite-difference solutions to the parabolic equation, JOURNAL OF THE ACOUSTICAL SOCIETY OF AMERICA 70(3) (1981), 795-800.

13. Lee, D., G. H. Knightly, and D. F. St. Mary, A higher order parabolic wave equation (submitted).

14. Lee, D., and J. S. Papadakis, Numerical solution for the parabolic wave equation: An ordinary-differential-equation approach, JOURNAL OF THE ACOUSTICAL SOCIETY OF AMERICA 68 (1980), 1482-1488.

15. McDaniel, S. T., Propagation of normal mode in the parabolic approximation, JOURNAL OF THE ACOUSTICAL SOCIETY OF AMERICA 57(2) (1975), 307-311.

16. McDaniel, S. T., Parabolic approximations for underwater sound propagation, JOURNAL OF THE ACOUSTICAL SOCIETY OF AMERICA 58(8) (1975), 1178-1185.

17.   St. Mary, D. F., Analysis of an implicit finite difference
      scheme for very wide angle underwater acoustic propagation,
      PROCEEDINGS 11TH IMACS WORLD CONGRESS ON SYSTEM SIMULATION
      AND SCIENTIFIC COMPUTATION 2 (1985), 153-156.

18.   St. Mary, D. F., D. Lee, and G. Botseas, A modified wide angle
      wave equation, JOURNAL OF COMPUTATIONAL PHYSICS (to appear).

19.   Tappert, F. D., The parabolic approximation method, in WAVE
      PROPAGATION AND ACOUSTICS, J. B. Keller and J. Papadakis, eds.,
      Springer-Verlag, New York, 1977.